Lecture Notes on Data Engineering and Communications Technologies

Volume 236

The aim of the book series is to present cutting edge engineering approaches to data technologies and communications. It will publish latest advances on the engineering task of building and deploying distributed, scalable and reliable data infrastructures and communication systems.

The series will have a prominent applied focus on data technologies and communications with aim to promote the bridging from fundamental research on data science and networking to data engineering and communications that lead to industry products, business knowledge and standardisation.

Indexed by SCOPUS, INSPEC, EI Compendex.

All books published in the series are submitted for consideration in Web of Science.

Alejandro C. Frery · Rajkumar Buyya ·
Ram Mohan Rao Kovvur · T. Hitendra Sarma
Editors

Computational Intelligence and Data Analytics

Proceedings of ICCIDA 2024

Springer

Editors
Alejandro C. Frery
Victoria University of Wellington
Wellington, New Zealand

Rajkumar Buyya
The University of Melbourne
Melbourne, VIC, Australia

Ram Mohan Rao Kovvur
Department of Information Technology
Vasavi College of Engineering
Hyderabad, Telangana, India

T. Hitendra Sarma
Department of Information Technology
Vasavi College of Engineering
Hyderabad, Telangana, India

ISSN 2367-4512 ISSN 2367-4520 (electronic)
Lecture Notes on Data Engineering and Communications Technologies
ISBN 978-981-96-0450-0 ISBN 978-981-96-0451-7 (eBook)
https://doi.org/10.1007/978-981-96-0451-7

This Springer imprint is published by the registered company Springer Nature Singapore Pte Ltd.
The registered company address is: 152 Beach Road, #21-01/04 Gateway East, Singapore 189721, Singapore

Organizing Committee ICCIDA 2024

Chief Patrons
Sri P. Ramamohan Rao, President, Vasavi Academy of Education
Sri M. Krishna Murthy, Secretary, Vasavi Academy of Education

Patrons
Sri P. Balaji, CEO, Vasavi Academy of Education
Dr. S. V. Ramana, Professor & Principal, Vasavi College of Engineering

Honorary Chair
Prof. Ramachandram, Brane, Hyderabad

General Chair
Dr. K. Ram Mohan Rao, VCE, Hyderabad, India

Technical Programme Committee
Dr. Susana Muñoz Hernández, Director TEDECO, Madrid
Dr. Rammohan Mallipeddi, Kyungpook National University, South Korea
Dr. K. G. Srinivasa, IIIT Naya Raipur, India

Tutorial Chair
Dr. K. Raghavendra, ADRIN, ISRO, Hyderabad, India
Mr. Amit Gupta, Microsoft, Hyderabad, India
Mr. Amar Sharma, CEO WOIR, Hyderabad, India

Publication Chair
Prof. Alejandro C. Frery, University of Willington, New Zealand
Prof. Rajkumar Buyya, University of Melbourne, Australia
Dr. T. Hitendra Sarma, VCE, Hyderabad, India

Local Organizing Chairs
Dr. Tilottama Goswami, VCE, Hyderabad, India
Dr. Burgula Kezia Rani, VCE, Hyderabad, India
S. Aruna, VCE, Hyderabad, India
Dr. N. Anil Kumar, VCE, Hyderabad, India
Dr. Neelakhantappa, VCE, Hyderabad, India
Dr. S. K. Prashanth, VCE, Hyderabad, India

Sponsorship Committee
K. Srinivasa Chakravarthy, VCE, Hyderabad, India
R. Dharma Reddy, VCE, Hyderabad, India
L. Divya, VCE, Hyderabad, India

Finance Committee
C. Sireesha, VCE, Hyderabad, India
M. Sathya Devi, VCE, Hyderabad, India
Dr. Anjana Devi, VCE, Hyderabad, India

Publicity Committee
Dr. S. Sree Lakshmi, VCE, Hyderabad, India
DRL Prasanna, VCE, Hyderabad, India
Sruthi Anand, VCE, Hyderabad, India
Haseeba Yaseen, VCE, Hyderabad, India

Registration Committee
Dr. S. K. Chaya Devi, VCE, Hyderabad, India
S. Rajya Laxmi, VCE, Hyderabad, India
B. Leelavathy, VCE, Hyderabad, India
S. Renuka, VCE, Hyderabad, India
G. Radha, VCE, Hyderabad, India
Soume Sanyal, VCE, Hyderabad, India

Travel and Accommodation
Dr. Rama Krishna, VCE, Hyderabad, India
G. Rajashekhar, VCE, Hyderabad, India
N. David Raju, VCE, Hyderabad, India

Program Committee

Abhay Paroha, SLB, USA
Adeyinka Adedigba, Federal University of Technology, Minna, Nigeria
Amit Gupta, Microsoft, India
Anil Kumar, Vasavi College of Engineering, India
Anila Glory H., SASTRA Deemed University, India
Aubain Nzokem, York University, Canada
Bala Peddigari, TCS, India
Chandan M. C., The National Institute of Engineering Mysuru, India
Chebolu N. A. B. Sankar, RCI, India
Daison Darlan, Kyungpook National University, South Korea
Dharma Reddy, Vasavi College of Engineering, India
Dr. Swastika Chakraborty, Sikkim Manipal Institute of Technology, India
Dr. Viswanath P., IIIT Sri City, Chittore, India
Dr. Sudipta Basu Pal, University of Engineering & Management, Kolkata, India
Hammed Obasekore, Kyungpook National University, South Korea
Harshavardhan Awar, VNR Vignana Jyothi Institute of Engineering and Technology, India
Hitendra Sarma, Vasavi College of Engineering Hyderabad, India
Huda Abdel-Hamed, Sudan University of science and technology, Sudan
J. Somasekar, JAIN (Deemed-to-be University), Bangalore
Jaya Sreevalsan-Nair, Graphics-Visualization-Computing Lab, IIIT Bangalore, India
Kapil Juneja, Bennett University, India
Kaushik Sarkar, Narula Institute of Technology, India
Kavitha Devireddy, Osmania University, India
Kezia Rani B., Vasavi College of Engineering, India
Kune Raghavendra, ADRIN, DOS, India
Lakshmi Narayana, Defence Electronics Research Laboratory (DLRL/DRDO) (Retired), India
M. Naga Sree, G. Narayanamma Institute of Technology and science for Women, India

Madhavilata M., G. Narayanamma Institute of Technology and science for Women, India
Madugundu Neelakantappa, Vasavi College of Engineering, Hyderabad, India, India
Manoj Patil, BRANE, Hyderabad
Mirza Zainab, M. H. Saboo Siddik COE, Mumbai, India
Mohammad Halim Deedar, Universidad Politécnica de Madrid, Spain
K. Mrudula, GNITS Hyderabad, India
Mukesh Kumar Tripathi, Vardaman Engineering College, Hyderabad, India
Murali Kanthi, CMR Technical Campus, Kandlakoya, Hyderabad
N. Anjani Devi, Vasavi College of Engineering Hyderabad, India
Nadella Sandhya, VNRVJIET, India
Nagaraju Devarakonda, VIT AP University, Amaravathi, Andhra Pradesh, India
Naresh M., NRSC, Hyderabad, India
Oladayo Solomon, Kyungpook National University, South Korea
Pooja Agarwal, PES, Bangalore, India
Pradeep Kumar, Maulana Azad National Urdu University Hyderabad, India
Prashanth S. K., Vasavi College of Engineering Hyderabad, India
Prashil Negandhi, Dwarkadas J. Sanghvi College of Engineering, India
Rajitha Rani, VLSI Engineer, India
Ram Mohan Rao Kovvur, Vasavi College of Engineering, India
Raman Dugyala, Vardhaman College of engineering, India
Ramesh Babu Arasamudi, Amrita Vishwa Vidyapeetham, India
Rammohan Mallipeddi, Kyungpook National University, South Korea
Ramnaresh Yadav, JNTU Hyderabad, India
Rukma Rekha N., University of Hyderabad, India
Salman Abdul Moiz, School of Computer & Information Sciences, University of Hyderabad, India
Samir Goswami, TMI, Hyderabad, India
Samson Chepuri, MVSR Engineering College, Hyderabad, India
Sandeep Kumar Mogha, Chandigarh University Mohali, India
Sandhya Banda, Maturi Venkata Subba Rao (MVSR) Engineering College, India
Sarat Chandra, Kerry Group, Ireland
Saravanan Parthasarathy, Medyaan Healthcare Private Limited, Chennai, India
Siddesh G. M., M. S. Ramaiah Institute of Technology, Bangalore, India
Srinivasa K. G., Indian Institute of Information Technology, Naya Raipur, India
Suresh Veluru, AI Lead at ServisBOT, Republic of Ireland
Surya Mukhi, Osmania University, India
Susana Muñozhernández, Universidad Politécnica de Madrid, Spain
Swamy Das Marri, CBIT, Hyderabad, India
Swapna Agarwal, Indian Statistical Institute, India
Thangaramya K., Vellore Institute of Technology, Vellore, India
Tilottama Goswami, Vasavi College of Engineering, Hyderabad, India
Tolu Aboyeji Esther, Kyungpook National University, South Korea
Usha Rani Nelakuditi, School of Electrical, Electronics and Communication VFSTR, India

Usigbe Member Joy, Kyungpook National University, Daegu., South Korea
Veera Prakash, Srinivasa Ramanujan Institute of Technology, India
Venu Gopal, Vardhaman College of Engineering, India
Vijayakumar Kadappa, BMS College of Engineering, India
Viji Gripsy, PSGR Krishnammal College for Women, Coimbatore, India
Y. C. A. Padmanabha Reddy, CBIT, Hyderabad, India
Zeuban Ivan, Kyungpook National University, South Korea
Vaishnavi Vats, Kyungpook National University, South Korea

Preface

The International Conference on Computational Intelligence and Data Analytics (ICCIDA 2024) has been recognized as a very successful conference hosted by the Department of Information Technology, Vasavi College of Engineering, and sponsored by Vasavi Academy of Education, during 28–29 June 2024 at Vasavi College of Engineering (Autonomous), Hyderabad, Telangana, India.

ICCIDA aims to provide an exciting platform for the young researchers to exchange their innovations and explore future prospects of research with the global research community in academia and industry, in the areas of computational intelligence, data analytics and its allied fields.

It is worth mentioning that ICCIDA 2024 has made its significance by attracting 368 papers with an acceptance ratio of 9.78%. The submissions are from different countries including France, Mexico, Nigeria, Algeria, Bangladesh, Canada, Cyprus, Czechia, Pakistan, South Korea, UAE, Croatia, United States, Indonesia, Saudi Arabia, Iran and India. All the articles are critically peer-reviewed by the program committee of 77 experts from different areas covering all the themes of the conference, from all parts of the globe. Finally, 36 papers were accepted and 30 presented by the authors during the conference.

The technical program committee, chaired by Prof. Rammohan Mallipeddi (Kyungpook National University (KNU), South Korea), Dr. Susana Muñoz Hernández (Technical University of Madrid, Director TEDECO, Madrid, Spain), and Prof. K. G. Srinivasa (Indian Institute of Information technology, Naya Raipur, India), have put their collective efforts to make this conference a technically enriched successful event. The Technical Program started with a tutorial talk on "Principles, Tools and Practices in Data Visualization" and a workshop on "Harnessing the power of AI to Education". The two--day conference schedule comprises 9 keynote talks by eminent speakers in the emerging research areas viz., Evolutionary Computing, Quantum Computing, Cloud and Edge Computing, Statistical Machine Learning, Soft Computing, Deep Learning, Data Science and IoT.

We sincerely thank the esteemed speakers:

- Prof. Rajkumar Buyya (Director, Cloud Computing and Distributed Systems Lab, University of Melbourne, Australia)
- Dr. M. Tanveer (INSA Associate Fellow and SERB Ramanujan Fellow, OPTIMAL Research Lab, Department of Mathematics, IIT Indore)
- Prof. P. N. Suganthan (IEEE Fellow, KINDI Center for Computing Research, Qatar University)
- Dr. P. Sateesh Kumar (Associate Professor, Computer Science and Engineering, IIT, Roorkee)
- Prof. Vadlamani Ravi (Professor Institute for Development and Research in Banking Technology Hyderabad, India)
- Dr. Raghavendra Kune (Advanced Data Processing Research Institute, Department of Space, ISRO, Hyderabad)
- Prof. Swagatam Das (Professor, Electronics and Communication Sciences Unit, Indian Statistical Institute, Kolkata)
- Dr. Susana Muñoz Hernández (Technical University of Madrid, Director TEDECO, Madrid, Spain)
- Dr. Rammohan Mallipeddi (Kyungpook National University (KNU), South Korea)
- Mr. Amit Gupta, Senior Data Scientist, Microsoft, Hyderabad
- Prof. K. G. Srinivasa, Indian Institute of Information technology, Naya Raipur.

We sincerely thank Mr. Aninda Bose and his team of Springer Nature for their strong support towards publishing this volume in the series of Lecture Notes on Data Engineering and Communications Technologies—Indexed by Scopus, Inspec and EiCompendex.

We thank Dr. S. V. Ramana, Principal, VCE and the management of Vasai Academy of Education, Hyderabad, for their constant support to host ICCIDA 2024.

Dr. Ram Mohan Rao Kovvur
Conference Chair

Contents

About the Editors

Alejandro C. Frery received his Ph.D. in Applied Computing from the Instituto Nacional de Pesquisas Espaciais (INPE, São José dos Campos, Brazil, 1993) and his Postdoctoral from IMPA. In 2003, he became a Full Professor at the Universidade Federal de Alagoas, Brazil. He was University Vice-Chancellor for Research, Graduate Courses, and Innovation from Jan'2016 to Jan'2020. Currently, he is a Statistics and Data Science Professor, at Victoria University of Wellington, New Zealand. Prof. Frery was Editor-in-Chief of the IEEE Geoscience and Remote Sensing Letters (2014–18). He was a Distinguished Lecturer for IEEE GRSS (2015–19). In 2018, he received the IEEE GRSS Regional Leader Award. Prof. Frery served in various capacities in IEEE GRSS (Advisory Committee) AdCom, in the past 2 decades. Since 2023, he has been Vice-President of Publications.

Dr. Rajkumar Buyya is Redmond Barry Distinguished Professor and Director of the Cloud Computing and Distributed Systems (CLOUDS) Laboratory at the University of Melbourne, Australia. He has authored over 750 publications and seven textbooks. He is the World's Top-Cited(#1) Author and the World's Most Productive(#1) Author in Cloud Computing. He served as Founding Editor-in-Chief of the IEEE Transactions on Cloud Computing. Dr. Buyya has been recognized as Web of Science's "Highly Cited Researcher" for four consecutive years since 2016, IEEE Fellow, Scopus Researcher of the Year 2017 with Excellence in Innovative Research Award by Elsevier, and the "Best of the World", in Computing Systems field, by The Australian 2019 Research Review.

Dr. Ram Mohan Rao Kovvur received Ph.D. in Computer Science and Engineering from JNTUH in the year 2014 with a Research specialization in Grid Computing. He has over 27 years of teaching experience and served in various administrative capacities. Currently, he is Professor and Head of the Information Technology Department, at Vasavi College of Engineering, Hyderabad, India. He has published 35 technical papers in reputed Journals and Conferences. He is the Principal Investigator of the project titled "Drone/UAV Based Multi-Sensor High-Quality Precision

Photogrammetry Laser Mapping and Data Capture, Processing for Geo-Spatial Platforms" funded by DST, India (34.78 lakh). Dr. Rao has established a Deep Learning Lab under the MODROBS Scheme of AICTE-INDIA with a grant of ₹19.31 Lakhs.

Dr. T. Hitendra Sarma received Ph.D. in Computer Science and Engineering from JNT University Anantapur, India, in December 2013. He has more than 15 years of teaching and research experience. Currently, he is working as an Associate Professor in the Department of Information Technology, at Vasavi College of Engineering, Hyderabad, India. Dr. Sarma has published more than 36 research articles in various peer-reviewed International Journals and Conferences published by Springer, Elsevier, and IEEE. Dr. Sarma completed a project titled "Dimensionality Reduction and Pixel Classification in Hyperspectral Images", funded by the Science and Engineering Research Board (SERB), India (2019–2021). He served as Volume Editor for the proceedings of ICECIT2015 & ICECIT2018 (Springer-LNEE), ICCIDA2022 (Springer-LNDECT), and MIGARS-2023 (IEEE). Dr. Sarma is a Senior Member of IEEE, IEEE CIS, IEEE GRSS, ACM, IE(I), and CSI.

On-Demand Cold Start Frequency Reduction with Off-Policy Reinforcement Learning in Serverless Computing

Siddharth Agarwal, Maria A. Rodriguez, and Rajkumar Buyya

Abstract Function-as-a-Service (FaaS) is a cloud computing paradigm offering an event-driven execution model to applications. It features 'serverless' attributes by eliminating resource management responsibilities from developers, and offers transparent and on-demand scalability of applications. To provide seamless on-demand scalability, new function instances are prepared to serve the incoming workload in the absence or unavailability of function instances. However, FaaS platforms are known to suffer from *cold starts*, where this function provisioning process introduces a non-negligible delay in function response and reduces the end-user experience. Therefore, the presented work focuses on reducing the frequent, on-demand cold starts on the platform by using Reinforcement Learning (RL). The proposed approach uses model-free Q-learning that consider function metrics such as CPU utilisation, existing function instances and response failure rate, to proactively initialize functions, in advance, based on the expected demand. The proposed solution is implemented on Kubeless and evaluated using an open-source function invocation trace applied to a *matrix multiplication* function. The evaluation results demonstrate a favourable performance of the RL-based agent when compared to Kubeless' default policy and a function keep-alive policy by improving throughput by up to 8.81% and reducing computation load and resource wastage by up to 55% and 37%, respectively, that is a direct outcome of reduced cold starts.

Keywords Cold start · Function-as-a-Service · Reinforcement Learning · Serverless computing

S. Agarwal · M. A. Rodriguez · R. Buyya (✉)
Cloud Computing and Distributed Systems Laboratory, School of Computing and Information Systems, The University of Melbourne, Melbourne, VIC, Australia
e-mail: rbuyya@unimelb.edu.au

S. Agarwal
e-mail: siddhartha@student.unimelb.edu.au

M. A. Rodriguez
e-mail: maria.read@unimelb.edu.au

A. C. Frery et al. (eds.), *Computational Intelligence and Data Analytics*, Lecture Notes on Data Engineering and Communications Technologies 236,
https://doi.org/10.1007/978-981-96-0451-7_1

1 Introduction

In cloud computing, a serverless deployment model removes the burden of managing and provisioning resources from the developers, allowing them to focus solely on the application development process. The term *serverless*, interchangeably used with Function-as-a-Service (FaaS), does not imply an absence of servers, but instead, accentuates delegating the responsibility of complex resource management tasks to cloud service providers (CSP) [13, 16]. The FaaS paradigm puts forward an event-driven, serverless computing model with fine-grained pay-per-use pricing where resources are billed based on their actual service time. Functions (i.e. a fragment of code containing business logic) are designed to scale on demand; they are stateless, short-lived and run on lightweight containers or virtual machines (VMs) in response to a triggering event. Such an abstraction increases agility in application development, offering lower administrative and ownership costs. The FaaS model has attracted a wide range of applications such as IoT services, REST APIs, stream processing and prediction services, which have strict availability and quality of service requirements in terms of response time. Conceptually, the FaaS model is designed to spin a new function instance for each demand request and shut down the instance after service [13]. However, in practice, commercial FaaS offerings like AWS Lambda, Azure Functions and Google Cloud Function may choose to re-use a function instance or keep the instance running for a limited time to serve subsequent requests [33]. Some open source serverless frameworks like Kubeless [1] and Knative, have similar implementations to re-use an instance of a function to serve subsequent requests.

An increase in workload demand leads to an instantiation process involving the creation of new function containers and the initialisation of the function's environment within those containers, after which incoming requests are served. Such a process usually requires downloading the client code, setting up code dependencies and the runtime environment, setting up container networking, and finally executing the code to handle the incoming request. Hence, instantiating a function's container introduces a non-negligible time latency, known as *cold start*, and gives rise to a challenge for serverless platforms [10, 15, 27, 36]. Some application-specific factors such as programming language, runtime environment and code deployment size as well as function requirements like CPU and memory, affect the cold start of a function [20, 22, 27, 31]. To automate the process of creating new function instances and reusing existing ones, serverless frameworks usually rely on resource-based (CPU or memory) horizontal scaling, known as horizontal pod auto-scaling (HPA) in Kubernetes-based frameworks like Kubeless, to respond to incoming requests. Resource-based scaling policies implement a reactive approach and instantiate new functions only when resource usage rises above a pre-defined threshold, thus leading to cold start latencies and an increase in the number of unsuccessful requests.

Threshold-based scaling decisions fail to consider factors like varying application load and platform throughput and hence, pose an opportunity to explore dynamic techniques that analyse these factors to address cold starts. This work presents a model-free Q-learning agent to exploit resource utilisation, available function instances and platform response failure rate to reduce the number of cold starts for CPU-intensive serverless functions. We define a reward function for the RL agent to dynamically establish the required number of function instances for a given workload demand based on expected average CPU utilisation and response failure rate. The RL-based agent interacts with the serverless environment by performing scaling actions and learns through trial and error during multiple iterations. The agent receives delayed feedback, either positive or negative, based upon the observed state, and consequently learns the appropriate number of function instances to fit the workload demand. This strategy uses no prior knowledge about the environment, demand pattern or workload, and dynamically adjusts to the changes for preparing required functions in advance to reduce cold starts. The proposed work scales the number of function instances by proactively estimating the number of functions that are needed to serve incoming workload to reduce the frequent cold starts. It utilises a practical workload of matrix multiplication involved in an image processing task, serving as a sample real-world function request pattern [28], and formally presents the cold start as an optimisation problem. Also, we structure the Q-learning components around the function metrics such as average CPU utilisation and response failure rate and evaluate our approach against the default resource-based policy and commercially accepted function keep-alive technique.

In summary, the key contributions of our work are

1. We analyse function resource metrics such as CPU utilisation, available instances and the proportion of unsuccessful responses to propose a Q-learning model that dynamically analyses the application request pattern and improves function throughput by reducing frequent cold starts on the platform.
2. We present a brief overview of explored solutions to address function cold starts and highlight the differences between contrasting approaches to the proposed agent.
3. We perform our experiments on a real-world system setup and evaluate the proposed RL-based agent against the default resource-based policy and a baseline keep-alive technique.

The rest of the paper is organised as follows. Section 2 highlights related research studies. In Sect. 3, we present the system model and formulate the problem statement. Section 4 outlines the proposed agentâŁ™s workflow and describes the implementation hypothesis and assumptions. In Sect. 5, we evaluate our technique with the baseline approach and highlight training results and discuss about performance in Sect. 6. Section 7 highlights future research direction and Sect. 8 summarises and concludes the paper.

2 Related Work

In this section, we briefly discuss about the Function-as-a-Service paradigm in serverless computing and elaborate on the current function cold start mitigation techniques and approaches.

2.1 Serverless Computing or Function-as-a-Service

Serverless computing offers a cloud service model wherein the server management or resource management responsibility lies with the CSP. In [13], the authors discussed the potential of this new, less complex computing model introduced by Amazon in 2014. They explain a function-based, serverless commercial offering of AWS Lambda, i.e. the Function-as-a-Service platform. They highlighted three primary differences between traditional cloud computing and serverless computing as follows: decoupled computation and storage, code execution without resource management, and paying in proportion to the resources used. The research posits that the serverless or FaaS model promotes business growth, making the use of the cloud easier.

Baldini et al. [8] introduced the emerging paradigm of FaaS as an application development architecture that allows the execution of a piece of code in the cloud without control over underlying resources. They identified containers and the emergence of microservices architecture as the promoter of FaaS model in serverless. They used FaaS and serverless interchangeably and defines it as a 'stripped down' programming model that executes stateless functions as its deployment unit.

Since the inception of serverless computing, there have been many commercial and open source offerings such as AWS Lambda, Microsoft Azure Functions, Google Cloud Functions, Fission and OpenWhisk. These platforms represent FaaS as an emerging technology but Hellerstein et al. [12] put together gaps that furnish serverless as a bad fit for cloud innovations. The authors criticized the current developments of cloud computing and state that the potential of cloud resources is yet to be harnessed. On the contrary, the work in [27] argued that serverless offerings are economical and affordable as they remove the responsibility of resource management and complexity of deployments from consumers. It presented the opportunities offered by multiple FaaS offerings and gives an overview of other existing challenges and indicates potential approaches for future work.

A Microsoft work [24] estimated that there will be near 500 million new applications in the subsequent 5 years, and it would be difficult for the current development models to support such large expansions. Another recent study by Datadog [2], publishes that over 70% organisations using AWS cloud services, 50% organisations using Microsoft Azure services and Google Cloud platform have adopted serverless computing into their architectures. FaaS is designed to increase development agility, reduce the cost of ownership and decrease overheads related to servers and other cloud resources. The term 'serverless' has been in the industry since the introduction of Backend-as-a-Service (BaaS). Despite the serverless benefits, FaaS experiences two major challenges, which are categorized as (i) system-level and (ii) programming and DevOps challenges [8, 13, 24]. The former identifies the cost of services, security, resource limits and cold start while scaling, and the latter focuses on tools and IDEs, deployment, statelessness and code granularity in the serverless model.

2.2 Function Cold Start and Mitigation

Researchers in [33] described function cold start as the time taken to execute a function. This process involves assigning a container to a function, accessing the code package and copying the function image, loading the image into memory, unpacking it and executing the function handler. It broadly classified the approaches to deal with function cold start in, environment optimisation and pinging. The former approach acts either by reducing container preparation time or decreasing the delay in loading function libraries, while the latter technique continuously monitors the functions and periodically pings them to keep the instances warm or running.

An adaptive container warm-up technique to reduce the cold start latency and a container pool strategy to reduce resource wastage is introduced in [36]. The proposed solution leverages a Long-Short Term Memory (LSTM) network to predict function invocation times and non-first functions in a chain to keep a warm queue of function containers ready. Although both the discussed techniques work in synchronisation, the first function in the chain suffered from a cold start.

The research in [15] explained platform-dependent overheads like pod provisioning and application implementation-dependent overheads. It presented a pool-based pre-warmed container technique, marked with selector 'app-label' to deal with the function cold start problem. To tackle the incoming demand, a container pool is checked first for existing pre-warmed containers, or the platform requests new containers as per the demand.

Another study [18] exploited the data similarity to reduce the function cold start. It criticized the current container deployment technique of pulling new container images from the storage bucket and introduced a live container migration over a peer-to-peer network. Similarly, [10] aimed to reduce the number of cold start occurrences by utilizing the function composition knowledge. It presented an application-side solution based on lightweight middleware. This middleware enable the developers to control the frequency of cold start by treating the FaaS platform as a black box.

Based on the investigation in [22], network creation and initialisation were found to be the prime contributors to the cold start latency. The study expressed that cold starts are caused due to work and wait times involved in various set-up processes like initializing networking elements. The study explained the stages of the container lifecycle and states that the clean-up stage demands cycles from the underlying containerisation daemon, hindering other processes. Therefore, a paused container pool manager is proposed to pre-create a network for function containers and attach the new function containers to configured IP and network when required.

Some studies [12, 20, 27] have identified significant factors that affect the cold start of a function. These include runtime environment, CPU and memory requirements, code dependency setting, workload concurrency and container networking requirements. Most works [14, 17, 21, 29, 35] focus on commercial FaaS platforms like AWS Lambda, Azure Functions, Google Cloud Functions and fall short to evaluate open source serverless platforms like OpenLambda, Fission, Kubeless, etc. Very few studies [23, 25, 31] have successfully performed analysis on an open-source serverless platform and provided possible solution by targeting the container level fine-grained control of the platform.

Recent research works [5, 9, 26, 34] introduce the paradigm of RL to the FaaS platforms in different ways. Schuler et al. [26] focuses on request-based provisioning of VMs or containers on the Knative platform. The authors demonstrated a correlation between latency and throughput with function concurrency levels and thus propose a Q-Learning model to determine the optimal concurrency level of a function for a single workload. Vahidinia et al. [34] proposed a two-layer adaptive approach, an RL algorithm to predict the best idle-container window, and an LSTM network to predict future invocation times to keep the pre-warmed containers ready.

The study demonstrated the advantages of the proposed solution on the OpenWhisk platform using a simple HTTP-based workload and a synthetic demand pattern. Another research [9] focused on resource-based scaling configuration (CPU utilisation) of OpenFaaS and adjusts the HPA settings using an RL-based agent. They assumed a serverless-edge application scenario and a synthetic demand pattern for the experimentation and present their preliminary findings based on latency as SLA.

Agarwal et al. [5] introduced the idea of Q-learning to ascertain the appropriate amount of resources to reduce frequent cold starts. The authors shared the preliminary training results with an attempt to show the applicability of reinforcement learning to the serverless environment. They utilised the platform exposed resource metrics to experiment with a synthetic workload trace, i.e. Fibonacci series calculation, to simulate a compute-intensive application and predict the required resources.

Our proposed work introduces a Q-Learning strategy to reduce frequent cold starts in the FaaS environment. Contrasting existing solutions, we apply the model-free Q-Learning to determine required number of function instances for the workload demand that eventually reduces number of on-demand cold starts. Furthermore, the existing solutions take advantage of either continuous pinging, pool-based approaches, container migration and network building or exploit platform-specific implementations like provisioned concurrency while failing to experiment with CPU-intensive real-world application workloads. Similar to [5], our work utilises available resource-based metrics and response failure rate to accomplish the learning, but improves over the discussed approach. Contrasting to their model, we formulate the problem of cold starts as an optimisation approach to proactively spawn the required function instances and minimize frequent, on-demand cold starts.

As part of their Q-learning model, the study used fixed-value constants in the reward modelling and we address this issue by carefully analysing the problem and curate it as a threshold-based reward system. Additionally, we experiment with *matrix multiplication* function, that can be used as part of image processing pipeline, to train and evaluate our agent and utilise the open-sourced function invocation trace [28] by Azure. Further we describe our design decisions and utilise constants based upon the trial-error analyses. The successful learning of the agent resulted in the preparation of near to optimal function instances in a timeframe to reduce the on-demand function creation or cold starts and improve the platform's throughput. A summary of related works is presented in Table 1.

3 System Model and Problem Formulation

FaaS is an event-driven cloud service model that allows stateless function deployment. It delivers high scalability and scale-to-zero feature being economical to infrequent demand patterns. New functions n_i, where $1 \leq n_i \leq N$ and N is the maximum scale, are instantiated on-demand to serve the incoming load (scale up) and removed (scale down) when not in use after a certain time span or below a configured, resource-based threshold metric value for every i iteration window. The preparation time of function containers, i.e. cold start C_t, adds to the execution time of a request. These frequent on-demand cold starts result in an increased computation pressure on existing resources, neglecting expected average CPU utilisation (ϕ_o), and expected request failure rate (τ_o). Therefore, an intelligent, learning-based solution is proposed to address them.

In this study, we consider Kubeless, an open-source Kubernetes-native serverless platform that leverages Kubernetes primitives to provide serverless abstraction. It wraps function code inside a docker container with pre-defined resource requirements, i.e. $RR_f = (cpu_f, mem_f, tout_f)$ and schedules them on worker nodes. Similar to commercial FaaS providers, Kubeless has an idle-container window of 5 min to re-use functions and scales down to a minimum of one function if the collected metrics (default 15 s window) are below the set threshold. We take into account the general illustration of FaaS platform and consider a stochastic incoming request pattern $D = \{d_1, d_2, \ldots, d_i\}$ with d_i requests in ith iteration window. We analyse the request pattern for a timeframe T divided in i iteration windows of duration t_i. The system model of the examined scenario is depicted in Fig. 1. The workflow of the potential cold start is explained in Fig. 2.

3.1 Problem Formulation

We formulate the function cold start as an optimisation problem aimed at minimizing the number of cold starts (Eq. 1) by preparing required instances, beforehand and aid the agent in learning a policy to reduce the request failure rate while maintaining average CPU utilisation.

Table 1 Related work summary

Work	Name	Platform	Solution focus	Strategy	Application type
Vahidinia et al. [33]	–	AWS Lambda	Cold start latency	Optimising environments & function pinging	Concurrent & sequential CPU & I/O intensive
Xu et al. [36]	AWU & ACPS	Kubernetes	Cold start latency & resource wastage	Invocation prediction (LSTM) & Container pool	Function chain model
Lin and Glikson [15]	–	Knative	Cold start frequency	Container pool & pod migration	Single function model
Bermbach et al. [10]	Naïve, Extended & Global Approach	AWS Lambda, Apache Open Whisk	Cold start frequency	Orchestration middleware	Function chain model
Mohan et al. [22]	Pause Container Pool Manager	Apache OpenWhisk	Cold start latency	Container Pool	Function chain model
Solaiman and Adnan [31]	WLEC	OpenLambda	Cold start latency	Container Pool	Single function model
Mahajan et al. [18]	–	AWS	Cold start latency	Container migration & content similarity	Single function model
Schuler et al. [26]	–	Knative	Cold start frequency	AI-based container concurrency	Emulated CPU & I/O intensive
Silva et al. [30]	Prebaking	OpenFaas	Cold start latency	CRIU process snapshot	Single function model
Vahidinia et al. [34]	–	OpenWhisk	Cold start frequency	RL-based idle window & LSTM based container pre-warming	Single function model
Benedetti et al. [9]	–	OpenFaas	Function Scaling	RL & SLA-based configuration	Single function model
Our work	–	Kubeless	Cold start frequency	AI-based function & throughput metrics	Single function model

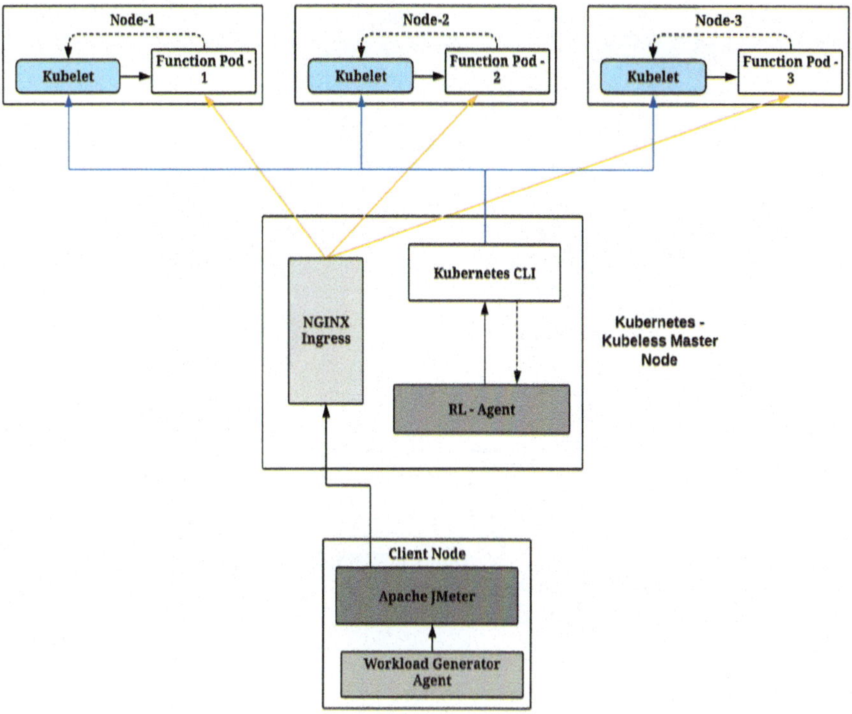

Fig. 1 System model

$$\min_{\phi,\tau,d_i} \quad (n_i) \tag{1}$$

such that

$$\tau_{d_i} < \tau_o; \ \phi_{d_i} < \phi_o \tag{2}$$

A cold start happens when there are no function instances available on the platform to deal with the incoming request and a new function instance is requested from the platform. FaaS services scale horizontally as per resource-based thresholds to be agile, usually considering the function's average CPU utilisation. Therefore, the goal of optimisation is to assess the incoming request pattern d_i for an application task in ith iteration window and configure a policy to prepare functions beforehand, considering actual and expected average CPU utilisation ($\phi_{d_i}\&\phi_o$) and request failure rate ($\tau_{d_i}\&\tau_o$). Since the preparation time, C_t remains similar for individual function containers, we focus on optimizing the frequency of cold start n_i for an individual iteration window.

With easy to implement and economical service model, enterprises are accommodating critical tasks like user verification, media processing, parallel scientific

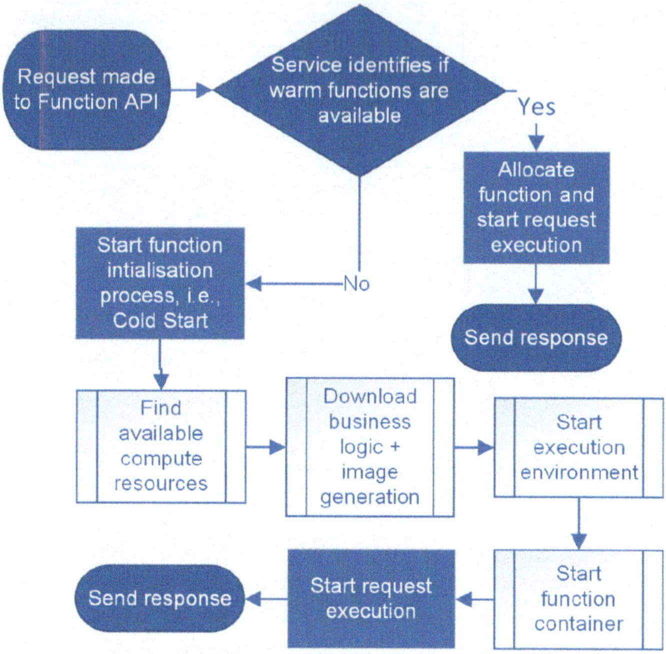

Fig. 2 Function warm start and cold start workflow

computations, anomaly detection and event-driven video streaming into the serverless paradigm. To assess the necessity of a dynamic solution, we consider *matrix multiplication* as workload, which is an important task in image processing workflow.

Reinforcement Learning Model In a model-free Q-Learning process, the agent learns by exploring the environment and exploiting the acquired information. The core components of the environment are state, action, reward and agent. The environment state represents the current visibility of the agent and is defined as a Markov Decision Process (MDP) [7, 32] where future environment state is independent of past states, given the present state information. Actions are the possible set of operations that the agent can perform in a particular state. Additionally, rewards are the guiding signals that lead the agent towards the desired goal by performing actions and transitioning between environment states. The agent maintains a Q-value table to assess the quality of action through obtained reward for the respective state and utilise it for future learning. Therefore, we propose a modelling scheme for the RL environment that is leveraged by a Q-Learning agent to learn a policy for function preparation.

We model the RL environment's state as $s_i = (\hat{n}_i, \phi_{d_i}, \tau_{d_i})$ where ϕ_{d_i} represents the average CPU utilisation of the available $\hat{n}_i$ function instances, τ_{d_i} represents the response failure rate, and i is the iteration window during a timeframe T. The agent's task is to prepare the estimated number of function instances in the upcoming iteration window either by exploring or exploiting the suitable actions. These actions of adding or adjusting the number of function instances, compensate for any expected cold starts from the incoming demand and help to improve the throughput of the system. Therefore, we define the agent's action as the number of function instances, n_i, to be added or removed from currently available functions $\hat{n}_{i-1}$ and represent it as a set $a_i = \{n_i | 1 \leq (\hat{n}_{i-1} + a_i) \leq N\}$. This heuristic helps the agent to control the degree of exploration by maintaining the number of functions within the threshold N, that is adapted based on deployed infrastructure capacity. Hence, we map the function resources and relevant metrics to RL environment primitives.

The motive of the RL-based agent is to learn an optimal policy, and we structure the rewards over resource-based metrics ϕ_{d_i}, function response failure rate τ_{d_i}, and expected threshold values (ϕ_o and τ_o). It evaluates the quality of action a_i in state s_i by keeping a value-based table, i.e. Q-table, that captures this information for every (s_i, a_i) pair. After executing the action, the agent waits for the duration of the iteration window and receives a delayed reward r_i, expressed based on the difference between the expected and actual utilisation and failure rate values, as shown in Eq. 3.

$$r_i = \frac{(\phi_o - \phi_{d_i}) + (\tau_o - \tau_{d_i})}{\hat{n}_i} \tag{3}$$

and the Q-table is represented as a matrix (Eq. 4) of dimension $S \times A$.

$$Q_{(S_n \times A_m)} = \begin{bmatrix} s_1, a_1 & \dots & s_1, a_m \\ \vdots & \ddots & \vdots \\ s_n, a_1 & \dots & s_n, a_m \end{bmatrix} \tag{4}$$

4 Q-Learning for Cold Start Reduction

In this work, we apply model-free Q-learning algorithm in FaaS paradigm to reduce frequent on-demand function cold starts. We select this algorithm due to its simple and easy implementation, model interpretability, strong theoretical convergence guarantees, ability to process the perceived information quickly using the Bellman equation and its adaptability to other advanced algorithms like Deep Q-Learning (DQN). As discussed in Sect. 3, we model the process of creating required function instances as an MDP and map the serverless computing primitives to RL agent's environment, state and actions. We explore and exploit an off-policy RL algorithm to reduce the on-demand function cold starts and determine the required function instances with the intuition of it being easy, simple to implement, less complex with stable convergence in a discrete action space. The proposed approach has two phases: an agent training phase and a testing phase. *Algorithm 1* demonstrates the agent training workflow. The environment setup process precedes the agent training, where the agent interacts with the environment and obtains information. After initial setup, the agent is trained for multiple epochs or timeframes where it assesses the function demand d_i over individual iteration windows i and ascertains appropriate function instances. During an iteration window i, the agent observes the environment state s_i, selects an action a_i according to ϵ-greedy policy. This greedy policy helps the agent to control its exploration and selects a random action with ϵ probability, otherwise exploiting the obtained information. This exploration rate is a dynamic value and decays with ongoing learning to prioritise the acquired information.

Algorithm 1 Q-Learning for Cold Start Reduction

Require: Initialise Environment variables
Ensure: Initialise Q-Table, $decayRate$, ϵ, $epoch$

$\epsilon = 0.01 + 0.99e^{(-decayRate \times epoch)}$
Repeat for each $TrainingEpoch$
$epoch \leftarrow epoch + 1$
while $t_{elapsed} < T$ **do**
  $s_i \leftarrow currentState(\hat{n}_i, \phi_{d_i}, \tau_{d_i}, i)$
  $a_i \leftarrow$ choose using ϵ-greedy policy from Q-Table
  Scale & wait for ith iteration window
  $r_i \leftarrow calculateReward(\phi_{d_i}, \tau_{d_i})$
  $s_{i+1} \leftarrow getNewState(\hat{n}_{i+1}, \phi_{d_{i+1}}, \tau_{d_{i+1}}, i+1)$
  $Q(s_i, a_i) \leftarrow (1-\alpha)Q(s_i, a_i + \alpha(r_i + \gamma \max_a Q(s_{i+1}, a_i))$
  $t_{elapsed} = t_{elapsed} + t_i$
end while

After performing the selected action, the agent waits for duration t_i of an iteration window to obtain the delayed reward r_i, calculated using the relevant resource-based metrics ϕ_{d_i} and function failure rate τ_{d_i}. This reward helps the agent in action quality assessment, and it combines the acquired knowledge over previous iterations using the Bellman Equation (Eq. 5). It is the core component in learning as it aids Q-value or Q-table updates and improves the agent's value-based decision-making capability. The equation uses two hyper-parameters learning rate, α and discount factor, γ. The learning rate signifies the speed of learning and accumulating new information, and the discount factor balances the importance of immediate and future rewards.

$$Q(s_i, a_i) = (1-\alpha)Q(s_i, a_i) + \alpha(r_i + \gamma \max_a Q(s_{i+1}, a_i)) \tag{5}$$

The agent then evaluates and adjusts the Q-value in Q-Table based upon the delayed reward for the corresponding (s_i, a_i) pair. The agent continues to analyse the demand over multiple iteration windows, selecting and performing actions, evaluating delayed rewards, assessing the quality of action and accumulating the information in Q-table, and repeating this process over multiple epochs for learning. Once the agent is trained for sufficient epochs and the exploration rate has decayed significantly, we can exploit obtained knowledge in the testing phase.

In the testing phase, the agent is evaluated using a demand pattern for the *matrix multiplication* function and the Q-table values guide the agent in taking informed actions. The agent determines the current environment state and obtains the best possible action, i.e. action with the highest Q-value for the corresponding state, and adjusts the required number of functions based on its understanding of the demand. We hypothesise that there exists a relationship between throughput and function availability to serve incoming requests. Therefore, we evaluate the agent's performance by considering metrics such as system throughput, function resource utilisation and available function instances. We further hypothesise that the RL-based agent learns to prepare and adjust required number of functions beforehand and improve the throughput while keeping the function's resource utilisation below the expected threshold.

5 Performance Evaluation

In this section, we provide the experimental setup and parameters, and perform an analysis of our agent compared to other complementary solutions.

5.1 System Setup

We set up our experimental test-bed as discussed in Sect. 3, using NeCTAR (Australian National Research Cloud Infrastructure) services on the Melbourne Research Cloud. We configure Kubernetes (v1.18.6) and Kubeless (v1.0.6) on a service cluster of 4 nodes, each with Ubuntu (18.04 LTS) OS image, 4 vCPUs, 16 GB RAM and 30 GB of disk storage to perform the relevant experiments. Typical serverless applications expect high scalability for their changing demands and can be compute-intensive, demanding a considerable amount of resources such as CPU, memory, or time to execute. These factors add to frequent cold starts on the platform by keeping the available functions or resources busy while requesting new functions for the subsequent workload demand. We use Python-based matrix multiplication (1024 pixels x 1024 pixels) to mimic the image processing task as our latency-critical application to deploy serverless functions.

The experimental setup mimics real-time application demand experienced in commercial FaaS platforms [11, 28]. We consider a single function invocation trace from the open-source Azure function data [28] and downsize it according to our resource capacity. We deploy the Apache JMeter load testing tool to generate the HTTP-based requests and randomize its request ramp-up period to guarantee the changing demand pattern for our workload. Also, we collect the relevant resource-based metrics and throughput information via Kubernetes APIs. Table 2 summarises the parameters used for system set-up.

Table 2 System setup parameter values

Parameter name	Value
Kubernetes version	v1.18.6
Kubeless version	v1.0.6
Nodes	4
OS	Ubuntu 18.04 LTS
vCPU	4
RAM	16 GB
Workload	Matrix Multiplication ($m \times m$)
m	1024

5.2 RL Environment Setup

To initialise the proposed RL-based environment, we first analyse and set up the function requirements according to deployed resource limits. After preliminary analysis, we configure the function requirements as 1 vCPU, 128 MB memory and 60 s function timeout, where timeout represents the maximum execution period for a function until failure. To experiment we assume a timeframe of 10 min to analyse the demand pattern of 100 requests during 5 iteration windows of 2 min. Based on the resource analysis and underlying Kubernetes assets we assume the function limit $N = 7$. These constraints allow us to put a considerable load or pressure on the different techniques discussed and effectively evaluate them against each other.

As discussed in Sect. 3, the RL-environment components depend upon resource metrics (average CPU utilisation), response failure rate, number of available functions and expected threshold values, summarized in Table 3. Since the proposed agent maintains a Q-table, these considerations help to minimise the risk of state-space explosion related to Q-Learning. The actions signify the addition or removal of functions based upon the function limit and the reward is modelled around the expected threshold values. We configure the Bellman Equation hyper-parameters: learning rate and discount factor as 0.9 and 0.99, respectively, based on the results of hyper-parameter tuning in [5]. The agent is structured to explore the environment and exploit the acquired knowledge. We use ϵ-greedy action selection policy to randomly select an action with initial $\epsilon = 1$ probability and exploit this information with a decay rate of 0.0025. These RL system parameter values were chosen after careful consideration of discussed workload and invocation pattern, according to the underlying resource capacity, and to showcase the applicability of RL-based agent in a serverless environment.

Table 3 RL-Environment parameter values

Parameter	Value
$cpu_f, mem_f, tout_f$	1, 128M, 60 s
N	7
T	10 min
i	5
t_i	2 min
ϕ_o	75%
τ_o	20%
α	0.9
γ	0.99
ϵ	1
$decayRate$	0.0025

5.3 Q-Learning Agent Evaluation

We train the RL-based agent for a timeframe of 10 min over 500 epochs to analyse an application demand and learn the ideal number of functions to reduce frequent cold starts. The agent is structured according to the RL-based environment design explained in Sect. 3 and around the implementation constraints. The quality of the RL-based agent is evaluated during a 2 h period to reduce the effect of any bias and performance bottlenecks.

We assess the effectiveness of our approach against the default scaling policy and commercially used function keep-alive policy on the serverless platform. Kubeless leverages the default resource-based scaling (HPA) implemented as a control loop that checks for the specific target metrics to adjust the function replicas. HPA has a default query period of 15 s to check and control the deployment based on the target metrics like average CPU utilisation. Therefore, the HPA controller fetches the specific metrics from the underlying API and calculates the average metric values for the available function instances. The controller adjusts the desired number of instances based on threshold violation but is unaware of the demand and only scales after a 15 s metric collection window. The expected threshold for function average CPU utilisation is set to be 75% with maximum scaling up to 7 instances. Therefore, whenever the average CPU utilisation of the function violates the threshold, new function instances are provisioned in real-time, representing a potential cold start in the system.

Also, HPA has a 5 min down-scaling window and during that period resources are bound to the platform irrespective of incoming demand which represents potential resource wastage. Therefore, it is worthwhile to analyse the performance of the RL-based agent against the function queuing or keep-alive approach that keeps enough resources bound to itself for an idle-container window.

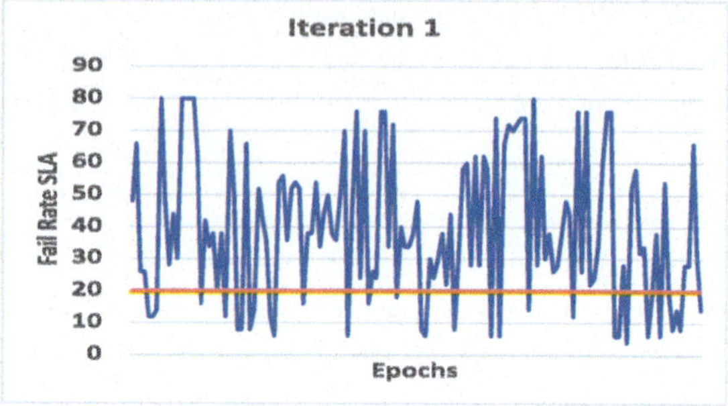

Fig. 3 Training iteration 1

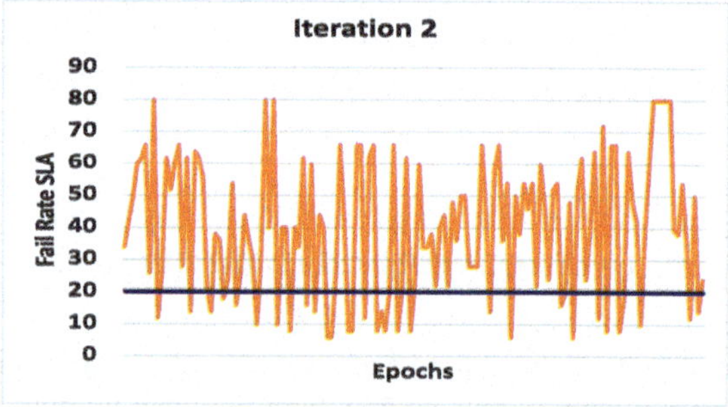

Fig. 4 Training iteration 2

Fig. 5 Training iteration 3

Figures 3, 4, 5, 6 and 7 illustrate the learning curve of the agent over multiple epochs and we observe that the agent continuously attempts to meet the expected thresholds. This highlights the agent's capability to obtain positive rewards and move towards the desired configuration. We compare the RL-based agent with HPA and successfully demonstrate the agent's ability to improve the function throughput, i.e. reduce the failure rate by up to 8.81%, Fig. 8. The RL-based agent further targets to maintain the expected CPU utilisation thresholds, Fig. 9, by reducing CPU stress up to 55% while determining the required function instances in Fig. 10. For example, in Fig. 10 during iteration windows 1 and 2, the HPA scales functions based on CPU utilisation threshold, unaware of the actual requirement for upcoming iteration and

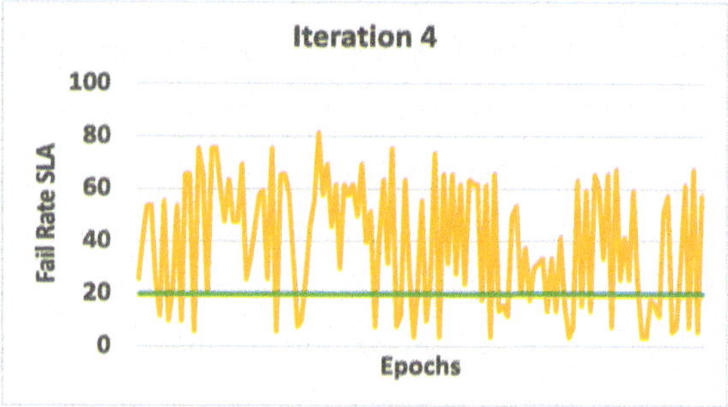

Fig. 6 Training iteration 4

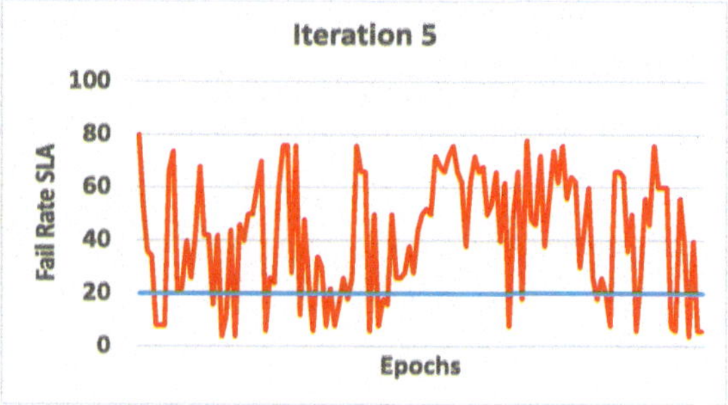

Fig. 7 Training iteration 5

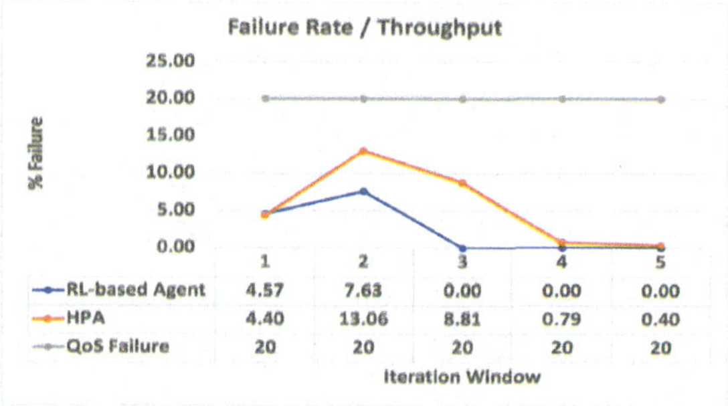

Fig. 8 RL agent v/s HPA: failure rate

results in resource wastage. Similarly, Fig. 10 illustrates the resource wastage by HPA during iterations 3 and 4.

Similar results are observed against function queuing or keep-alive policy, where we evaluate two queues with $N = 4$ and $N = 7$. The RL-based agent scales and prepares the function according to demand needs while the queue results in resource wastage of up to 37%, as shown in Fig. 11. Although the queuing policy manages to reduce the request failure rate to zero, it is due to extra resources available, as depicted in Fig. 12, but can not be precisely captured by HPA metrics and shows over CPU utilisation of up to 50% in Fig. 13. The proposed agent analysed the demand pattern by consuming sufficient function resources, preparing the ideal number of functions

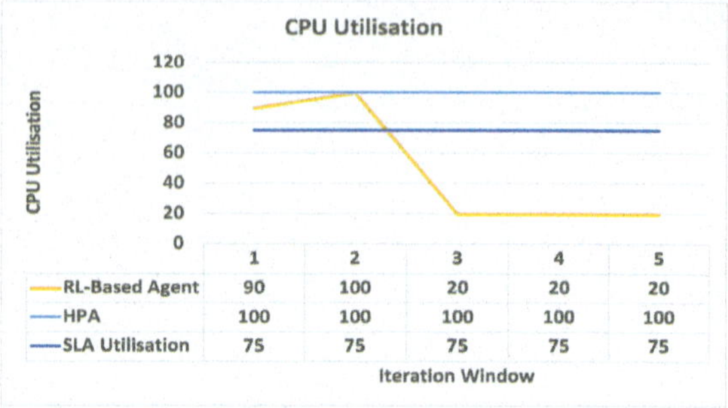

Fig. 9 RL agent v/s HPA: CPU utilisation

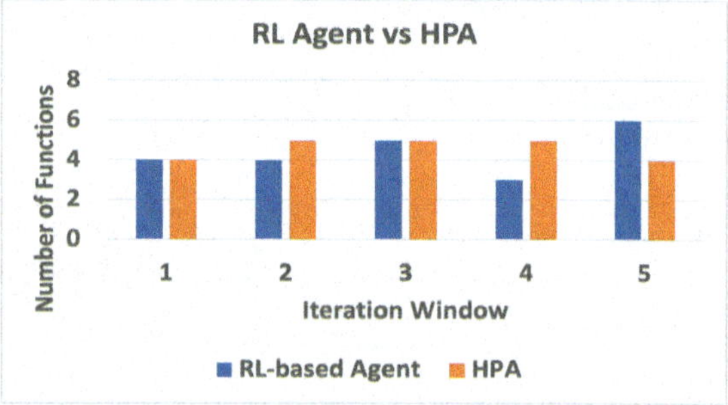

Fig. 10 HPA: function provision

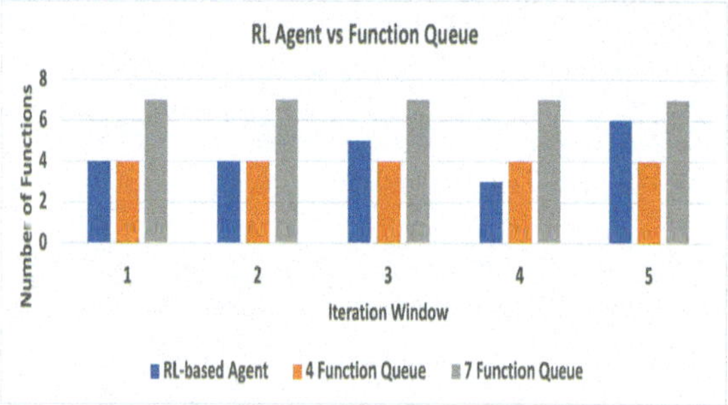

Fig. 11 Function queue: function provision

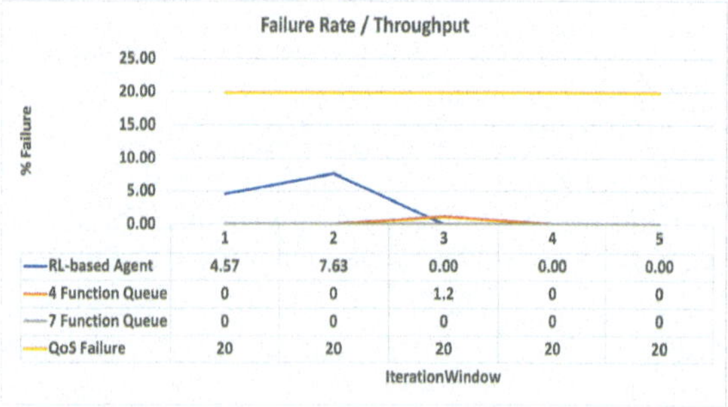

Fig. 12 RL agent v/s function queue: failure rate

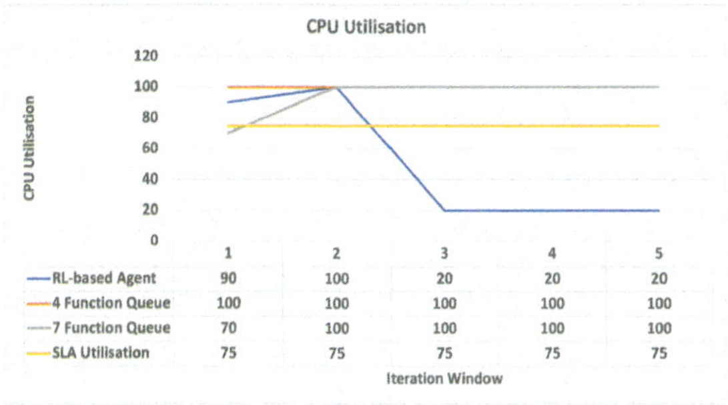

Fig. 13 RL agent v/s function queue: CPU utilisation

and trying to keep the desired CPU utilisation under control. Hence, the learning and testing analysis support our hypothesis that reducing on-demand cold starts can be directly linked to the throughput improvement.

6 Discussion

Function cold start is an inherent shortcoming of the serverless execution model. Thus, we have proposed an RL-based technique to investigate the demand pattern of the application and attempt to reduce the frequency of function cold starts. The proposed agent performs better than the baseline approaches under a controlled experimental environment. But there are certain points to recollect associated with the real-time appropriateness of the proposed solution.

We leverage the RL environment modelling, specifically Q-Learning constraints [32, 37], and in general, these algorithms are expensive in terms of data and time. The agent interacts with the modelled environment to acquire relevant information over multiple epochs that signify a higher degree of exploration. Hence, as evidenced in the proposed work, for an RL-based agent to outperform a baseline technique, a training period of 500 epochs is exploited for satisfactorily analysing the workload demand for a timeframe (10 min). Therefore, RL-based approaches are considerably expensive in practical applications with stringent optimisation requirements.

A classical Q-Learning approach is applied to discrete environment variables [32]. To constrain the serverless environment within the requirements of the Q-Learning algorithm, we consider the discrete variables to model cold starts. The size of the Q-table is large and is a function of state space and action space. But with the expansion of state space or the action space, the size of the Q-table grows exponentially [32, 37]. Therefore, Q-Learning experiences state explosion, making it infeasible to perform updates on Q-values and degraded space and time complexity.

The proposed agent analyses individual application demand, so the learning canâŁ™t be generalised for other demand patterns and requires respective training to be commissioned. Furthermore, the agent is trained for 500 iterations and evaluated, but the chance of exploring every state is bleak with limited iterations of training. Therefore, the agent expects to be guided by certain approximations to avoid acting randomly. The agent utilises resource-based metrics that affect the cold starts, so the availability of relevant tools and techniques to collect instantaneous metrics is essential. Also, the respective platform implementation of a serverless environment, such as metrics collection frequency, function concurrency policy and request queuing, can extend support to the analyses.

The difference between the approaches can be attributed to the following characteristics of the proposed RL-based agent-

1. The process of elimination of invalid states during the RL environment setup and lazy loading of Python, helps the agent to productively use the acquired information about the environment.
2. Although the RL-based agent outperforms HPA and function queue policy, there is a lack of function container concurrency policy. The CPU-intensive function workload is configured with an execution time of 60 s and thus affected by the concurrency control of the instance.
3. The composition of state space and reward function incorporates the effect of failures during the training, and therefore, the agent tries to compensate for the failures in consequent steps of learning by exploiting the acquired knowledge.

On the account of the performance evaluation results, we can adequately conclude that the proposed agent successfully outperforms competing policies for the given workload and experiment settings. We strengthen this claim by analysing the training and testing outcomes of the RL-based agent, focused on examining the workload pattern to reduce request failure which is a direct consequence of appropriate function instances representing reduced function cold starts.

7 Future Research Directions

As part of the future research in broader serverless computing domain, various aspects of resource management needs to be addressed. These could be broadly categorised under workload estimation and characterisation, resource scheduling, and resource scaling. Following the increased adoption of FaaS, challenges such as function cold start delay, co-located function interference, lack of QoS guarantees, user pricing model, runtime limitations and support for specialised hardware, resource efficiency and workload management emerge as concerning factors for the success of serverless computing [19]. With the changing dynamics of application workload, its resource demand and introduction of AI/ML models in real-time systems, there is need for adaptive and proactive methods to tackle inefficient resource management in serverless computing. Additionally, there is a need for autonomous resource management system to completely offload the function scheduling, resource allocation and resource scaling tasks to offer a true serverless experience for developers and utilise the provider resources to the maximum. However, a guarantee of QoS objectives is still warranted in terms of service latency, throughput, fault tolerance and cost incurred by the user. Other aspects of FaaS that require meticulous thinking with adoption in edge and fog computing environments are data privacy and data locality concerns that further reduces the flexibility of FaaS application model.

Specifically, to address the issue of function cold start delays, an impact of other important aspects such as function memory allocation, language runtime, deployment size, programming convention and function characteristics in conjunction with different techniques can be explored as discussed below:

- In the industrial works like [3, 4], an insight is provided for the serverless AWS Lambda functions and how an improvement can be made to its cold start latencies and operating costs with careful function initialisation phase consideration.
- An in-depth analysis of function fusion or monolithic function development on resource requirements and cold start of function workflows can also be explored with respect to the performance and run-time costs of the function.
- Furthermore, a trade-off analysis of techniques like function pre-baking, provisioned concurrency and reserved concurrency can also be valuable in performance optimisation of serverless function and associated application workflows.
- In addition to CPU-intensive functions, impact of cold start on different classes of functions like memory-intensive, I/O-intensive and network-intensive functions, utilised for application domains such as AI/ML model training and inference, and media processing, can also be explored.
- Consequently, a cost and performance analysis of RL-based agents can also be made for the function cold start reduction in these application cases. Similar to Q-Learning, the application of other policy-based techniques such as SARSA, which is known to converge faster than Q-Learning, can also be experimented with in the domain of the cold start problem. As an adaptation of Q-Learning, the proposed solution includes discrete values over continuous values for state representation. In this context, to avoid the problem of state space explosion, function approximation techniques such as DQNs, Proximal Policy Optimisation (PPO) and other complex deep learning methods such as Soft-Actor-Critic (SAC) or Recurrent models [6] can also be leveraged to estimate the information about optimal actions.
- To further explore the evolving technological space, an integration of Generative Adversarial Networks for synthetic training data generation or Large-Language-Models can be explored for pro-active cold start reduction using the historical data.

8 Summary and Conclusions

FaaS model executes the piece of code inside a container, known as a function and prepares new function containers on demand. New function containers undergo an initialisation process that puts together all the essential components before executing the function handler. This bootstrapping process consumes time in the order of a few seconds, known as function cold start and introduces a delay in the response of the function container.

This work visits the problem of function cold start by addressing frequent cold starts and analysing the application demand through an RL technique. We leverage the services of Apache JMeter to produce varying incoming request patterns and a CPU-intensive function workload to complement the invocation pattern and observe relevant cold starts. The system is set up using the Kubeless framework and the RL environment is modelled for the agent to examine the necessary metrics to make guided decisions in provisioning an appropriate number of function instances.

We present an evidence of leveraging Q-Learning to address cold starts on FaaS platforms and verify it with improved platform throughput, reduced resource wastage while maintaining expected thresholds, during the iterations. We evaluate the performance of our proposed agent against the HPA policy and function queue policy. We successfully observe the RL-based agent outperforming comparing techniques after a training of 500 epochs that verifies our hypothesis of strong association between success rate and reduced number of cold starts on the platform. After the test analyses, the Q-Learning agent successfully improves throughput by up to 8.81% and reduces resource wastage by up to 37% while preparing sufficient functions to reduce cold starts.

References

1. Kubeless-kubernetes native serverless. https://kubeless.io/docs/
2. Datadog, the state of serverless (2022). https://www.datadoghq.com/state-of-serverless-2022/. (June 2022)
3. Operating lambda: Performance optimization-part 1 (2021). https://aws.amazon.com/blogs/compute/operating-lambda-performance-optimization-part-1/. (May 2021)
4. Operating lambda: Performance optimization-part 2 (2021). https://aws.amazon.com/blogs/compute/operating-lambda-performance-optimization-part-2/. (May 2021)
5. Agarwal S, Rodriguez MA, Buyya R (2021) A reinforcement learning approach to reduce serverless function cold start frequency. In: Proceedings of the 21st IEEE/ACM international symposium on cluster, cloud and internet computing (CCGrid). IEEE, pp 797–803
6. Agarwal S, Rodriguez MA, Buyya R (2024) A deep recurrent-reinforcement learning method for intelligent autoscaling of serverless functions. IEEE Trans Serv Comput 1–12. https://doi.org/10.1109/TSC.2024.3387661
7. Ashraf M (2018) Reinforcement learning demystified: Markov decision processes. Towards Data Science
8. Baldini I, Castro P, Chang K, Cheng P, Fink S, Ishakian V, Mitchell N, Muthusamy V, Rabbah R, Slominski A et al (2017) Serverless computing: Current trends and open problems. In: Research advances in cloud computing. Springer, pp 1–20
9. Benedetti P, Femminella M, Reali G, Steenhaut K (2022) Reinforcement learning applicability for resource-based auto-scaling in serverless edge applications. In: Proceedings of the IEEE international conference on pervasive computing and communications workshops and other affiliated events (PerCom Workshops). IEEE, pp 674–679
10. Bermbach D, Karakaya AS, Buchholz S (2020) Using application knowledge to reduce cold starts in faas services. In: Proceedings of the 35th annual ACM symposium on applied computing, pp 134–143

11. Galstyan A, Czajkowski K, Lerman K (2004) Resource allocation in the grid using reinforcement learning. In: Proceedings of the third international joint conference on autonomous agents and multiagent systems, 2004. AAMAS 2004, vol 1. IEEE Computer Society, pp 1314–1315
12. Hellerstein JM, Faleiro J, Gonzalez JE, Schleier-Smith J, Sreekanti V, Tumanov A, Wu C (2018) Serverless computing: One step forward, two steps back. arXiv:1812.03651
13. Jonas E, Schleier-Smith J, Sreekanti V, Tsai CC, Khandelwal A, Pu Q, Shankar V, Carreira J, Krauth K, Yadwadkar N et al (2019) Cloud programming simplified: a berkeley view on serverless computing. arXiv:1902.03383
14. Lee H, Satyam K, Fox G (2018) Evaluation of production serverless computing environments. In: Proceedings of the IEEE 11th international conference on cloud computing (CLOUD). IEEE, pp. 442–450
15. Lin PM, Glikson A (2019) Mitigating cold starts in serverless platforms: a pool-based approach. arXiv:1903.12221
16. Lloyd W, Ramesh S, Chinthalapati S, Ly L, Pallickara S (2018) Serverless computing: an investigation of factors influencing microservice performance. In: Proceedings of the IEEE international conference on cloud engineering (IC2E). IEEE, pp 159–169
17. Lynn T, Rosati P, Lejeune A, Emeakaroha V (2017) A preliminary review of enterprise serverless cloud computing (function-as-a-service) platforms. In: Proceedings of the IEEE international conference on cloud computing technology and science (CloudCom). IEEE, pp 162–169
18. Mahajan K, Mahajan S, Misra V, Rubenstein D (2019) Exploiting content similarity to address cold start in container deployments. In: Proceedings of the 15th international conference on emerging networking experiments and technologies, pp 37–39
19. Mampage A, Karunasekera S, Buyya R (2022) A holistic view on resource management in serverless computing environments: taxonomy and future directions. ACM Comput Surv (CSUR) 54(11s):1–36
20. Manner J, Endreß M, Heckel T, Wirtz G (2018) Cold start influencing factors in function as a service. In: Proceedings of the IEEE/ACM international conference on utility and cloud computing companion (UCC Companion). IEEE, pp 181–188
21. McGrath G, Brenner PR (2017) Serverless computing: design, implementation, and performance. In: Proceedings of the 37th IEEE international conference on distributed computing systems workshops (ICDCSW). IEEE, pp 405–410
22. Mohan A, Sane H, Doshi K, Edupuganti S, Nayak N, Sukhomlinov V (2019) Agile cold starts for scalable serverless. In: Proceedings of the 11th USENIX workshop on hot topics in cloud computing (HotCloud 19)
23. Mohanty SK, Premsankar G, Di Francesco M et al (2018) An evaluation of open source serverless computing frameworks. in Proceedings of the IEEE international conference on cloud computing technology and science (CloudCom), pp 115–120
24. Rosenbaum E (2020) Next frontier in microsoft, google, amazon cloud battle is over a world without code. https://www.cnbc.com/2020/04/01/new-microsoft-google-amazon-cloud-battle-over-world-without-code.html. (Apr 2020)
25. Santos J, Wauters T, Volckaert B, De Turck F (2019) Towards network-aware resource provisioning in kubernetes for fog computing applications. In: Proceedings of the IEEE conference on network softwarization (NetSoft). IEEE, pp 351–359
26. Schuler L, Jamil S, Kühl N (2021) Ai-based resource allocation: Reinforcement learning for adaptive auto-scaling in serverless environments. In: Proceedings of the IEEE/ACM 21st international symposium on cluster, cloud and internet computing (CCGrid). IEEE, pp 804–811
27. Shafiei H, Khonsari A, Mousavi P (2019) Serverless computing: a survey of opportunities, challenges, and applications. ACM Comput Surv (CSUR)
28. Shahrad M, Fonseca R, Goiri Í, Chaudhry G, Batum P, Cooke J, Laureano E, Tresness C, Russinovich M, Bianchini R (2020) Serverless in the wild: characterizing and optimizing the serverless workload at a large cloud provider. In: Proceedings of the USENIX annual technical conference (USENIX ATC 20), pp 205–218
29. Shilkov M (2021) Comparison of cold starts in serverless functions across aws, azure, and gcp. https://mikhail.io/serverless/coldstarts/big3/ (Jan 2021)

30. Silva P, Fireman D, Pereira TE (2020) Prebaking functions to warm the serverless cold start. In: Proceedings of the 21st international middleware conference, pp 1–13
31. Solaiman K, Adnan MA (2020) Wlec: a not so cold architecture to mitigate cold start problem in serverless computing. In: Proceedings of the IEEE international conference on cloud engineering (IC2E). IEEE, pp 144–153
32. Sutton RS (1998) Reinforcement learning: past, present and future. In: Asia-Pacific conference on simulated evolution and learning. Springer, pp 195–197
33. Vahidinia P, Farahani B, Aliee FS (2020) Cold start in serverless computing: current trends and mitigation strategies. In: Proceedings of the international conference on omni-layer intelligent systems (COINS). IEEE, pp 1–7
34. Vahidinia P, Farahani B, Aliee FS (2022) Mitigating cold start problem in serverless computing: a reinforcement learning approach. Proc IEEE Internet of Things J
35. Wang L, Li M, Zhang Y, Ristenpart T, Swift M (2018) Peeking behind the curtains of serverless platforms. In: Proceedings of the USENIX annual technical conference (USENIX ATC 18), pp 133–146
36. Xu Z, Zhang H, Geng X, Wu Q, Ma H (2019) Adaptive function launching acceleration in serverless computing platforms. In: Proceedings of the IEEE 25th international conference on parallel and distributed systems (ICPADS). IEEE, pp 9–16
37. Zychlinski S (2019) Qrash course: reinforcement learning 101 deep q networks in 10 minutes. https://towardsdatascience.com/qrash-course-deep-q-networks-from-the-ground-up-1bbda41d3677. (10 Jan 2019)

Benchmarking Covariance Matrix Evolution Strategies on Autonomous Parking Navigation

Esther Aboyeji, Daison Darlan, Oladayo Solomon Ajani, and Rammohan Mallipeddi

Abstract This paper offers a comparative study of different Covariance Matrix Adaptation Evolution Strategy (CMA-ES) techniques, evaluating their performance in autonomous parking navigation and maneuvering scenarios. The two-stage trajectory optimization framework is employed, where the global path is first generated using the A-star algorithm, followed by local optimization using various CMA-ES-based algorithms. The study evaluates the performance of these algorithms under four parking mission cases, classified as either simple or difficult, with a planning horizon of four waypoints. The results indicate that while CMA-ES algorithms are effective, certain variants trade off performance for computational efficiency. The findings suggest that hybrid or ensemble versions of CMA-ES might offer improved solutions for optimization-based autonomous parking tasks. The analysis provides insights into the suitability of different CMA-ES variants, which can inform choices for specific autonomous driving applications.

Keywords Autonomous parking · CMA-ES · Trajectory optimization · Evolution strategy · Autonomous vehicles

1 Introduction

Two-stage Trajectory optimization for Autonomous Ground Vehicles (AGVs) parking navigation and maneuvering is gaining a lot of traction in the literature as a solution approach to the problem of autonomous parking [2, 5, 12]. This is because the framework addresses the problem of autonomous parking in a human-like scenario thus leading to increased accuracy, fast path or trajectory planning, and potential towards collision avoidance in the presence of dynamic or moving obstacles. Generally in the two-stage optimization framework, a global planner such as the well-known A star algorithm is first used to generate a feasible global path or waypoints

E. Aboyeji · D. Darlan · O. S. Ajani · R. Mallipeddi (✉)
Department of Artificial Intelligence, School of Electronics Engineering, Kyungpook National University, Daegu, South Korea
e-mail: mallipeddiram@gmail.com

A. C. Frery et al. (eds.), *Computational Intelligence and Data Analytics*, Lecture Notes on Data Engineering and Communications Technologies 236,
https://doi.org/10.1007/978-981-96-0451-7_2

from the initial point to the desired parking position. Consequently, the problem is transformed into a local planning issue, which will be systematically addressed in the subsequent stage. Specifically, the generated path serves as the initial input for the optimization process in the second stage [5, 6]. In some cases, the waypoints generated in the first stage are divided into several segments depending on the predefined planning horizon, and the beginning and end of each consecutive waypoint are used as the reference path to be optimized in the second stage [2, 12].

Although several works from the literature have demonstrated the suitability of using the A star algorithm in the first stage of the optimization, researchers have employed different algorithms in the second stage [1]. On a broad scale algorithms employed in the second stage can be classified as gradient and gradient-based algorithms. Although gradient-based techniques, including the Interior Point Method (IPM), Sequential Quadratic Programming (SQP), and Trust Region methods, have demonstrated success in various applications, they are known to encounter difficulties under certain conditions. For instance, [2] demonstrates that non-gradient-based methods encounter challenges when applied to long planning horizons. In this context, a long planning horizon denotes the quantity of waypoints used to establish the objective for each local motion planning task. For practical implementation, shorter planning horizons are often adequate for smaller vehicles, whereas larger vehicles may require extended planning horizons. As a result, non-gradient methods with proven scalability to different planning horizons are been promoted in the literature.

In the class of the non-gradient-based methods, algorithms such as PSO, GA, ES, etc have been employed in the literature. Recently [2] demonstrated the suitability of the covariance Matrix evolution strategy (CMA-ES) algorithm with an ensemble of mutations for autonomous parking navigation and maneuvering. intuitively the superior performance of the algorithm under varying planning horizons as demonstrated in the work is due to its highly sophisticated and well-crafted mechanisms to guide the evolution process. Furthermore, the performance of other algorithms, such as Differential Evolution (DE) and Particle Swarm Optimization (PSO), deteriorates as the planning horizon increases. The degradation arises because extending the planning horizon increases the problem's dimensionality, making it challenging for global search algorithms to refine the initial solutions offered by the global planner. Specifically, DE and PSO, by treating the problem as a global optimization task, can cause the initial solutions from the first stage of optimization to deviate further from the optimal solution, thereby making the search process more complex. Although CMA-ES-based algorithms are suitable for the second stage of the optimization process and different variants of the algorithm exist where each variant is either aimed at improved performance in terms of the underlying optimization, or reduction of computational complexity. However, there is no work that presents an analysis of CMA-ES variants on the autonomous parking navigation task. This is important to isolate the contribution of the different components featured in different CM-ES variants and to provide a basis for the choice of different variants of the algorithms under different user preferences in autonomous driving tasks.

Therefore this work presents a comparative analysis of different CMA-ES-based algorithms for Autonomous Parking Navigation and Maneuver. To perform the analysis, we leverage the same framework in [2] and employ four different parking mission cases where two can be classified as simple mission cases and the other two as difficult mission cases. In terms of the algorithm, we employ the original CMA-ES algorithm as well as four variants from literature [2, 9, 10]. All the analyses were performed for mission cases with a planning horizon of 4 which is believed to be relatively difficult for most optimization algorithms to solve however it is important in the context of long vehicles as well as predictive controllers as such Model Predictive Control (MPC). The results show that some improvement might be necessary to efficiently deploy CMA-ES for optimization-based autonomous parking. Furthermore, some of the algorithms trade-off performance for speed and hybrid or ensemble versions of CMA-ES might be more beneficial in autonomous parking and real-world optimization problems in general.

The subsequent sections of this paper are arranged as follows: Section 2 outlines the formulation of the parking trajectory optimization problem. Section 3 offers an overview of the two-stage optimization framework employed in this study. In Sect. 4, background on CMA-ES and some of its variants are presented while Sect. 5 presents the experimental setup and results. Finally, the conclusion and future works are presented in Sect. 6.

2 Autonomous Parking as an Optimization Problem

This Section presents a step-by-step formulation of the autonomous parking problem as an optimization problem. Additionally, the discretized form of the problem, necessary for solving it using any optimization algorithm, is thoroughly described.

2.1 *Optimal Control Formulation for Parking Maneuver Planning*

The underlying problem can be described as a typical Optimal Control Problem (OCP) as detailed in [12] where the dynamics of the system are expressed in terms of the dynamics of the vehicle under consideration.

Model of Vehicle The motion of the vehicle utilized in this study is characterized by a car-like kinematic model, as illustrated in Fig. 1. In this figure, the coordinates (x, y) represent the position of the rear wheel's central point. The parameters d, r, f, v, a, ψ, and δ correspond to the inter-axial distance, rear overhang, front overhang, velocity, acceleration, angular velocity, and steering angle, respectively. The vehicle's state can be encapsulated in a six-dimensional state vector $\dot{q}$, which is governed by a set of differential equations as presented in (1). Additionally, the mathematical relationship between two essential vehicle-dependent parameters, namely the turn radius ρ and the inter-axial distance, is defined as shown in (2).

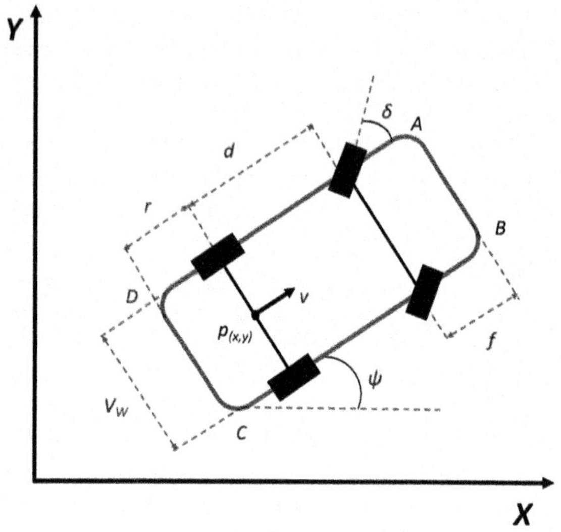

Fig. 1 Schematic diagram of a front-wheel steering vehicle

$$\dot{q} = \begin{bmatrix} \dot{x} \\ \dot{y} \\ \dot{\psi} \\ \dot{v} \\ \dot{\delta} \\ \dot{a} \end{bmatrix} = f(q, u) = \begin{bmatrix} v\,cos(\psi) \\ v\,sin(\psi) \\ \frac{v}{L}\,tan(\delta) \\ a \\ \frac{\delta_{ref}-\delta}{t_d} \\ \frac{a_{ref}-a}{t_a} \end{bmatrix} \tag{1}$$

$$\rho = \frac{d}{tan(\delta)} \tag{2}$$

As defined in Eq. (1), the actual control input vector u consists of the reference steering angle δ_{ref} and the reference longitudinal acceleration a_{ref}. This two-dimensional input vector is constrained within a set of bounds, defined by the upper limits $\delta_{ref,max}$ and $a_{ref,max}$, and the lower limits $\delta_{ref,min}$ and $a_{ref,min}$, as follows:

$$u = \begin{bmatrix} \delta_{ref} \\ a_{ref} \end{bmatrix}, \quad \begin{bmatrix} \delta_{ref,min} \\ a_{ref,min} \end{bmatrix} \leq u \leq \begin{bmatrix} \delta_{ref,max} \\ a_{ref,max} \end{bmatrix} \tag{3}$$

Although the The sideslip dynamics of the vehicle model are not considered in this analysis, the dynamics of the actuators are represented by first-order transfer functions with time constants t_d and t_a.

Objective Function To address the underlying OCP problem as an optimization problem, a fitness or objective function to be optimized needs to be defined. In this work, the fitness function is expressed in terms of steering the vehicle from an initial point to a desired reference state or location (q_{ref}). This can be expressed mathematically as follows:

$$h(q(T;\ t_0, q_0, u(\cdot))) = \left\| q_{ref} - q(T;\ t_0, q_0, u(\cdot)) \right\|_W \tag{4}$$

In the formulation, the initial time t_0 and state q_0 are specified with respect to a measurable control $u(\cdot)$. While Eq. (4) might suggest a convex nature of the fitness function with respect to $q(T)$, it is, in fact, a non-linear and non-convex function. This is because q is mapped to u by a non-linear model with respect to the minimization variable, which is the control history $u(t)$.

Parking Constraints Similar to several engineering problems, autonomous parking missions require some constraints to be satisfied based on the associated parking missions and vehicle dynamic model. As presented in [12], constraints related to Collision Avoidance, Comfort, and Safety as well as Performance need to be handled in this work. In the literature, these constraints are typically formulated as algebraic equalities or inequalities. Consequently, the collision avoidance constraints are expressed as

$$J_{col}(q(t)) = 0 \tag{5}$$

The constraints related to safety and comfort are defined as follows:

$$\begin{aligned} -5.0 \leq v(t) \leq 5.0 \quad & \forall\, t \in [0, T] \\ -1.0 \leq a_{long}(t) = a(t) \leq 1.0 \quad & \forall\, t \in [0, T] \\ -0.8 \leq a_{lat}(t) = \frac{v(t)^2}{\frac{L}{tan(\delta(t))}} \leq 0.8 \quad & \forall\, t \in [0, T] \\ -0.7 \leq j_{long}(t) = \frac{a_{ref}(t) - a(t)}{t_a} \leq 0.7 \quad & \forall\, t \in [0, T] \\ -0.3 \leq j_{lat}(t) = a_{\dot{lat}}(t) \leq 0.3 \quad & \forall\, t \in [0, T] \end{aligned} \tag{6}$$

where a_{long}, a_{lat}, j_{long}, and j_{lat} represent the longitudinal acceleration, lateral acceleration, longitudinal jerk, and lateral jerk, respectively. Finally, the performance criteria to be met are defined as follows:

$$\begin{aligned} \left\| x_{ref} - x(T) \right\| &\leq 0.1 \\ \left\| y_{ref} - y(T) \right\| &\leq 0.1 \\ \left\| \psi_{ref} - \psi(T) \right\| &\leq 0.1 \\ \left\| v_{ref} - v(T) \right\| &\leq 0.1 \\ \left\| a_{ref} - a(T) \right\| &\leq 0.1 \end{aligned} \tag{7}$$

where x_{ref}, y_{ref}, ψ_{ref}, v_{ref}, and a_{ref} denote the components of the reference state vector q_{ref}.

The comprehensive optimization problem, including the objective function and its corresponding constraints, can be summarized as follows:

$$\begin{aligned}&\text{minimize } J = h(q(T;\ t_0, q_0, u(\cdot)))\\&\text{subject to } \forall h(q) \in [t_0, q_0, T]\\&\qquad\text{Eq. (5) specifies the constraints for collision avoidance}\\&\qquad\text{Eq. (6) details the constraints concerning safety and comfort}\\&\qquad\text{Eq. (7) describes the constraints for performance-related objectives}\end{aligned} \tag{8}$$

2.2 *Discrete Representation of the Optimal Parking Model*

To solve the optimization problem presented in (8), it is necessary to discretize the dynamic model of the vehicle, the fitness function, and the constraints in relation to time, as it cannot be addressed directly in its continuous form [6]. This is usually done by breaking down the time interval $[t_0, T]$ into N_k discrete time segments. Consequently, the optimization task is reformulated to finding the optimal control values across these discrete time steps or segments to minimize the objective function (final state error) and ensure that the constraints are satisfied. The state of the vehicle is computed through numerical integration of the system dynamics. Thus, the problem can be re-expressed as follows:

$$\begin{aligned}&\text{minimize } J = h(q_{N_k}(T_{N_k};\ t_{N_k}, q_{N_k}, u(\cdot)))\\&\text{subject to } \forall h(q_k), k \in 0, 1, 2, ..., N_k - 1\\&\qquad\text{Eq. (5) specifies the constraints for collision avoidance}\\&\qquad\text{Eq. (6) details the constraints concerning safety and comfort}\\&\qquad\text{Eq. (7) describes the constraints for performance-related objectives}\end{aligned} \tag{9}$$

where the expressions for Eqs. (5), (6), and (7) are redefined in terms of N_K [12].

3 Two-Stage Optimization Framework

The two-stage optimization approach employed to solve the OCP is presented in this Section. In the motion planning process, two main stages are identified: parking navigation and parking maneuver. The initial stage addresses the movement from the vehicle's starting position to the region near the parking slot, while the subsequent stage involves the movement from this location to the designated parking spot or reference state. This separation of the planning process into distinct stages allows for tailored optimization of each stage, thereby enhancing the effectiveness of the overall parking strategy.

3.1 *Initial Waypoint Generation*

As outlined earlier, the initial phase of the optimization process involves generating waypoints, which are then used to reduce the problem to a local planning task. Previous studies have demonstrated that initial waypoints for the navigation and maneuver stages of the parking mission can be created separately through a path-planning algorithm. In our approach, we adopted the waypoint generation method from [12], which utilizes the A* search algorithm. These waypoints are subsequently employed to define the local planning problem for the second phase.

Depending on the chosen planning horizons, the generated waypoints are organized into several segments to define the local planning problem. The objective of this local planning task is articulated in terms of the waypoints, where the initial state for each segment is set as the final state of the preceding segment, and the terminal state is defined as the last waypoint within the current segment. The overall objectives and constraints for each local planning problem are formulated in accordance with the initial and final states of each segment, as outlined in [12]. This approach mirrors human strategies for parking, where, despite the final goal being the parking spot, a step-by-step method is employed to achieve each intermediate objective until the overall parking goal is reached.

4 Covariance Matrix Evolution Strategy and Variants

A Comprehensive tutorial detailing all the features of the standard CMA-ES algorithms can be found in [8]. Here we summarize the entire evolution process of CMA-ES as shown in Algorithm 1.

The process begins with setting up the strategy parameters and distribution means (L1), followed by generating λ candidate solutions through sampling from a Gaussian distribution with covariance matrix C and step-size σ (L3-8). In the selection and recombination stage, the best-ranked μ solution candidates referred to as elites are selected from the offspring and recombined based on the recombination weights to obtain a new distribution mean to drive the next generation (L9-11). Consequently, information regarding the nature of the problem landscape obtained from the offspring in the current generation is used to facilitate the step-size adaption process (L12-14) and the covariance matrix adaption process (L15-18) to obtain both C and σ for the next generation. From the above, it can be concluded that the success of the CMA-ES algorithm relies on the way new offspring are generated (mutation) and the well-crafted adaption of the strategy parameters which are relatively computationally expensive. Specifically, the eigen decomposition process required in the adaption of

Algorithm 1: Pseudocode of the CMA-Evolution Strategy algorithm

Initialize: Set parameters λ, $w_{i=1...\lambda}$, c_m, c_σ, d_σ, c_c, c_1, $and c_\mu$, $\boldsymbol{p}_\sigma \leftarrow 0$, $\boldsymbol{p}_c \leftarrow 0$, $\boldsymbol{C} \leftarrow \boldsymbol{I}$, and $g \leftarrow 0$, distribution mean $m \in \mathbb{R}^n$ and step-size $\sigma \in \mathbb{R}_{>0}$.

while *not done* **do**

- $g \leftarrow g + 1$
- **generate new offspring population**
- **for** $k = 1, ..., \lambda$ **do**
  - $z_k \sim N(0, I)$
  - $y_k = \boldsymbol{B}\boldsymbol{D}z_k \sim N(\boldsymbol{0}, \boldsymbol{C})$
  - $\boldsymbol{x}_k = \boldsymbol{m} + \sigma \boldsymbol{y}_k \sim N(\boldsymbol{m}, \sigma^2 \boldsymbol{C})$
- **end**
- **Elite selection and combination**
- $\langle \boldsymbol{y} \rangle_w = \sum_{i=1}^{\mu} w_i \boldsymbol{y}_{i:\lambda}$ where $\sum_{i=1}^{\mu} w_i = 1, w_i > 0 \;\; for \;\; i \in 1...\mu$
- $\boldsymbol{m} \leftarrow \boldsymbol{m} + c_m \sigma \langle \boldsymbol{y} \rangle_w$ equals $\sum_{i=1}^{\mu} w_i \boldsymbol{x}_{i:\lambda} \;\; if \;\; c_m = 1$
- **Mutation strength adaptation**
- $\boldsymbol{p}_\sigma \leftarrow (1 - c_\sigma)\boldsymbol{p}_\sigma + \sqrt{c_\sigma(2 - c_\sigma)\mu_{eff}}\boldsymbol{C}^{-\frac{1}{2}} \langle \boldsymbol{y} \rangle_w$
- $\sigma \leftarrow \sigma \times exp\left(\frac{c_\sigma}{d_\sigma}\left(\frac{\|\boldsymbol{p}_\sigma\|}{\boldsymbol{E}\|N(\boldsymbol{0},\boldsymbol{I})\|} - 1\right)\right)$
- **Covariance matrix adaptation**
- $\boldsymbol{p}_c \leftarrow (1 - c_c)\boldsymbol{p}_c + h_\sigma \sqrt{c_c(2 - c_c)\mu_{eff}} \langle \boldsymbol{y} \rangle_w$
- $w_i^\circ = w_i \times (1 \;\; if \;\; w_i \geq 0 \;\; else \;\; n/\left\|\boldsymbol{B}\boldsymbol{D}^{-1}\boldsymbol{B}^T \boldsymbol{y}_{i:\lambda}\right\|^2)$
- $C \leftarrow (1 + \underbrace{c_1\delta(h_\sigma) - c_1 - c_\mu \sum w_j}_{\text{usually equals to 0}})C + c_1 p_c p_c^T + c_\mu \sum_{i=1}^{\lambda} w_i^\circ \boldsymbol{y}_{i:\lambda} \boldsymbol{y}_{i:\lambda}^T$

end

the C matrix is asymptotically of the order $O(n^3)$. As a result, researchers have proposed different variants aimed at either reducing the overall computation complexity of the CMA-ES algorithm or improving the mutation mechanism to facilitate a better blend of exploration and exploitation.

In the context of this work, four CMA-ES variants have been selected. On one hand, the selection of the algorithms is because they cover the different main reasons why CMA-ES variants are being developed as mentioned earlier. On the other hand, most of the algorithms have featured in two-stage optimization of autonomous vehicle parking missions from literature [2]. Specifically, we select cCMA-ES [2], mCMA-ES [2], fMA-ES [10] and sCMA-ES [9].

cCMA-ES [2] is a variant of the standard CMA-ES where the only difference is that instead of generating new solution offspring based on Gaussian Mutation, the solution offspring are selected based on Cuahcy Distribution. From the literature, it is proven that the Cauchy distribution comes with the advantage of a better exploration compared to the Gaussian distribution [3].

mCMA-ES proposed in [2] differs from the original CMA-ES algorithm in that it combines the exploration and exploitation capability of Cauchy and Gaussian distribution respectively to realize an ensemble of mutation-based CMA-ES algorithm. In other words, the offspring are generated using both Cauchy and Gaussian distribution concurrently.

fMAES [10] is a CMA-ES variant based on d matrix adaptation evolution strategy (MA-ES) [4] where the covariance matrix adaption is reduced to a matrix adaption process thus reducing the computational complicity of the resulting algorithm. As an extension to MA-ES, fMA-ES features a fast (f) efficient M-update.

In sCMAES [9], a Reference Vector (RV)-based ranking strategy to calculate the mean and update the covariance matrix in CMA-ES is proposed. The proposed algorithm is designed specifically for constrained optimization problems and includes a repair mechanism that utilizes the Levenberg-Marquardt (LM) algorithm combined with a Broyden update to generate feasible alternatives for solutions that are initially infeasible.

Apart from sCMA-ES which features a dedicated constraint handling mechanism all the other algorithms are unconstrained. To apply them to the autonomous parking problem, we incorporate the superior of feasible constraints holding method [7, 11] to handle the underlying constraints.

5 Experimental Setup and Results

5.1 Experimental Configuration and Evaluation Metrics

Figure 2 presents a layout of 24 different parking missions of which 4 are employed in this work. In the figure, the numbers represent the featured parking spaces corresponding to each parking mission. As earlier mentioned, the analysis presented in this work is limited to four parking missions (1, 2, 15, and 24). The choice of these parking missions is motivated by the results of previous work where it was discovered that the problems could be classified as either easy or hard problems [2]. Hence we employ two easy and two hard problems for the analysis respectively. In Table 1, the values of the mission-related parameters such as the parking space length (P_{SL}), the parking space width (P_{SW}), etc are presented. Furthermore, the Table provides the bounds for the mission-related variables.

In the evaluation of optimization-based autonomous parking algorithms, performance is typically measured using metrics such as motion length, objective value, collision rate, and the satisfaction of mission performance constraints [5]. Additionally, this study includes algorithm runtime as a performance metric. The effectiveness of each algorithm is assessed by calculating the mean and standard deviation for each metric across six independent runs. For the collision rate metric, a result of zero (0) indicates no collisions, while one (1) signifies that collisions occurred. In contrast, the Mission Performance Constraints Satisfaction (MPCS) metric uses zero (0) to indicate that constraints were not met, and one (1) for when constraints are satisfied, with satisfaction determined as an objective function value of 0.1 or less, as outlined

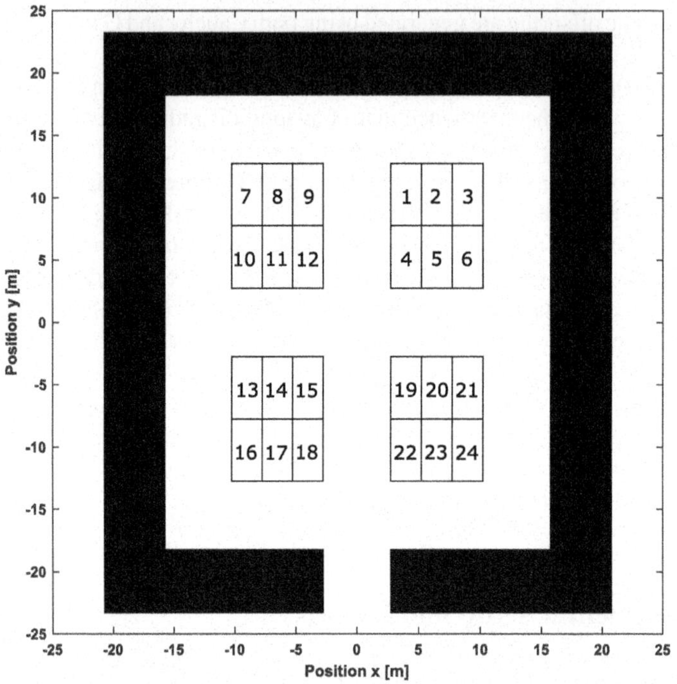

Fig. 2 The layout of the parking environment and 24 parking scenarios are illustrated, showcasing parking space dimensions of 2.5 m by 5.0 m and a lane width of 5.5 m

Table 1 Ranges of variables

Variables	Ranges	Variables	Ranges
P_{SL}, m	5	v, m/s	[–5.0, 5.0]
P_{SW}, m	2.5	a_{long}, m/s^2	[–1.0, 1.0]
P_{LW}, m	5.5	a_{lat}, m/s^2	[–0.8, 0.8]
p_x, m	[–15, 15]	δ, deg	[–37, 37]
p_y, m	[–25, 25]		

in [12]. All experiments were conducted in MATLAB on a 64-bit computer with an Intel(R) Core(TM) i9-11900 K processor at 3.50 GHz and 32 GB of RAM .

5.2 *Experimental Results*

Table 2 presents the summary of optimization results obtained from the five featured algorithms on the selected mission cases. From the results, it can be seen that both the standard CMA-ES and sCMA-ES fail totally on all four mission cases in terms of

Table 2 Results of CMA-ES and its variants on 4 selected autonomous parking mission cases in terms of average run time, objective value, collisions, and MPCS

Case	Metrics	Algorithms									
		CMA-ES		cCMA-ES		fCMA-ES		sCMA-ES		mCMA-ES	
		Mean	Std	Mean	Std	Mean	Std	Mean	Std	Mean	Std
1	Run Time	9.67E+02	2.39E+02	7.08E+02	1.53E+01	8.50E+02	2.04E+02	1.33E+03	1.75E+01	1.47E+03	3.68E+02
	Objective Value	2.07E-01	7.05E-02	2.32E-01	6.60E-02	1.57E+00	3.00E+00	1.02E+00	5.82E-01	**8.25E-03**	**1.25E-03**
	Collisions	1	0	0.833333	0.345033	5.00E-01	4.63E-01	1	0	**0**	**0**
	MPCS	0	0	0	0	1.67E-01	3.45E-01	0	0	**1**	**0**
2	Run Time	8.77E+02	1.13E+02	7.71E+02	1.09E+00	1.12E+03	2.16E+02	1.46E+03	2.96E+01	1.59E+03	3.94E+02
	Objective Value	1.57E-01	4.38E-02	5.16E-01	3.62E-01	1.90E-01	3.48E-01	2.50E+00	1.37E+00	**7.67E-03**	**9.44E-04**
	Collisions	1	0	1	0	8.33E-01	3.45E-01	1	0	**0**	**0**
	MPCS	0	0	0	0	1.67E-01	3.45E-01	0	0	**1**	**0**
15	Run Time	5.61E+02	6.07E+01	5.21E+02	6.41E-01	9.44E+02	2.15E+02	1.12E+03	3.71E+02	1.71E+03	3.20E+02
	Objective Value	2.68E-01	1.09E-01	2.52E-01	1.05E-01	1.60E-01	2.29E-01	8.66E-01	9.80E-01	**9.43E-03**	**4.73E-04**
	Collisions	1	0	0.666667	0.436436	5.00E-01	4.63E-01	1	0	**0**	**0**
	MPCS	0	0	0.333333	0.436436	3.33E-01	4.36E-01	0	0	**1**	**0**
24	Run Time	3.76E+02	6.26E+01	3.27E+02	5.09E-01	2.95E+02	2.79E+01	6.02E+02	2.32E+00	5.05E+02	2.29E+02
	Objective Value	1.69E+00	6.73E-01	1.57E+00	9.57E-01	8.28E-01	6.01E-01	5.23E+00	2.17E+00	**8.42E-03**	**1.26E-03**
	Collisions	1	0	1	0	8.33E-01	3.45E-01	1	0	**0**	**0**
	MPCS	0	0	0	0	1.67E-01	3.45E-01	0	0	**1**	**0**

all the metrics. For cCMA-ES it was successful in satisfying the mission constraints (collision and performance) in one run out of the 6 runs on mission case 1 and in two runs out of the 6 runs on mission case 15. It is important to note that both mission cases 1 and 15 are classified as easy mission cases. The performance of fCMA-ES is relatively better compared with CMA-ES, sCMA-ES, and cCMA-ES. Specifically, fCMA-ES archives a 50% success rate on mission cases 1 and 15 which are classified as easy mission cases. On mission cases 2 and 24 (which are classified as more difficult problems), fCMA-ES only succeeds in one out of the 6 independent runs. This slightly better performance of fCMA-ES could be attributed to its slightly loose matrix adaption instead of the covariance matrix adaption featured in the standard algorithm. mCMA-ES on the other hand satisfied all the mission constraints in all runs for all the mission cases. The good performance of mCMA-ES can be attributed to its featured efficient blend of mutation strategies resulting in an efficient blend of exploration and exploitation.

In terms of algorithm run-time, fCMA-ES outperforms all the other algorithms. This is expected as it features a dedicated mechanism to reduce the algorithm run-time complexity. Generally, although fCMA-ES as well as other algorithms such as CMA-ES, cCMA-ES, and sCMA-ES have better runtime than mCMA-ES this is not important because the performance of the algorithms is not good enough for any real implementation.

Figure 3d presents a graphical performance of each of the understudied algorithms on parking mission 24. From the Figure, it can be seen that each algorithm performs differently on the parking mission with sCMA-ES having the worst performance. The poor performance of sCMAES can be attributed to its inability to fine-tune the

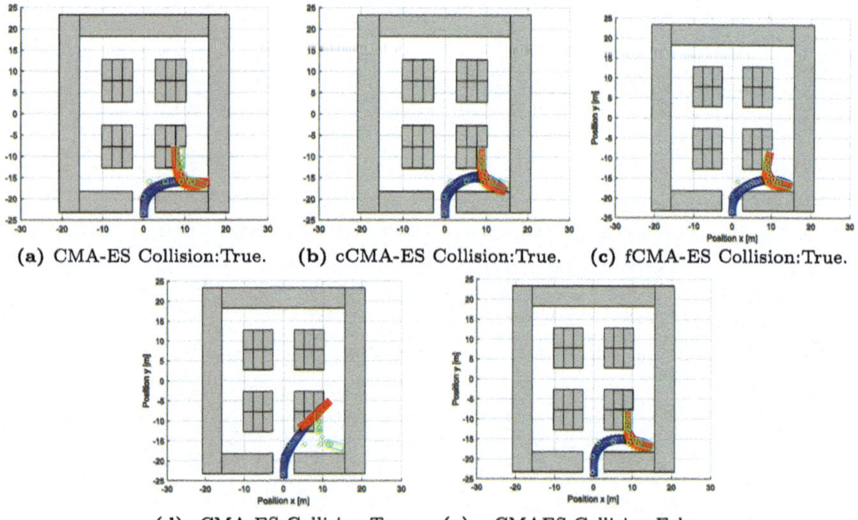

(a) CMA-ES Collision:True. (b) cCMA-ES Collision:True. (c) fCMA-ES Collision:True.

(d) sCMA-ES Collision:True. (e) mCMAES Collision:False.

Fig. 3 Performance comparison of different CMA-ES based algorithms on parking mission 24

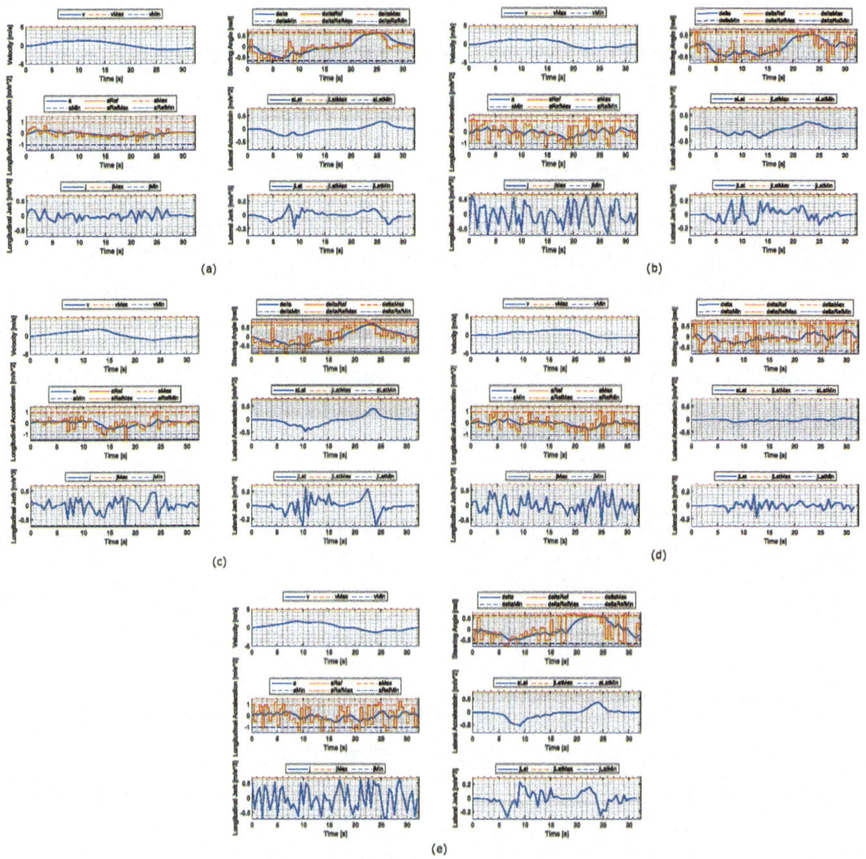

Fig. 4 Trajectories for parking mission 24 are compared across different algorithms using metrics such as velocity, steering angle, longitudinal acceleration, lateral acceleration, longitudinal jerk, and lateral jerk. The algorithms compared are: **a** CMA-ES, **b** cCMA-ES, **c** fCMA-ES, **d** sCMA-ES, and **e** mCMA-ES

initial waypoint generated in the first stage of the optimization rather it moved the initial solutions further away from the parking lot. Although other algorithms such as CMA-ES, cCMA-ES, and fCMA-ES were able to maintain the path generated in the first stage they were not able to satisfy the mission constraints even though they feature the same constraint handling technique as mCMA-ES.

In Fig. 4, the plot of trajectories obtained by each of the underlying algorithms are presented in terms of velocity, steering angle, lateral and longitudinal acceleration, and lateral and longitudinal jerk. The results supports the notation from previous works [2] that CMA-ES as well as other variants such as cCMA-ES, fCMA-ES and sCMA-ES features more of either exploration or exploitation rather than an efficient blend of exploration and exploitation as observed in mCMA-ES.

6 Conclusion

In this work, the standard CMA-ES algorithm as well as 4 other variants are evaluated on 4 different two-stage autonomous parking missions. The 4 parking missions can be classified as either hard or easy problems based on previous analysis from literature. The results of the analysis shows that an efficient blend of exploration and exploitation facilitated through ensemble of mutations featured in mCMA-ES provides a significant performance edge over other the standard CMA-ES as well as other simplified version of CMA-ES designed specifically to reduce its computational complexity.

Acknowledgements This research was supported by the Core Research Institute Basic Science Research Program through the National Research Foundation of Korea(NRF) funded by the Ministry of Education Korea (2021R1A6A1A03043144).

References

1. Aboyeji E, Ajani OS, Mallipeddi R (2023) On the scalability of parking trajectory optimization of autonomous ground vehicles. In: 2023 14th international conference on information and communication technology convergence (ICTC). IEEE, pp 344–349
2. Aboyeji ET, Ajani OS, Mallipeddi R (2024) Covariance matrix adaptation evolution strategy based on ensemble of mutations for parking navigation and maneuver of autonomous vehicles. Expert Syst Appl 249:123565
3. Ajani OS, Mallipeddi R (2022) Adaptive evolution strategy with ensemble of mutations for reinforcement learning. Knowl-Based Syst 245:108624
4. Beyer HG, Sendhoff B (2017) Simplify your covariance matrix adaptation evolution strategy. IEEE Trans Evol Comput 21(5):746–759. https://doi.org/10.1109/TEVC.2017.2680320
5. Chai R, Tsourdos A, Savvaris A, Chai S, Xia Y (2018) Two-stage trajectory optimization for autonomous ground vehicles parking maneuver. IEEE Trans Industr Inf 15(7):3899–3909
6. Chai R, Tsourdos A, Savvaris A, Chai S, Xia Y, Chen CP (2020) Multiobjective overtaking maneuver planning for autonomous ground vehicles. IEEE Trans Cybern 51(8):4035–4049
7. Deb K (2000) An efficient constraint handling method for genetic algorithms. Comput Methods Appl Mech Eng 186:311–338
8. Hansen N (2016) The cma evolution strategy: a tutorial. arXiv:abs/1604.00772, https://api.semanticscholar.org/CorpusID:15038271
9. Kumar A, Das S, Mallipeddi R (2022) A reference vector-based simplified covariance matrix adaptation evolution strategy for constrained global optimization. IEEE Trans Cybern 52(5):3696–3709. https://doi.org/10.1109/TCYB.2020.3013950
10. Loshchilov I, Glasmachers T, Beyer HG (2019) Large scale black-box optimization by limited-memory matrix adaptation. IEEE Trans Evol Comput 23(2):353–358. https://doi.org/10.1109/TEVC.2018.2855049
11. Mallipeddi R, Suganthan PN (2010) Ensemble of constraint handling techniques. IEEE Trans Evol Comput 14(4):561–579. https://doi.org/10.1109/TEVC.2009.2033582
12. Rocosa RC (2018) Motion planning for non-holonomic autonomous vehicles in parking spaces: an optimal control problem approach. Master's thesis, Delft University of Technology, Netherlands

Easy Accuracy Improvement of Fuzzy Modeling Using Available Datasets

Susana Muñoz Hernández

Abstract In recent years, we have witnessed the rise of brute force methods in the development of artificial intelligence applications. For decades, efforts in artificial intelligence have sought to emulate human reasoning to create "intelligent" reasoning models akin to human cognition. However, experts often struggle to identify the mental rules guiding their decisions, and not all the rules they believe they use are reliable. Conversely, testing all possible factor combinations to find correlations between concepts can be inefficient and may result in testing numerous absurd combinations. In this paper, we present a prototype proposal aimed at improving the accuracy of fuzzy modeling in two key aspects by analyzing available datasets. Firstly, it enhances the truth values of the definition of fuzzy concepts based on the distribution of available data. Secondly, it provides a more accurate credibility assessment for defined fuzzy rules according to their success with respect to previous datasets. Both improvements contribute to achieve more reliable and accurate predictions.

Keywords Fuzzy logic · Logic programming · Fuzzy modeling · Credibility · Data mining

1 Introduction

Fuzzy logic has been instrumental in managing uncertainty and imprecision across various domains. Fuzzy logic, first introduced by Zadeh in 1965 [1], offers a mathematical framework to handle the concept of partial truth, where the truth value may range between completely true and completely false [2]. This approach has been particularly useful in fields such as control systems, artificial intelligence, and

Work partially funded by the Grant PID2019-104735RB-C44 (SAFER project) by the Spanish Ministry of Science and Innovation.

S. Muñoz Hernández (✉)
Technical University of Madrid, Campus de Montegancedo, 28660 Madrid, Spain
e-mail: susana.munoz@upm.es
URL: https://babel.ls.fi.upm.es/~susana

A. C. Frery et al. (eds.), *Computational Intelligence and Data Analytics*, Lecture Notes on Data Engineering and Communications Technologies 236,
https://doi.org/10.1007/978-981-96-0451-7_3

decision-making processes, where binary logic is insufficient [3, 4]. In this research line, we also can find the concept of *rough sets* [5, 6]. This fact leads us to the fact of programming [7]. In this context, Prolog [8], a logical programming language, has been extended to incorporate fuzzy logic, leading to the development of frameworks like Fuzzy Prolog [9] and RFuzzy [10].

Fuzzy Prolog emerged as a powerful tool for integrating fuzzy logic into the traditional Prolog environment, enabling the handling of uncertainty directly within the logic programming paradigm [9, 11]. For instance, in medical diagnosis systems, fuzzy logic allows for the representation of symptoms and their degrees of severity, leading to more accurate diagnoses (i.e., [12]). Subsequent research by Muñoz-Hernández and Wiguna explored its application in dynamic environments like RoboCup soccer, demonstrating its capability to handle real-time decision-making under uncertainty [13].

In Fuzzy Prolog, a truth value is represented as a union of intervals of real numbers within the [0, 1] range (Borel algebra), whereas in RFuzzy [9], it is more simply represented as a single real number within the [0, 1] range. RFuzzy, developed by Muñoz-Hernández and colleagues, extends Fuzzy Prolog by introducing multi-adjoint logic [14], allowing for more sophisticated fuzzy reasoning capabilities as constructive negation [15] as priorities in the rule execution [16] and combining crisp and fuzzy logic [11]. Besides, it is simple to transform any dataset into a Prolog database[17] with the goal of easy fuzzy querying using fuzzy functions and fuzzy rule definitions [18–21].

1.1 Case Study: Diabetes Dataset

We use one of the datasets available in [22]. The dataset [23], originally compiled by the National Institute of Diabetes and Digestive and Kidney Diseases, is designed to diagnostically predict diabetes based on specific diagnostic measurements it contains. Several constraints were placed on the selection of these instances from a larger database. In particular, all patients here are females at least 21 years old of Pima Indian heritage.

The dataset consists of several medical predictor variables and one target variable, *outcome* (value 1 if diabetes, 0 otherwise). Predictor variables include the number of *pregnancies* the patient has had, level of *glucose*, *blood pressure*, *skin thickness*, *insulin* level, their *BMI*[1], *diabetes pedigree*[2], and *age*. The *csv* dataset can be easily converted into a list of Prolog facts by being uploaded in UFleSe [18–21].

Taking advantage of the web interface of UFleSe it is possible to define the limits of ranges of the definition of a list of Fuzzy functions that can be later used for diagnosis of diabetes for future patients. A set of fuzzy functions can be defined just identifying the break points of the function definition over a continues domain. Some examples of increasing functions are *many_pregnancies*, *high_glucose*, *high_insulin*, *obese*

[1] BMI value is multiplied by 10 to use integer numbers without loosing accuracy.

[2] Diabetes pedigree value is multiplied by 1000 to use integer numbers without loosing accuracy.

(defined over the values of BMI), and *old*. Some examples of decreasing functions are *few_pregnancies*, *low_glucose*, *low_insulin*, *low_bmi*, and *young* (defined over the values of age). Some examples of functions that represent middle values are *high_bmi* (but not obese) and *middle_age*. We show the simple syntax of some of their definitions:

```
many_pregnancies(diabetes) :~
   function(pregnancies(diabetes), [(3, 0), (5, 1)]).

few_pregnancies(diabetes) :~
   function(pregnancies(diabetes), [(3, 1), (5, 0)]).

obese(diabetes) :~
   function(bmi(diabetes), [(250, 0), (300, 1)]).

high_bmi(diabetes) :~
   function(bmi(diabetes), [(245, 0), (250, 1), (299, 1), (304,0)]).

low_bmi(diabetes) :~
   function(bmi(diabetes), [(184, 1), (187, 0)]).
```

Using the intuitive fuzzy criteria defined above it is possible to define rules for diabetes diagnosis taking into account one fuzzy criteria as the examples below. The problem is that we have no idea about the real credibility of each rule so we suppose they are completely reliable (credibility truth value 1 that we aggregate to the truth value of the body of the rule with the product as operator aggregator). We provide some examples to show the simple syntax:

```
ruleP(diabetes) :~     rule(many_pregnancies(diabetes))
                       with_credibility (prod, 1).

ruleG(diabetes) :~     rule(high_glucose(diabetes))
                       with_credibility (prod, 1).

ruleI(diabetes) :~     rule(high_insulin(diabetes))
                       with_credibility (prod, 1).

ruleOb(diabetes) :~    rule(obese(diabetes))
                       with_credibility (prod, 1).

ruleB(diabetes) :~     rule(high_bmi(diabetes))
                       with_credibility (prod, 1).

ruleOl(diabetes) :~    rule(old(diabetes))
                       with_credibility (prod, 1).
```

In the same way, we can define rules based on the combination of two, three, or more criteria with an aggregation operator (we have chosen the maximum operator) and we assign them as well total credibility. Any aggregation operator can be used (minimum, maximum, product, Łukasiewicz operator or any other that satisfies the properties of a lattice according to the multi-adjoint logic[14]). We provide below some examples:

```
rulePB(diabetes) :~   rule(max, (many_pregnancies(diabetes),
                                 high_bmi(diabetes)))
                      with_credibility (prod, 1).

ruleGOb(diabetes) :~   rule(max, (high_glucose(diabetes),
                                  obese(diabetes)))
                      with_credibility (prod, 1).

rulePBG(diabetes) :~  rule(max, (many_pregnancies(diabetes),
                                 high_bmi(diabetes),
                                 high_glucose(diabetes)))
                      with_credibility (prod, 1).

rulePObG(diabetes) :~ rule(max, (many_pregnancies(diabetes),
                                 obese(diabetes),
                                 high_glucose(diabetes)))
                      with_credibility (prod, 1).
```

2 Improving a Fuzzy Model Definition

Expert systems have long utilized fuzzy logic to handle complex decision-making processes where binary logic falls short. These systems emulate the decision-making ability of a human expert, using rules and knowledge bases to solve complex problems. By incorporating fuzzy rules, these systems can make more nuanced decisions that reflect real-world ambiguities.

Despite the advancements in Fuzzy Prolog and RFuzzy to provide a simple syntax that allows to be able to provide a fuzzy model easily, one of the persistent challenges is ensuring the precision of the definition of fuzzy functions and the accuracy of fuzzy rules. The complexity and variability inherent in real-world data can lead to inaccuracies in fuzzy modeling. Traditional methods often rely on the experts' knowledge to define fuzzy rules and membership functions, but this approach can be limited by the subjective nature of human expertise and the difficulty in capturing all relevant nuances. The proposal is to fix this subjectivity by fixing the accuracy of the fuzzy criteria definition and obtaining the credibility of the fuzzy rules by utilizing available information, aiming to improve their reliability and accuracy in prediction for practical applications.

Assigning credibility to fuzzy rules involves evaluating their performance over time and adjusting their influence accordingly. Historical data provides a rich source of information for this purpose, allowing the system to learn from past diagnosis and outcomes.

The main idea of our proposal is, in a first step, to use expert knowledge to provide an initial definition of fuzzy functions and fuzzy rules (with the simple syntax showed at the example in Sect. 1.1) that can help to model a fuzzy problem but pruning the amount of wrong or absurd options that could be tested without the guide of a professional. In a second step, it is possible to improve the accuracy of the fuzzy function definition by comparison with a normal distribution of available data, see Sect. 3. The third step would be to assign the corresponding credibility to the created fuzzy rules, according to its satisfaction with the available datasets, see Sect. 4.

3 Improving the Accuracy of Fuzzy Functions

The improvement of fuzzy function definition and the credibility of fuzzy rules can be done by leveraging historical data and incorporating techniques from machine learning and optimization; furthermore, it is possible to refine fuzzy models and enhance their predictive capabilities. The accuracy problem in fuzzy systems is particularly critical in applications where high precision is required, such as medical diagnosis and financial forecasting.

Improving the definitions of fuzzy functions involves refining the membership functions that represent fuzzy sets. Historical data plays a crucial role in this process, as it provides empirical evidence that can be used to adjust the shapes and boundaries of membership functions.

Different techniques can be used for improving the accuracy of fuzzy criteria definition. One of the simplest ways to improve the definition of functions that represent big, medium, or small values is the use of the normal distribution or the Gaussian bell. According to the set of values that we have in a dataset for a variable, e.g., price, size, etc., we can define the ranges of the domain of the variables that we desire to use to define the fuzzy functions that represent big, medium, or small values according to a normal distribution of the available values (see Fig. 1).

According to a normal distribution with average μ and typical deviation σ, a reasonable approximation is to consider high values the ones higher than $\mu + \sigma$ and not high at all the ones lower than $\mu + \sigma/2$ (see Fig. 2).

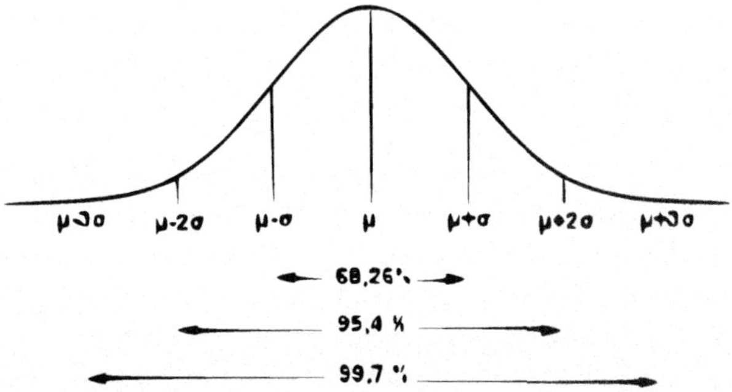

Fig. 1 Normal distribution representation of a dataset with average μ and typical deviation σ

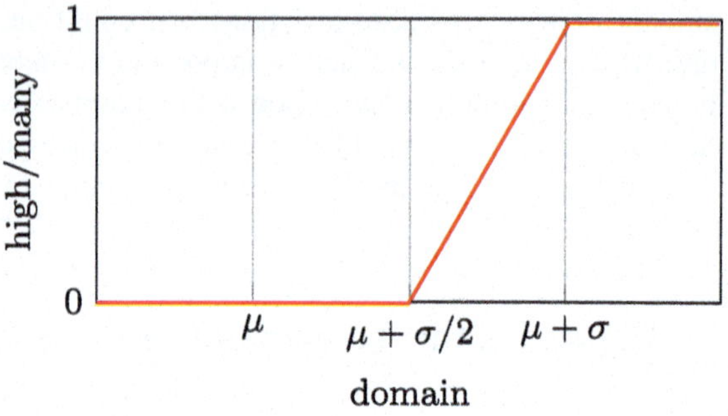

Fig. 2 Fuzzy function representing a high/many concept related to μ and σ values of a normal distribution

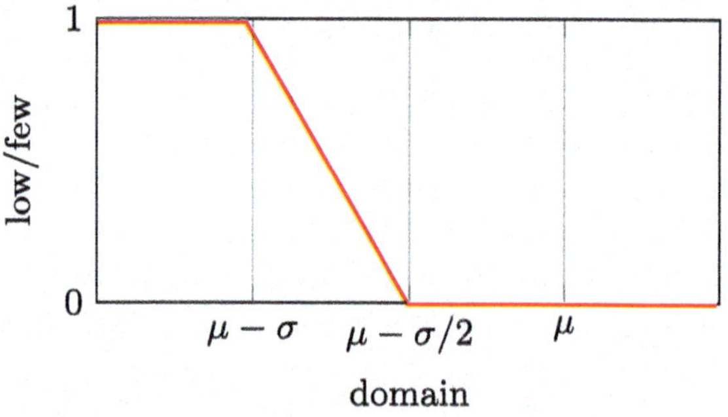

Fig. 3 Fuzzy function representing a low/few concept related to μ and σ values of a normal distribution

Symmetrically, we consider low values the ones lower than $\mu - \sigma$ and not low at all the ones higher than $\mu - \sigma/2$ (see Fig. 3).

3.1 Case Study: Improving Accuracy of Fuzzy Functions of a Diabetes Dataset

To continue with the case study presented in Sect. 1.1, we are going to improve the accuracy of some fuzzy function definitions. The breakpoints for defining '*obese*,' '*high_bmi*,' and '*low_bmi*' are standard values established by specialists. However, determining the limits for functions like '*many_pregnancies*' or '*few_pregnancies*' is more challenging, as these concepts depend on the specific cohort under consideration. Thanks to the dataset [23] provided by [22] there are available data of 768 cases of diabetes diagnosis in the particular cohort that we are modeling. We have used the normal distribution of data related to the variable that provides the number of pregnancies to adjust the break points of the definition of the fuzzy criteria defined over it by calculating its average or mean, μ, and its deviation, σ. According to Sect. 3, we obtain more accurate function definitions where the break points are more precise (according to a normal distribution of the dataset values):

```
many_pregnancies(diabetes) :~
```

```
    function(pregnancies(diabetes), [ (5.53, 0), (7.22, 1) ]).

few_pregnancies(diabetes) :~
    function(pregnancies(diabetes), [ (0.48, 1), (2.16, 0) ]).
```

In this case, the average, μ, is 3.85 pregnancies and the deviation, σ, is 3.37, so that the accuracy of the *many_pregnancies* function is improved by considering that less than 5.53 pregnancies ($\mu + \sigma/2$) are not many at all and more than 7.22 pregnancies ($\mu + \sigma$) are definitely many (for this population cohort). In a similar way, the accuracy of the *few_pregnancies* function is improved by considering that only no pregnancies are considered few pregnancies ($\mu - \sigma$) more than 2 pregnancies ($\mu - \sigma/2$) are definitely not few (for this population cohort).

4 Assigning Credibility to Fuzzy Rules Definitions

One of the significant challenges in expert systems is the need to prune the vast number of potential combination of data to arrive at the most reliable rules that are able to model the problem. Fuzzy logic helps in this process by assigning degrees of truth to various criteria, thereby narrowing down the options more effectively. Enhancing the credibility of fuzzy rules through historical data analysis further refines these systems, making their outputs more trustworthy and effective. This approach aligns with the ongoing efforts to integrate advanced fuzzy logic techniques into expert systems for better performance and reliability.

The problem that arises at this point, after dealing with the precision of the definition of fuzzy functions, is the credibility of the proposed fuzzy rules. We have implemented a simple algorithm in Prolog to compare two sets of normalized data (knowing that one contains the truth values of real data and another one contains the truth values that are assigned to the data using the fuzzy function definition) to see how the truth values deviate from the real data using MAE (Mean Absolute Error), a common metric used in machine learning.

In summary, we applied a criterion such as MAE to compare the results of the fuzzy function definitions in the modeling of a problem with the complete amount of real values that we have in the available corresponding dataset. We assign to each rule a value of credibility that is inverse to the MAE of its application to the whole available dataset.

4.1 Case Study: Credibility Assignment to Fuzzy Rules of the Diabetes Dataset

In the examples of rules of Sect. 1.1, we define a set of fuzzy rules for diabetes diagnosis based on one, two, or three fuzzy criteria. But, how good is this fuzzy

Table 1 Values of credibility of the rules defined for the diabetes data (I part)

Rule name	Aggregated fuzzy functions	Credibility
ruleP	*many_pregnancies*	0.6501864521696253
ruleG	*high_glucose*	0.5634765625
ruleI	*high_insulin*	0.6520833333333333
ruleOb	*obese*	0.5967968749999999
ruleB	*high_bmi*	0.5151041666666667
ruleOl	*old*	0.640625
ruleM	*middle_age*	0.6739583333333333

Table 2 Values of credibility of the rules defined for the diabetes data (II part)

Rule name	Aggregated fuzzy functions	Credibility
rulePG	*many_pregnancies, high_glucose*	0.564226993959566
rulePI	*many_pregnancies, high_insulin*	0.651228118836292
rulePOb	*many_pregnancies, obese*	0.5983433493589744
rulePB	*many_pregnancies, high_bmi*	0.518415618836292
rulePOl	*many_pregnancies, old*	0.6412020771696253
rulePM	*many_pregnancies, middle_age*	0.6708988227317555
ruleGI	*high_glucose, high_insulin*	0.5645182291666667
ruleGOb	*high_glucose, obese*	0.5216276041666666
ruleGB	*high_glucose, high_bmi*	0.45657552083333347
ruleGOl	*high_glucose, old*	0.5552083333333334
ruleGM	*high_glucose, middle_age*	0.5946614583333333
ruleIOb	*high_insulin, obese*	0.5973697916666665
ruleIB	*high_insulin, high_bmi*	0.5151041666666667
ruleIOl	*high_insulin, old*	0.6416666666666667
ruleIM	*high_insulin, middle_age*	0.6739583333333333
ruleObOl	*obese, old*	0.5902343749999999
ruleObM	*obese, middle_age*	0.6205989583333332
ruleBOl	*high_bmi, old*	00.5098958333333333
ruleBM	*high_bmi, middle_age*	0.53515625

model of the problem of diagnosis of diabetes? We were assuming credibility 1 for all the rules and we could aggregate this value with the truth value obtained for the rules using the maximum aggregation operator for example.

The accuracy of the model could be better if, besides improving the fuzzy functions definition (see Sect. 3), we assign an appropriate credibility of each rule. We have focus in the diagnosis of diabetes. We have defined in Sect. 1.1 a list of fuzzy functions defined over the available fields of a dataset of patients. Combining one (see Table 1), two (see Table 2), or three (see Table 3) of these functions we define model of

Table 3 Values of credibility of the rules defined for the diabetes data (III part)

Rule name	Aggregated fuzzy functions	Credibility
rulePBG	*many_pregnancies, high_bmi, high_glucose*	0.4636695636094675
rulePObG	*many_pregnancies, obese, high_glucose*	0.5266677453155818
rulePIM	*many_pregnancies, hight_insulin, middle_age*	0.6708988227317555

diagnosis based on a set of rules and using the simple algorithm described in Sect. 4 we calculate their credibility.

5 Conclusions

It is common to have an interest in modeling real-world problems that involve human fuzzy criteria. The problem is that people able to implement mathematical or programming models do not use to be the same people that are experts in the particular problem field. On the other hand, simple fuzzy syntax, as, e.g., RFuzzy syntax, can be easy for modeling but the problem is the model accuracy. In this paper, we have provided a case study very simple to show a significant improvement of the accuracy and reliability of a fuzzy model by using the data of the available dataset. We have shown an improvement of the accuracy of the fuzzy function definitions by using a Gauss bell distribution and we have assigned an appropriate credibility to the fuzzy rules provided for diabetes diagnosis according to their success with the dataset. The credibility of the rule is the inverse to the MAE of applying the rule to the elements of the dataset. Present work is related to include these improvements in a web interface tool to facilitate its use by researches and future work is related to test more advanced machine learning and AI techniques for the improvement of the accuracy of fuzzy definitions.

References

1. Zadeh LA (1965) Fuzzy sets. Inf. Control 8(3):338–353
2. Zadeh LA (2008) Is there a need for fuzzy logic? Inf Sci 178(13):2751–2779
3. Shapiro EY (1983) Logic programs with uncertainties: a tool for implementing rule-based systems. In: Proceedings of the eighth international joint conference on artificial intelligence, IJCAI'83, vol 1. Morgan Kaufmann Publishers Inc., San Francisco, CA, USA, pp 529–532
4. Skowron A, Dutta S (2018) Rough sets: past, present, and future. Nat Comput 17. https://doi.org/10.1007/s11047-018-9700-3
5. Skowron A, Dutta S (2018) Rough sets: past, present, and future. Nat Comput 17. https://doi.org/10.1007/s11047-018-9700-3
6. Vojtáš P (2001) Fuzzy logic programming. Fuzzy Set Syst 124(3):361–370
7. Vojtáš P (2001) Fuzzy logic programming. Fuzzy Set Syst 124(3):361–370

8. Li D, Liu D (1990) A fuzzy prolog database system. John Wiley & Sons Inc, New York, NY, USA
9. Guadarrama S, Muñoz-Hernández S, Vaucheret C (2004) Fuzzy prolog: a new approach using soft constraints propagation. Fuzzy Set Syst 144(1):127–150
10. Muñoz-Hernández S, Pablos-Ceruelo V, Strass H (2011) RFuzzy: Syntax, semantics and implementation details of a simple and expressive fuzzy tool over prolog. Inf Sci 181(10):1951–1970
11. Muñoz-Hernández S, Pablos-Ceruelo V, Strass H (2011) RFuzzy: Syntax, semantics and implementation details of a simple and expressive fuzzy tool over prolog. Inf Sci 181(10):1951–1970
12. Pablos-Ceruelo V, Muñoz-Hernández S (2011) Introducing priorities in RFuzzy: syntax and semantics. In: CMMSE 2011: proceedings of the 11th international conference on mathematical methods in science and engineering, vol 3. Benidorm (Alicante), Spain, pp 918–929, June 2011
13. Deedar MH, Muñoz-Hernández S (2018) Extending a flexible searching tool for multiple database formats. In: Emerging trends in electrical, communications, and information technologies. Springer
14. Deedar MH, Muñoz-Hernández S (2020) User-friendly interface for introducing fuzzy criteria into expressive searches. In: Intelligent systems and applications. Springer International Publishing, Cham, pp 982–997
15. Deedar MH, Muñoz-Hernández S (2019) Allowing users to create similarity relations for their flexible searches over databases. In: Artificial intelligence and soft computing. Springer International Publishing, Cham, pp 526–541
16. Deedar MH, Muñoz-Hernández S (2020) Uflese: user-friendly parametric framework for expressive flexible searches. Can J Electr Comput Eng 43(4):235–250
17. Li D, Liu D (1990) A fuzzy prolog database system. John Wiley & Sons Inc, New York, NY, USA
18. Vaucheret C, Guadarrama S, Muñoz-Hernández S (2002) LPAR, "Fuzzy prolog: a simple general implementation using CLP(R)". In: Baaz M, Voronkov A (eds) (LPAR). Lecture notes in artificial intelligence, vol 2514. Springer, pp 450–464
19. Deedar MH, Muñoz-Hernández S (2020) User-friendly interface for introducing fuzzy criteria into expressive searches. Intelligent systems and applications. Springer International Publishing, Cham, pp 982–997
20. Deedar MH, Muñoz-Hernández S (2019) Allowing users to create similarity relations for their flexible searches over databases. Artificial intelligence and soft computing. Springer International Publishing, Cham, pp 526–541
21. Deedar MH, Muñoz-Hernández S (2020) Uflese: user-friendly parametric framework for expressive flexible searches. Can J Electr Comput Eng 43(4):235–250
22. Dua D, Graff C (2019) UCI machine learning repository: pima Indians diabetes database. https://www.kaggle.com/datasets/uciml/pima-indians-diabetes-database?resource=download
23. Kaggle Datasets. https://www.kaggle.com/datasets

Machine Learning—Applications

An Efficient Approach to Forest Fire Segmentation in Satellite Images Using Superpixels Based Fuzzy Clustering and CIELch Color Model

P. Ganesan, L. M. I. Leo Joseph, B. S. Sathish, V. Elamaran, M. Rajendra Prasad, G. Sajiv, and S. Jency

Abstract Forest fires are a serious worry because of the havoc they do to the ecosystem. In addition to causing the destruction of invaluable plants and animals, forest fires seriously endanger both human life and property. Consequently, there is an increasing demand for precise and effective techniques for identifying and classifying forest fires in satellite imagery. Recent developments in technology have made it possible to separate forest fire zones in satellite images more precisely and efficiently by utilizing sophisticated image processing techniques. In this study, we offer an effective method for segmenting forest fires in satellite images that makes use of the advantages of fuzzy clustering, superpixels, and the CIELch color model. Superpixels make it possible to choose homogenous areas of the image, which can greatly increase the segmentation process' accuracy. This method, when integrated

P. Ganesan (✉) · M. R. Prasad
Department of ECE, Vidya Jyothi Institute of Technology, Hyderabad, India
e-mail: gganeshnathan@gmail.com

M. R. Prasad
e-mail: rajendraprasad@vjit.ac.in

L. M. I. L. Joseph
Department of ECE, SR University, Warangal, Telangana, India
e-mail: leojoseph.lmi@sru.edu.in

B. S. Sathish
Department of CSE AI, Sri Venkateshwara College of Engineering, Bengaluru, India
e-mail: sathish.bs_ca@svcengg.edu.in

V. Elamaran
Department of ECE, School of EEE, SASTRA Deemed University, Thanjavur, India
e-mail: elamaran@ece.sastra.edu

G. Sajiv · S. Jency
Saveetha School of Engineering, SIMATS, Chennai, India
e-mail: sajivg2003.sse@saveetha.com

S. Jency
e-mail: jencys9026.sse@saveetha.com

A. C. Frery et al. (eds.), *Computational Intelligence and Data Analytics*, Lecture Notes on Data Engineering and Communications Technologies 236,
https://doi.org/10.1007/978-981-96-0451-7_4

with fuzzy clustering, which enables the representation of partial memberships of pixels to various clusters, may efficiently capture the minute variations in the fire-affected regions, even in intricate and ever-changing environments such as forests. Furthermore, compared to conventional RGB or HSV models, the CIELch color model offers a more perceptually uniform color space, which makes it better suited for capturing the color features of forest fires in satellite images. As a result, segmentation outcomes are more precise and dependable since color information is essential for determining the boundaries and extent of the regions damaged by the fire. We have assessed the suggested method using a variety of forest fire satellite images. The results of the experiment show that, even under difficult environmental circumstances like varying illumination and the presence of smoke, it is successful in precisely detecting and demarcating the area of forest fire zones.

Keywords Forest fire · Segmentation · Superpixels · CIELch · SLIC · Clustering · MFCM

1 Introduction

Wildfires, also referred to as forest fires, are naturally occurring occurrences in regions of forests and wild lands that are highlighted by the fast and uncontrollable burning of vegetation. These fires have a significant impact on biodiversity, nutrient cycles, and vegetation patterns within ecosystems. While some wildfires are triggered by natural events like lightning strikes, many are ignited by human activity such as arson, industrial accidents, burning of agricultural land, and cigarette waste. Forest fires may have catastrophic effects on human settlements and the environment. In addition to destroying vegetation and causing soil erosion, wildfires often result in the loss of wildlife habitat and deterioration in air quality from smoke and particulate matter emissions. Furthermore, there are serious threats to people's health and safety from wildfires, especially when they threaten inhabited areas or vital infrastructure.

A major environmental problem facing the entire world, forest fires pose serious risks to human settlements, wildlife, and ecosystems. Due to human activity and climate change, wildfires are occurring more frequently and with greater severity. As a result, prompt and accurate identification of forest fire occurrences is now essential for efficient fire mitigation and management measures. With its wide coverage and regular updates, satellite imagery has proven to be an invaluable tool for tracking and identifying forest fires over vast geographic areas. However, because of a number of reasons, including the complexity of natural landscapes, the diversity of fire patterns, and the existence of confusing elements like smoke and shadows, the automated segmentation of forest fire zones in satellite pictures remains a tough challenge.

Conventional approaches to forest fire detection frequently depend on labor-intensive, error-prone manual thresholding procedures or the interpretation of satellite information. This problem has a possible solution in the form of automated segmentation techniques, which allow for quick analysis of satellite data and prompt identification of regions damaged by fires. In recent years, superpixels based image representation has become more and more common as a useful preprocessing step for a variety of computer vision applications, such as object detection and segmentation. Superpixels provide more accurate and efficient segmentation by reducing the dimensionality of the picture while retaining critical spatial information by clustering pixels into small and perceptually homogenous groups. By enabling the flexible assignment of data points to many groups, fuzzy clustering, on the other hand, expands on conventional clustering algorithms. For segmentation tasks, where pixels may belong to numerous classes or display unclear membership, this flexibility is very beneficial. Fuzzy clustering captures complicated spatial and spectral correlations by capturing the inherent uncertainty in image data, which yields more reliable segmentation results. We present the usage of the CIELch color model for forest fire segmentation in addition to superpixels based fuzzy clustering. Compared to more conventional color spaces like RGB or HSV, the CIELch color space, which relies on perceptual uniformity, offers superior color discrimination and greater consistency across various devices and viewing situations. Our goal is to improve the segmentation results by improving the separability of regions impacted by fire from the background through the transformation of satellite images into the CIELch space.

The rest of the article is structured as follows: An overview of relevant research on forest fire segmentation is given in Sect. 2. The approach, including the procedures for creating superpixels, fuzzy clustering, and color space in the translation, is presented in depth in Sect. 3. Experiments on benchmark datasets are presented in Sect. 4, and Sect. 5 discusses the main conclusions and revelations.

2 Related Work

Globally, forest fires are a major threat to biodiversity and ecosystems. Efficient fire control and mitigation tactics depend on the early identification and precise segmentation of forest fire zones in satellite images. Using advances in machine learning, image processing, and remote sensing, scholars have suggested a number of methods for segmenting and detecting forest fires in recent years. This study gives a broad overview of the body of research on the segmentation of forest fires, emphasizing the benefits and drawbacks of various strategies while establishing the background for the suggested methodology that is discussed in Table 1.

3 Materials and Methods

The work utilized simple linear iterative clustering (SLIC) [1, 2] for the selection of superpixels. This algorithm clusters all pixels into groups with the same attributes such as color [3]. The example for superpixels based segmentation depicted in Fig. 1. The input image and boundaries according to the amount of superpixels displayed on

Table 1 Methodologies for forest fire segmentation

Technique	Principle	Advantages	Limitations	Research findings
Deep learning (CNNs)	Learns complex features from labeled image data	High accuracy, handles variations	Large datasets, high computational cost	• Achieves high accuracy in benchmark datasets • Requires large-scale training for real-world scenarios
Machine learning (random forests)	Ensemble learning using decision trees	Efficient training, handles high-dimensional data	Lower accuracy compared to deep learning	• Offers good accuracy with lower computational demands • May struggle with highly complex fire patterns
Thresholding techniques	Classifies pixels based on a predefined threshold	Simple to implement, fast processing	Prone to noise and illumination variation	• Suitable for real-time applications • Limited accuracy in complex backgrounds
Spectral indices based methods	Utilizes spectral reflectance properties of fire	Targets unique spectral signature of fire	Limited accuracy in complex backgrounds	• Effective for initial fire detection • Prone to false positives in areas with similar spectral signatures
Fuzzy logic based methods	Incorporates uncertainty in pixel classification	Handles ambiguity in fire pixel characteristics	Requires expertise for defining fuzzy rules	• Can handle variations in fire pixel intensity • Subjectivity in defining fuzzy rules can impact accuracy

(continued)

Table 1 (continued)

Technique	Principle	Advantages	Limitations	Research findings
Level set segmentation	Evolves a curve to encompass the fire region	Accurate segmentation of irregular shapes	Requires careful parameter tuning	• Achieves good segmentation boundaries • Sensitive to parameter selection and initialization
Active contours based methods	Deformable curves guided by image features	Handles complex fire shapes	Sensitive to initial contour placement	• Can capture intricate fire boundaries • Initial contour placement can significantly impact segmentation
Texture Analysis Methods	Captures spatial distribution of pixel intensities	Identifies fire based on distinct texture patterns	Sensitive to image resolution and fire size	• Effective for capturing fire texture • Limited robustness to variations in fire size and resolution
Subtractive clustering techniques	Iteratively removes non-fire pixel clusters	Efficient for large datasets	May not capture complex fire characteristics	• Can handle large datasets efficiently • Might struggle with segmenting fires with non-uniform texture

Fig. 1a, b respectively. Figure 1c displays the outcome of the process after the color of every pixels modified based on the mean RGB values of the particular superpixel cluster. Superpixels are small and intuitively homogenous image areas created by clustering pixels according to related criteria such as hue or intensity [4]. Some interesting features of superpixels includes

- They share comparable features including color, intensity, and texture.
- They are spatially connected, resulting in smooth and cohesive structures [5].
- They are small as well as clearly established reducing intra-region variation while increasing contrast of inter-region [6].
- They minimize the amount of computation required for grouping pixels into homogenous groups, leading to improved processing effectiveness [7].

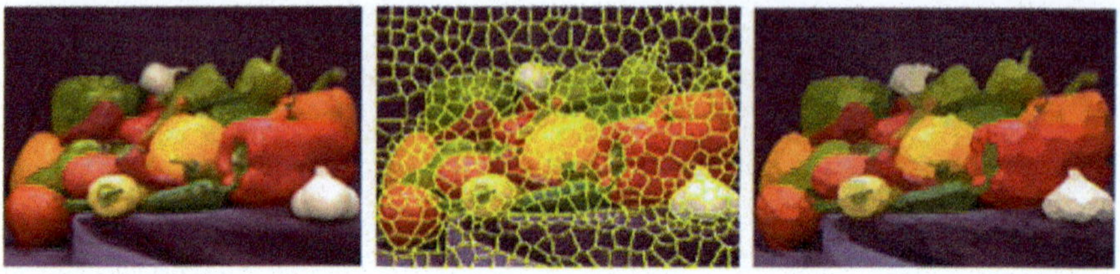

Fig. 1 An example for superpixels based segmentation

The amount of Superpixels utilized in the segmentation procedure has a noteworthy impact on the computational efficiency and quality of the segmentation outcome as shown in Table 2.

The planned methodology to segment forest fires in satellite images is illustrated in Fig. 2. The framework for forest fire segmentation is explained as follows. The work utilized a collection of satellite forest fire images acquired from https://earthobservatory.nasa.gov/images. This site is a huge collection of satellite images for scientific research. Preprocessing includes resizing, noise removal, correcting for atmospheric effects, and enhancement methods like radiometric and atmospheric correction. In this work, resizing is applied to improve the computational efficiency. Now the image in RGB should transform to CIELch model. This transformation is crucial for providing a precise, uniform, and intriguing color depiction across various devices and applications. It is essential for color management, analysis, and visualization. CIELch is a viable option for identifying fire regions in images owing to its perceived consistency and clear demarcation between intensity (L) and chrominance (c and h) components. The image first transforms into XYZ space using (1).

$$\begin{bmatrix} X \\ Y \\ Z \end{bmatrix} = \begin{bmatrix} 0.4125 & 0.3576 & 0.1805 \\ 0.2127 & 0.7152 & 0.7218 \\ 0.0194 & 0.1192 & 0.9504 \end{bmatrix} \begin{bmatrix} R \\ G \\ B \end{bmatrix} \quad (1)$$

Table 2 The impact of superpixels in segmentation process

No. of superpixels	Effect on segmentation	Advantages	Limitations
High (more superpixels)	High computational cost	Captures finer details within objects	May lead to over-segmentation, especially for smaller objects
Medium (balanced)	Good balance between efficiency and detail	Captures essential object features	May miss some intricate details
Low (less superpixels)	Low computational cost	Faster processing	May miss important details within objects and lead to under-segmentation

Then XYZ image was translated into CIELab space using (2) and then transformed to CIELch as shown in (3). Here we utilize D65 illuminant, i.e., the value of X_n, Y_n and Z_n are 0.9505, 1.0000 and 1.0888 respectively

$$\begin{aligned} L &= 116 * f(Y/Y_n) - 16 \\ a &= 500 * f(X/X_n) - f(Y/Y_n) \\ b &= 200 * f(Y/Y_n) - f(Z/Z_n) \end{aligned} \tag{2}$$

$$\begin{aligned} L &= L \\ C &= \sqrt{a^2 + b^2} \\ H &= \frac{180}{\pi}\left\{\pi + \arctan\left(\frac{b}{a}\right)\right\} \end{aligned} \tag{3}$$

Fig. 2 The planned methodology to segment forest fires

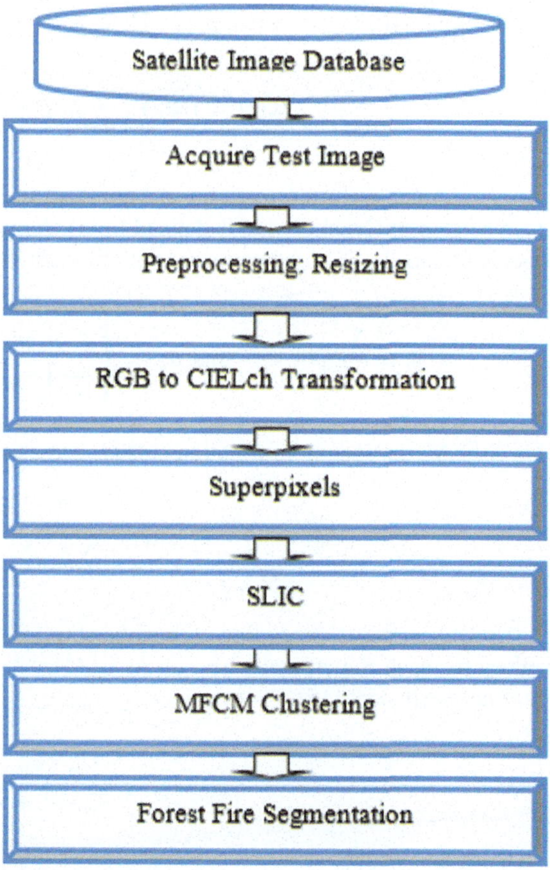

As a next vital step, the CIELch image clustered into superpixels. The amount of superpixels is the deciding factor for the efficiency and quality of segmentation result as discussed in Table 2. Superpixels combine nearby pixels with comparable properties, lowering computing complexity while preserving spatial information. This work employed SLIC technique to segment images into tiny, intuitively uniform regions. SLIC utilizes a grid structure to generate superpixels and iteratively progress superpixels based on color similarity and pixel proximity. The important characteristics from each Superpixel, including texture descriptions, color histograms, and spatial properties are retrieved and generate spectral and spatial feature vectors for fire-affected locations in each Superpixel.

Now, as a segmentation process employ the modified fuzzy c-means (MFCM) clustering technique to categorize superpixels according to the extracted characteristics. MFCM enables a superpixel to be partially classified into numerous clusters, representing the inherent ambiguity of fire pixel classification [8, 9]. It progressively optimizes an objective function (OF) that calculates the distance between superpixels and centroid [10, 11] as shown in (4). Here, m (weighting factor) is 2

$$J_m(U, V) = \sum_{i=1}^{c}\sum_{k=1}^{n} \mu_{ik}^{m} \|x_k - v_i\|^2 \tag{4}$$

The number of clusters (e.g., fire and non-fire) and a fuzzification value to manage membership uncertainty [12] is defined in (5). Assign fuzzy membership values to each Superpixel using predetermined initially centroids [13]. Repeat till convergence.

$$V_i = \frac{\sum_{k=1}^{n} \mu_{ik}^{m} X_k}{\sum_{k=1}^{n} \mu_{ik}^{m}} \text{ and } \mu_{ik} = \left\{ \sum_{j=1}^{c} \left\{ \frac{\|x_k - v_i\|}{\|x_k - v_j\|} \right\}^{2/(m-1)} \right\}^{-1} \tag{5}$$

For MFCM, the spatial information [14] is characterized as in (6)

$$S_{ij} = \sum_{k \in W(X_j)} U_{ik}\alpha_{k1} + \frac{\sum_{k \in (X_j)} U_{ik}\alpha_{k2}}{\sum_{t=1}^{c} \sum_{k \in W(X_j)} U_{tk}} \tag{6}$$

Update centroids depending on membership values and update the membership values for each Superpixel based on its distance from the cluster centers and the fuzziness exponent [15].

$$U_{ij(new)} = \frac{U_{ij}^{p} * S_{ij}^{q}}{\sum_{k=1}^{c} U_{kj}^{p} * S_{kj}^{q}} \tag{7}$$

The weightage for each cluster computed is depicted in (8) and OF modified as (9)

$$W_{ji} = \frac{1}{1 + e^{-\left\{\frac{\|x_j - v_i\|^2}{\sum_{j=1}^{n} \|x_j - v_i\|^2 \left(\frac{c}{n}\right)}\right\}}} \tag{8}$$

$$J_{Mod} = \sum_{k=1}^{n} \sum_{i=1}^{c} \left(U_{ik}^{m} W_{ji}^{m}\right) \|X_k - V_i\|^2 \tag{9}$$

4 Results and Discussion

The efficiency of the proposed method of forest fire region segmentation is evaluated using the satellite images collected from https://earthobservatory.nasa.gov/images. The forest fire image of south-central Chile shown in Fig. 3 is captured by MODIS on NASA Aqua satellite on 03.03.2023.

Now the image in RGB is transformed to CIELch model which is illustrated in Fig. 4a. This transformation is crucial for providing a precise, uniform, and intriguing color depiction across various devices. It is essential for color management, analysis, and visualization. CIELch is a viable option for identifying fire regions in images owing to its perceived consistency and clear demarcation between intensity (L) and chrominance (c and h) components. The work utilized SLIC for the selection of superpixels. The amount of Superpixels utilized in segmentation procedure has noteworthy impact on the computational efficiency and quality of the segmentation outcome. Initially, we opted with 10,000 Superpixels which is shown in Fig. 4b.

As a clustering process, MFCM is applied to categorize superpixels according to the extracted characteristics. MFCM enables a superpixel to be partially classified into numerous clusters, representing the inherent ambiguity of fire pixel classification. Here, the image is clustered into six regions for the clear demarcation of fire and non-fire regions. The clustering is based on color map shown in Fig. 5.

Fig. 3 The test image

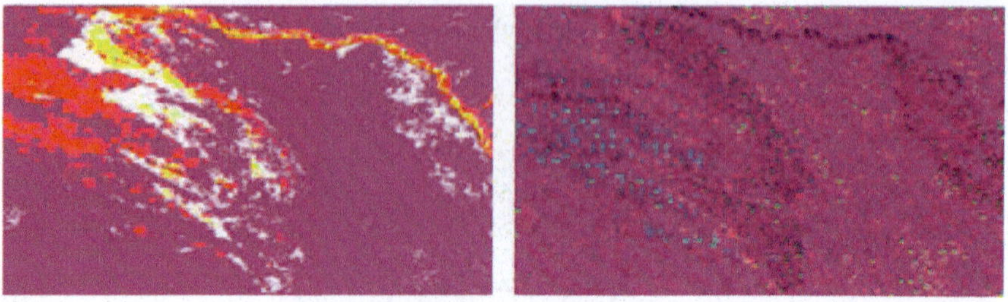

Fig. 4 The image in CIELch and number (10,000) of superpixels for segmentation

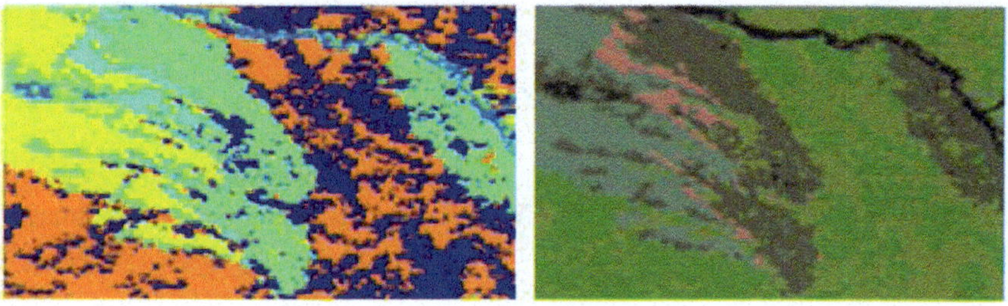

Fig. 5 The end result of proposed method for forest fire segmentation

The impact of the number of Superpixels for the proposed forest fire segmentation is illustrated in Fig. 6. Here, the outcome for 20,000 (the first row), 30,000 (the second row), 50,000 (the third row), and 100,000 (the fourth row). More number of pixels took more execution time but captured fine details of the image. However, this will lead to over-segmentation. For small amount of pixels, execution time is low. However, some important details may be lost which leads to under-segmentation.

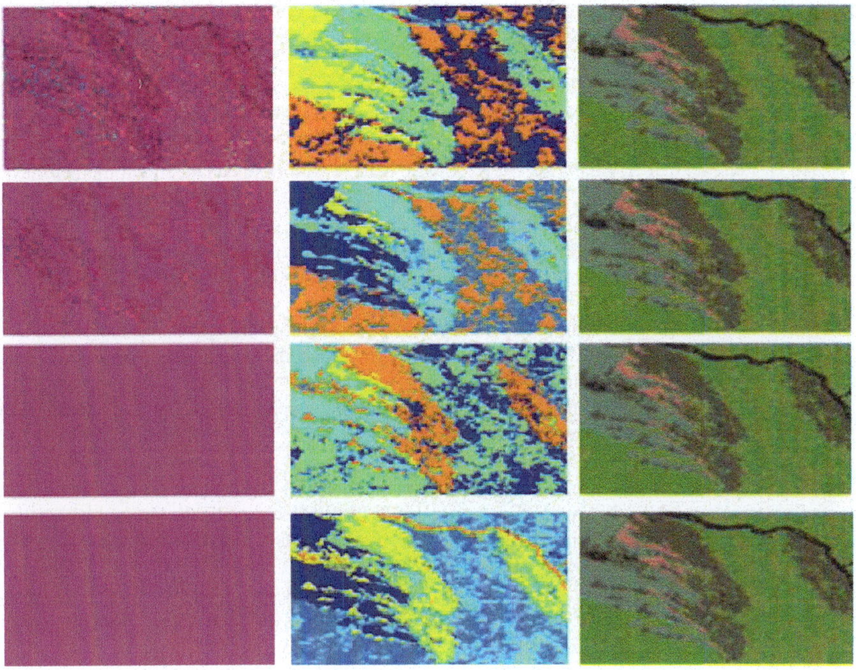

Fig. 6 The impact of number of superpixels for forest fire segmentation

5 Conclusion

After doing a thorough investigation of forest fire segmentation in satellite images, it is clear that using Superpixel-based fuzzy clustering with the CIELch color model is an efficient strategy. The proposed approach is crucial for effective fire-fighting and environmental management. The suggested approach significantly improves the accuracy and efficiency of segmentation. The utilization of superpixels reduces computing complexity while retaining spatial cohesiveness, and the CIELch model improves the separation of fire-affected regions from the satellite images. MFCM refines the process of segmentation by incorporating the degree of ambiguity and uncertainty in the image data. The experimental findings confirmed the efficacy of the suggested methodology, which outperformed conventional methods. Accuracy and efficiency are critical for early detection and fast reaction to forest fires, eventually helping to preserve natural ecosystems and safeguard human lives and infrastructure. Overall, the integration of superpixel-based fuzzy clustering with the CIELch model appears to be a viable approach for forest fire segmentation in satellite images, with prospective uses in real-time surveillance systems and decision support tools.

References

1. Wang Z, Lyu J, Tang X (2023) AutoSMIM: automatic superpixel-based masked image modeling for skin lesion segmentation. IEEE Trans Med Imaging 42(12):3501–3511. https://doi.org/10.1109/TMI.2023.3290700
2. Hu H, Feng C, Cui X, Zhang K, Bu X, Yang F (2023) A sample enhancement method based on simple linear iterative clustering superpixel segmentation applied to multibeam seabed classification. IEEE Trans Geosci Remote Sens 61:1–15
3. Sun L (2023) Multi-path minimum spanning tree and superpixel based cost aggregation for stereo matching. IEEE Access 11:121096–121108. https://doi.org/10.1109/ACCESS.2023.3328246
4. Wang Z, Lyu J, Luo W, Tang X (2022, March) Superpixel inpainting for self-supervised skin lesion segmentation from dermoscopic images. In: Proceedings of IEEE 19th international symposium on biomedical imaging (ISBI), pp 1–4
5. Shi C, Pun C-M (2020) Multiscale superpixel-based hyperspectral image classification using recurrent neural networks with stacked autoencoders. IEEE Trans Multimed 22(2):487–501. https://doi.org/10.1109/TMM.2019.2928491
6. Dalju HB, Rushdi MA, Morsy A (2021) Superpixel-based segmentation and classification of gastrointestinal landmarks and diseases. In: 2021 IEEE biomedical circuits and systems conference (BioCAS), Berlin, Germany, pp 1–5. https://doi.org/10.1109/BioCAS49922.2021.9645002
7. Lei T, Jia X, Zhang Y, Liu S, Meng H, Nandi AK (2019) Superpixel-based fast fuzzy C-means clustering for color image segmentation. IEEE Trans Fuzzy Syst 27(9):1753–1766. https://doi.org/10.1109/TFUZZ.2018.2889018
8. Kalist V (2015) Possiblistic-fuzzy C-means clustering approach for the segmentation of satellite images in HSL color space. Procedia Comput Sci 57:49–56
9. Ganesan P, Rajini V (2013) Segmentation and denoising of noisy satellite images based on modified fuzzy C means clustering and discrete wavelet transform for information retrieval. Int J Eng Technol 5(5):3858–3869
10. Sathish BS et al (2015) Color image segmentation based on genetic algorithm and histogram threshold. Int J Appl Eng Res 10(6):5205–5209
11. Zhang S, Lu T, Li S, Fu W (2022) Superpixel-based Brownian descriptor for hyperspectral image classification. IEEE Trans Geosci Remote Sens 60:1–12, Art no. 5520112. https://doi.org/10.1109/TGRS.2021.3133878
12. Sajiv G (2016) Comparative study of possiblistic fuzzy C-means clustering based image segmentation in RGB and CIELuv color space. Int J Pharm Technol 8(1):10899–10909
13. Zhang Y, Bai X, Fan R, Wang Z (2018) Deviation-sparse fuzzy C-means with neighbor information constraint. IEEE Trans Fuzzy Syst
14. Zhang H, Wang Q, Shi W, Hao M (2017) A novel adaptive fuzzy local information c-means clustering algorithm for remotely sensed imagery classification. IEEE Trans Geosci Remote Sens 55(9):5057–5068
15. Palanivel K, Sathish BS, Kalist V, Shaik KB (2015) Performance of fuzzy based clustering algorithms for the segmentation of satellite images—a comparative study. In: 2015 IEEE seventh national conference on computing, communication and information systems (NCCCIS), Coimbatore

An Improved XGBoost Classifier for Micro Expression Recognition Using Hybrid Optimization Algorithm

B. Sai Sarvagna, Vignesh A. Nair, Goutham Vijay, and Lekha S. Nair

Abstract In this work, we suggest a new approach to identify Micro Expressions (ME) in the facial region using Machine Learning, categorizing them into simulated and neutralized types. The study explores diverse techniques for feature extraction and pre-processing to emphasize unique patterns, aiming to improve model convergence, generalization, and mitigate overfitting. The proposed method recognizes challenges related to spontaneous subtle motion, environmental variations, and imbalanced datasets in Micro Expression Recognition. To tackle these issues, we incorporated the XGBoost classifier and a hybrid GA-PSO algorithm for optimizing feature selection, enhancing the model's ability to differentiate and classify various classes within a dataset. The research utilizes the CASME II dataset, emphasizing its relevance in addressing real-world scenarios and validating the proposed approach. The combination of XGBoost and GA-PSO presents a promising direction for future advancements in this field.

Keywords Micro expression recognition · Machine learning · Spontaneous Subtle Motion · Feature extraction · CASME II dataset · Generalization capabilities

1 Introduction

Facial expression recognition is a crucial aspect of computer vision, significantly contributing to the understanding of human emotions conveyed through facial expressions. The increasing reliance on digital platforms and interactive technology has heightened the need for systems capable of identifying and responding to human emotions [1] This demand is driven by applications in virtual communication, human-computer interfaces, and psychological studies, where accurate facial expression detection is invaluable [2].

B. S. Sarvagna · V. A. Nair · G. Vijay · L. S. Nair (✉)
Department of Computer Science and Engineering, Amrita School of Computing, Amrita Vishwa Vidyapeetham, Amritapuri, India
e-mail: lekhas@am.amrita.edu

A. C. Frery et al. (eds.), *Computational Intelligence and Data Analytics*, Lecture Notes on Data Engineering and Communications Technologies 236,
https://doi.org/10.1007/978-981-96-0451-7_5

Traditional computer vision algorithms have achieved notable success in emotion recognition; however, they often struggle with the subtle nuances of micro facial expressions. To address this challenge, we propose a novel approach that enhances micro facial expression recognition using a hybrid optimization algorithm. Our methodology involves extracting essential facial features-such as the nose, mouth, and eyes-from videos in the CASME II dataset.

Initially, a Genetic Algorithm (GA) is employed to select the most relevant features, overcoming the limitations of existing algorithms in capturing micro expressions. Subsequently, a Particle Swarm Optimization (PSO) algorithm is applied to refine and optimize these selected features, resulting in a robust hybrid optimization strategy. This approach aims to improve human-computer interaction, facilitate emotional computing, and enhance human experience sensing and interaction.

2 Related Works

Facial expression recognition has emerged as a significant research area in computer vision and affective computing, focusing on deciphering the intricate language of human emotions conveyed through facial expressions to better comprehend human behavior. A notable challenge in developing automatic micro-expression recognition systems is the scarcity of databases containing sufficient training samples of high quality. Yan [3] addressed this issue by meticulously selecting appropriate samples for database construction, resulting in the creation of the CASME II database, which has also been utilized in our research.

Zhao [4] proposed the MERSiamC3D model, which employs a two-stage learning approach for Micro Expressions (MEs) recognition. In the initial stage, MERSiamC3D extracts generic features of MEs, leveraging a convolutional neural network to extract spatial-temporal information from key frames, selectively discarding redundant frames while highlighting the movement of the apex frame. This method achieved an accuracy of approximately 80% on the CASME II dataset. Similarly, Sajjad [5] proposed a framework incorporating CNN as one of its modules. However, due to its high processing complexity and significant computational resource requirements, there is a need to explore and adopt lightweight deep learning technologies.

The transition of facial expression recognition (FER) towards real-world challenges underscores the increasing reliance on deep neural networks [6]. Preprocessing involves separating non-facial areas and identifying, aligning, and cropping the facial area to mitigate the impact of head movement [7]. Uzun [8] demonstrated notable improvement in accuracy through feature selection and cross-validation processes conducted with the PSO algorithm. Integrating GA within PSO, as explored by Mistry [9], benefits from GA's diversity-preserving mechanisms, enhancing the likelihood of finding globally optimal solutions.

Zong [10] proposed the FCAN-XGBoost methodology, which combines the FCAN feature attention network with the XGBoost algorithm. XGBoost effectively handles deep features, optimizing resource utilization without compromising classification performance. Furthermore, integrating the XGBoost classifier into a CNN model [11] significantly improves automated data analysis, achieving superior classification accuracy.

In this paper, we propose a new two-stage approach to micro facial expression recognition, analyzing facial expression recognition using the CASME II database to achieve improved recognition rates. XGBoost is employed in conjunction with both genetic algorithms (GA) and particle swarm optimization (PSO) to enhance the classification performance of the model. Combining GA with PSO enables a more effective balance between exploration and exploitation. While PSO excels at exploiting promising regions within the search space, GA's strength lies in exploring a wide range of areas through genetic operations such as crossover and mutation.

3 Dataset

The research utilizes the CASMEII dataset that we obtained through verification process and validating the proposed approach. Due to the limited size of existing spontaneous micro-expression databases and the inadequate video quality of current micro-expression datasets, which struggle to capture the nuanced details of micro-expressions, CASMEII has been chosen. The dataset provided consists of three distinct classes angry, happy and disgust, comprising a total of 255 videos recorded from 26 participants. Each video contains frames captured as images, with all images cropped to dimensions of 170×140 pixels. The dataset is categorized into its respective classes. The videos have a frame rate of 200 frames per second this high frame rate is chosen to capture detailed information on facial muscle movements, which is critical for recognizing micro-expressions that are very brief and subtle compared to existing databases like SMIC. Our methodology involves the extraction of essential facial features, including the nose, mouth, and eyes, from videos in the CASME II dataset [3].

4 Proposed Method

The proposed facial expression detection system uses cutting-edge computer vision and machine learning algorithms to effectively recognise and understand human emotions from photos or videos of faces. The system consists of linked modules with diverse functions to recognise facial expressions [12] (Fig. 1).

The initial phase involves users uploading CASME II dataset to application and then application will read all images, where each image undergoes normalization

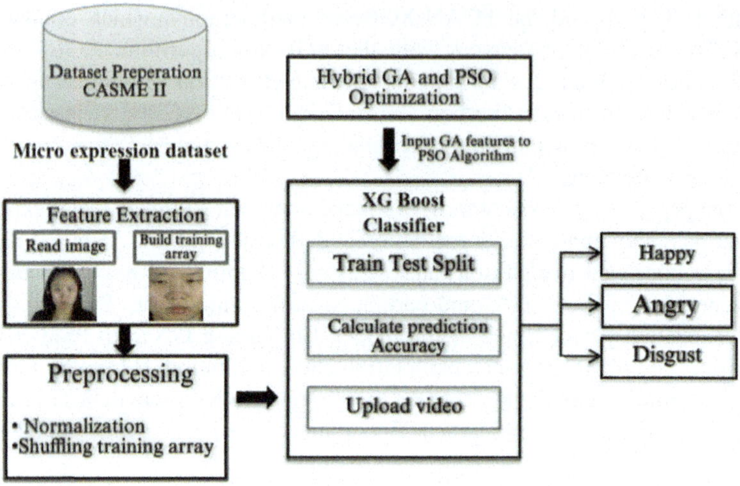

Fig. 1 Flow chart

to ensure uniformity in pixel values across the dataset. To ensure a balanced representation of classes, the training array is shuffled systematically to distribute to all classes. Subsequently, the preprocessed dataset is subjected to a dual-stage optimization process employing Genetic Algorithm (GA) and Particle Swarm Optimization (PSO).

Genetic Algorithm is responsible for feature selection, It operates on the provided dataset to select relevant features. Inputs to GA are data containing facial landmark data and corresponding labels. It optimizes the feature subset by maximizing the classification accuracy. Therefore it results in a binary array indicating which features are selected and which are not. This array is used to extract the selected features from the dataset. The Particle Sworm Optimisation part builds upon the the feature selection done by Genetic Algorithm. It aims to optimize the selected features to improve classification performance. Inputs to PSO is the dataset with the features selected by Genetic Algorithm then it ealuates the classification performance, optimizes the selected features and results in an another binary array indicating further refined selected features.

After the features are selected and followed optimization using GA and PSO, the dataset undergoes partitioning into two subsets: One for training the model and other for testing the performance. The training subset which is 80% of the dataset is used to train the machine learning model, allowing it to learn patterns from the data. The testing subset which is 20% of the dataset is used to evaluate the model's performance on unseen data, providing an estimate of how well the model generalizes to new, unseen samples. After the split, it displays information about the total number of images and features in the dataset, as well as the sizes of the training and testing subsets.

The features which are processed and optimized using Genetic Algorithm (GA) and Particle Swarm Optimization (PSO), they are used as input for training the XGBoost classifier, renowned for its robust performance in classification tasks [11]. During the training phase, the XGBoost model learns to categorize images using optimized features selected from the training dataset. Subsequently, the model undergoes evaluation using a separate testing dataset. Key performance metrics such as accuracy, precision, recall, and F1-score are computed to assess the model's proficiency in accurately identifying micro expressions. Additionally, a confusion matrix is generated to visualize the model's classification performance across different expression classes.

Although the primary dataset used for this research is CASME II, our methodology ensures rigorous evaluation to validate the proposed method's effectiveness. For future work, we plan to extend our evaluation to include other standard datasets such as SAMM and SMIC to further substantiate the robustness and generalizability of our approach. By focusing on comprehensive evaluation metrics and robust cross-validation, we aim to demonstrate the effectiveness of our method.

Once trained and evaluated, the XGBoost model is ready to make predictions on new data. The application enables users to upload videos containing micro-expressions for analysis it detects faces in a video stream and predict the corresponding expression label Angry, Disgust, or Happy -based on the learned features learned during training.

This comprehensive approach integrates preprocessing, optimization, model training, evaluation, and application, demonstrating a holistic methodology for micro-expression recognition.

4.1 Improved XGBoost Classifier

In our studies, we have chosen to employ an enhanced version of the XGBoost classifier for micro-expression recognition, inspired by its well-documented effectiveness in classification tasks, especially in scenarios characterized by intricate datasets with high dimensionality and noise.

The XGBoost algorithm has been improved in our research, encompassing various aspects, including Customised Feature Selection, Hybrid Optimization Approach, and Fine-tuning Algorithmic Parameters.

The application utilizes Genetic Algorithm (GA) and Particle Swarm Optimization (PSO) techniques to identify and select significant features from the dataset. This process aims to enhance the quality of input features supplied to the XGBoost model by adopting a hybrid optimization strategy to leverage the strengths of both algorithms. This approach improves the feature selection process, allowing the XGBoost model to focus on the most discriminative features while minimizing overfitting. Hyperparameters such as Learning Rate (eta), Maximum Depth (max_depth), and Regularization Parameters (lambda, alpha) have been fine-tuned.

The objective function optimized by XGBoost can be represented as:

$$\text{Obj} = \sum_{i=1}^{n} l(y_i, \hat{y}_i) + \sum_{k=1}^{K} \Omega(f_k)$$

Here, l is the loss function measuring the discrepancy between the predicted ($\hat{y}_i$) and actual values (y_i). $\Omega(f_k)$ represents the regularization term associated with each tree to control its complexity.

The regularization terms include: Decreasing the learning rate to 0.08 during training leads to more precise updates to the model's weights, potentially improving overall performance. Increasing the maximum depth of each tree to 4 compared to the default value, enables the model to create more intricate decision boundaries. However, it is important to note that deeper trees can potentially lead to overfitting.

$$\Omega(f_k) = \gamma \cdot |w|$$

Here, $|w|$ denotes the absolute value of the leaf weights, and γ is the regularization parameter.

$$\Omega(f_k) = 0.5 \cdot \lambda \cdot w^2$$

Here, w^2 represents the square of the leaf weights, and λ is the regularization parameter.

Incorporating an L1 regularization alpha term to 0.1 penalizes large weights in the model, promoting a simpler model structure and mitigating overfitting. Meanwhile, elevating the L2 regularization lambda term to 1.1 penalizes the squared magnitude of weights, promoting smoother decision boundaries and regulating model complexity by discouraging large weights.

4.2 *Hybrid Optimization*

In the Genetic Algorithm,

Exploration: This exploration phase allows the algorithm to search through the feature space effectively. Parameters such as population size (n_population) and the number of generations (n_generations). These parameters control the diversity of feature combinations explored during the optimization process.

Evaluation and Selection: GA evaluates each combination's performance based on its success in recognizing micro expressions. Combinations that perform well are selected for further consideration. Using cross-validation ($cv = 5$) and a specified scoring metric ($scoring =$ "$accuracy$"). Combinations with higher scores are retained, while others are discarded. The selector variable ultimately holds the selected features based on their performance.

Recombination and Mutation: The Genetic Algorithm applies recombination ($crossover\_proba = 0.5$) introduces random changes ($mutation\_proba = 0.2$) to generate new potential solutions. This step ensures the algorithm generate the new feature combinations and preventing premature convergence.

In the PSO algorithm,
Initialization: We begin by initializing a swarm of particles. These particles represent potential solutions or feature subsets. The features selected by the GA serve as the initial positions of these particles. Each particle's position corresponds to a specific combination of features.
Local and Global Optimization: The PSO algorithm iteratively adjusts the positions of particles based on their past performance and the overall best-known positions. This adjustment process incorporates both local and global information. $f\_per\_particle(m, alpha)$ function evaluates the performance of each particle (feature subset) based on a fitness criterion. The parameter m represents the binary vector indicating which features are selected. The fitness criterion considers both the classification accuracy (local information) and the proportion of selected features (global information).
Refinement: As the PSO algorithm progresses through iterations, it fine-tunes the positions of particles to explore the solution space more effectively. This refinement process aims to improve the overall effectiveness of the feature set by finding an optimal balance between feature relevance and computational efficiency.

5 Results and Discussions

The experimental findings substantiate the efficacy of the proposed research paper. Moving forward, we will delve into discussing these findings. A fundamental aspect of this research lies in the meticulous integration of Genetic Algorithm (GA) and Particle Swarm Optimization (PSO) for the optimization of features. The application of the XGBOOST algorithm, renowned for its ensemble learning capabilities, on the training dataset represents a critical step in the proposed research.

5.1 Uploading and Preprocessing Dataset

After uploading the CASME II Dataset, extracting and preprocessing face images is the most important phase. Several important procedures are performed in this step to prepare the data for analysis and modeling [3]. First, feature extraction methods are employed to extract useful information from facial images. This includes facial landmarks, textures, and forms.

Thereafter, the dataset underwent normalization to regularize and ensure consistency across all photos, addressing variations in lighting, discrepancy, and color distribution that could potentially hamper facial emotion identification. By accounting for such differences, the normalization process enhanced the quality and reliability of the dataset for subsequent analysis. Also, randomization of the dataset was enforced to eliminate biases arising from the order of data donation.

In essence, the thorough execution of these preprocessing steps ensured the optimization of the dataset for subsequent analysis and modeling.

5.2 GA and PSO Optimization

In the third phase of our micro-expression recognition process, Genetic Algorithm (GA) and Particle Swarm Optimization (PSO) are utilized for hybrid optimization. After applying both GA and PSO techniques, the number of features in each facial landmark is significantly reduced. Prior to the optimization process, each facial landmark contained a substantial number of features, with 784 features per landmark. However, after GA optimization, the feature count decreased to 355, indicating successful selection of relevant characteristics. Subsequently, PSO further refined these features, ultimately reducing the count to 230. Application illustrates the features found before and after optimization showcasing the significant reduction in feature count and underscoring the effectiveness of the hybrid optimization approach in enhancing the system's performance.

5.3 Train and Test Split

Following the fourth phase of preprocessing, wherein the dataset is meticulously optimized, the subsequent step involves its division into train and test subsets. The training dataset contains 80% of processed data, whereas the test dataset contains 20% [4]. This strategic partitioning strategy ensures a balanced representation of data across both subsets, mollifying biases and facilitating robust model training [13].

The total number of images found in the dataset is 4976, with each image containing 230 features. The training dataset consists of 3980 images, while the test dataset comprises 996 images. Subsequently, the train and test split datasets are utilized as input for the XGBoost classifier, The model uses 200 frames per second for detecting micro-expressions from the CASME II dataset.

5.4 Experimental Outputs

Utilizing the enhanced XGBoost algorithm on preprocessed and partitioned datasets. Once the features are prepared, the XGBoost classifier training phase involves The model learning the mapping between the input features and the target labels.XGBoost builds an ensemble of decision trees sequentially. Each new tree corrects the errors of the previous ones. During training, XGBoost evaluates the importance of each feature and focuses on the most informative ones. Standard measures like accuracy, precision, recall, and F1-score are used to assess model performance on the test dataset [12].

Also when comparing XGBoost and Attention Neural Networks (ANNs) for certain applications, XGBoost demonstrates clear advantages in terms of feature selection efficiency, regularization, and performance on complex datasets.In our research the use of L1 and L2 regularization helps in mitigating overfitting, which is a common challenge in deep learning models. XGBoost has demonstrated success across various complex datasets such as biomedical data, financial time series, and large-scale recommendation systems.

The accuracy metric indicates the overall correctness of the model's prediction with a value of 92.16%. A precision score of 92.65% indicates that when the model predicts a micro-expression as Angry, Disgust, or Happy, it is correct about 92.65% of the time. A recall score of 89.62% which is the proportion of true positive instances out of the total actual positive instances. F1 score of 91.11% which is mean of precision and recall, providing a single measure of a model's performance by balancing the trade-off between precision and recall.

5.5 Predict Expressions

Micro expressions are predicted from the uploaded videos. This approach involves analyzing each video frame, extracting relevant facial attributes, and applying the trained XGBoost model to categorize detected emotions into specified categories like happy, disgusted, or angry [11].

Figure 2 predicted a disgust class from a video frame at that particular instance. For each frame, It extracts the region of interest (ROI) containing the face, crops it, and resizes it to a fixed size, It predicts the micro facial expression using the pre-trained XGBoost model. It draws the predicted expression label on the frame, It displays the processed frame with the predicted expression in a window and after processing all frames, it releases the video capture object and destroys any OpenCV windows (Fig. 3).

The Tables 1 and 2 outline the frequency distribution of each micro-expression observed in the uploaded videos. Application displays the final bar plot showing the frequency of predicted expressions in the uploaded video. The visual depiction offers a crucial means for analyzing and comprehending the emotional nuances conveyed

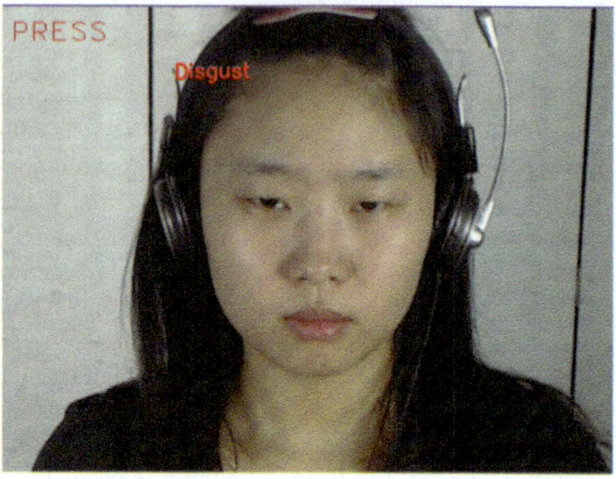

Fig. 2 Application predicted as disgust

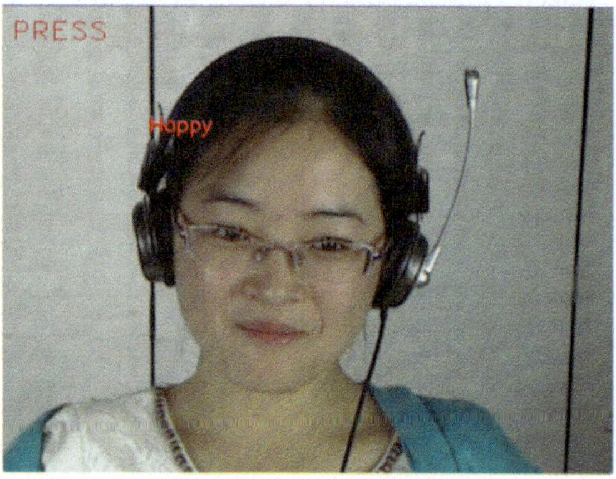

Fig. 3 Application predicted as happy

Table 1 Frequency of each micro-expression

Microexpression	Frequency
Anger	50
Happy	3
Disgust	89

by individuals in the video. By presenting the frequency distribution of predicted expressions, it facilitates deeper exploration and interpretation of the observed facial cues. This information allows us to discern the predominant expression, which can inform the determination of the final emotional assessment.

Table 2 Frequency of each micro-expression

Microexpression	Frequency
Anger	5
Happy	95
Disgust	4

6 Conclusion

Our micro facial expression recognition technology successfully interprets human emotions from uploaded videos and images. We have developed a robust system capable of classifying micro facial expressions into specified emotions. This framework utilizes feature extraction, hybrid optimization, and XGBoost machine learning. Real-time video analysis enables users to study and understand facial expressions in dynamic, real-world situations, enhancing system usability. Our research endeavors to optimize the accuracy and reliability of micro-expression classification, thereby advancing the efficacy of our study's method in this area.

By prioritizing feature extraction for micro expressions, this research addresses challenges in existing algorithms and provides a foundation for future investigations. Also Working with more classes of data, including neutral expressions, to improve robustness and applicability. Combining attention detection models with the circumplex model of emotion for micro-expression recognition holds significant potential for improving robustness in emotion recognition systems. However, one of the major challenges lies in the availability of datasets explicitly annotated with both micro-expressions and circumplex dimensions. To address this gap, future research could focus on re-annotating existing micro-expression datasets with arousal and valence scores, either through expert raters or automated methods. In this era of digital technology, our system leads emotion detection technology, paving the way for new ways to understand and empathize with human emotions.

References

1. Keshari T, Suja P (2019) Emotion recognition using feature-level fusion of facial expressions and body gestures. In: 2019 International conference on communication and electronics systems (ICCES)
2. Zhang Y, Wang C, Deng W (2021) Relative uncertainty learning for facial expression recognition. Adv Neural Inf Process Syst 34:17616–17627
3. Yan W-J et al (2014) CASME II: an improved spontaneous micro-expression database and the baseline evaluation. PLoS ONE 9(1):e86041
4. Zhao S et al (2021) A two-stage 3D CNN based learning method for spontaneous micro-expression recognition. Neurocomputing 448:276–289
5. Sajjad M et al (2020) Human behavior understanding in big multimedia data using CNN based facial expression recognition. Mob Netw Appl 25:1611–1621

6. Li S, Deng W (2020) Deep facial expression recognition: a survey. IEEE Trans Affect Comput 13(3):1195–1215
7. Esmaeili V, Shahdi SO (2020) Automatic micro-expression apex spotting using Cubic-LBP. Multimed Tools Appl 79(27):20221–20239
8. Uzun MZ, Celik Y, Basaran E (2022) Micro-expression recognition by using CNN features with PSO algorithm and SVM methods. Traitement du Signal 39(5):1685–1693
9. Mistry K et al (2017) A micro-GA embedded PSO feature selection approach to intelligent facial emotion recognition. IEEE Trans Cybern 47(6):1496–1509
10. Zong J et al (2023) FCAN-XGBoost: a novel hybrid model for EEG emotion recognition. Sensors 23(12):5680
11. Zivkovic M et al (2022) Hybrid CNN and XGBoost model tuned by modified arithmetic optimization algorithm for COVID-19 early diagnostics from X-ray images. Electronics 11(22):3798
12. Rosiani UD et al (2018) Micro expression: comparison of speed and marking accuracy in facial component detection. In: 2018 International conference on computer engineering, network and intelligent multimedia (CENIM), pp 221–226
13. Revina IM, Emmanuel WS (2021) A survey on human face expression recognition techniques. J King Saud Univ Comput Inf Sci 33(6):619–628

Detecting Equity Bubbles and Financial Crashes with Machine Learning

Vamsi Bodugu, Pukhraj Sharma, and Vijay Avuti

Abstract This research utilizes machine learning models to detect equity bubbles and predict financial crashes in the S&P 500 index. The imbalanced data problem is addressed, where non-bubbles significantly outnumber bubbles in the training data. Model features are selected based on economic theories and market fundamentals. The Random Forests model, with a Cross-Validation-tuned decision threshold, demonstrates the highest performance with a 93.0% balance accuracy rate. The model effectively balances sensitivity (100%) and specificity (86%). However, the Recurrent Neural Networks with Bidirectional Long-term Short Memory and focal loss function yield unsatisfactory results, likely due to limited data size. Key predictive features include market fundamental indicators, long-term market returns, and the psychology factor. Short-term market returns and macroeconomic indicators are less influential in crash prediction. The model focuses on long-term (over 3 months) and extreme market downturns. In conclusion, this research provides a machine learning approach to predict financial crashes and detect equity bubbles in the S&P 500 index. The Random Forests model effectively addresses imbalanced data, while highlighting the significance of market fundamentals. Limitations include the performance of Recurrent Neural Networks and the scope of long-term extreme market downturn predictions. These findings contribute to risk assessment and market forecasting in finance.

Keywords Asset bubbles · Market crashes · Machine learning · Deep learning · Random forests · Recurrent Neural Networks

V. Bodugu (✉) · P. Sharma · V. Avuti
Mumbai, India
e-mail: f20180842h@alumni.bits-pilani.ac.in

P. Sharma
e-mail: f20180534p@alumni.bits-pilani.ac.in

V. Avuti
e-mail: f20180840h@alumni.bits-pilani.ac.in

A. C. Frery et al. (eds.), *Computational Intelligence and Data Analytics*, Lecture Notes on Data Engineering and Communications Technologies 236,
https://doi.org/10.1007/978-981-96-0451-7_6

1 Introduction

Bubbles and market crashes are important themes of financial markets. Asset bubbles describe the situation where asset prices significantly deviate from their fundamental values. Notable historical bubble includes the Dutch tulip mania in 1637, the dot-com bubble in the 1990s, and the US housing bubble in the 2000s. Investors unaware of the potential risks of bubbles realize huge losses when bubbles burst and the market crashes. Some people believe the market crash in 1982 wiped out the cumulative profits of the history of American banking history [21].

More recently, despite the general economic downturn caused by the COVID-19 pandemic, stock markets are performing exceedingly well. The S&P 500 index was up by 18.23%, while the tech-heavy NASDAQ gained 43.6% in 2020. With the historical lesson of bubbles, we may wonder if we are experiencing one right now.

Given the catastrophic consequences caused by bubbles, it would be meaningful if we could identify bubbles and predict potential crashes. An accurate bubble detection tool may serve as an early warning system for policymakers or a risk management tool for investors. Building such a bubble detection tool is the primary motivation of this research. While bubbles and market crashes are broad phenomena, we focus on bubble detection in the stock markets in this paper. Specifically, the scope of the research is limited to bubble detection in the S&P 500 market index. The main research goal is to develop a method that can reliably detect equity bubbles and forecast market crashes in the S&P 500 index.

If we can realize the main research goal, we can answer certain further questions. Assuming that the prediction method is transparent enough, we can identify the most important factors for prediction. On the technical side, we can gain insights into selecting empirical methods that work best for equity bubble detection problems. Also, we can predict if we are in an equity bubble right now using our proposed prediction method.

Much previous research has been done on bubble detection and market crash prediction. Based on economic theories such as rational choices and fundamental value as discounted dividends, some researchers came up with statistical tests like variance bound test for bubble detection [18]. More recently, researchers have applied machine learning methods, which are powerful prediction tools for detecting complex patterns, to the study of predicting market crashes [5]. However, as we will see in the literature review section, previous studies are limited in two ways. First, theory-based studies rely highly on their theoretical assumptions and suffer from problems such as joint hypothesis testing. Second, model-free machine learning studies typically do not include enough input variables. They only include short-term trading variables and have unsatisfactory prediction power.

we attempt to solve these limitations in this paper. we adopted the bubble definition from observable data to avoid the theory reliance problem. As readers may have noticed, we have used the term "bubble" interchangeably with potential market crashes. That is precisely my definition of bubbles. While it may not be suitable for theoretical debates, e.g., "Do bubbles exist," it captures the practical side of the concept of bubbles for practitioners like policymakers and investors. With this definition, the bubble detection problem is transformed into a supervised machine learning classification problem, where we try to predict the output, the bubble, given the input variables. One of the distinguishing features of this research is that we include more input variables than many previous studies to boost the prediction power of models. We select variables suggested by economic theories as input for bubble detection, including macroeconomic data, long-term trading trends, and market fundamental indicators. It is worth noting that my research's validity would not depend on the correctness of theories because they merely serve as feature-selection tools.

We implemented three machine learning models as candidates for the best classifiers: logistic regression (logit), Random Forests (RF), and Recurrent Neural Networks with Bidirectional Long Term Short Memory (RNN-BiLSTM). Logistic regression is easy to implement and can be well interpreted. Yet, compared to the other two methods, it is limited because it restricts the function form to linearity after the logistic transformation. The RF model is good at detecting non-linear patterns and seems promising based on a previous study. RNN-BiLSTM has been found to have superior results with financial time series data compared to traditional methods. Besides, it is used in similar studies and has good performance.

One unique challenge of the modeling is that the dataset is imbalanced, meaning there are far more non-bubble instances than bubbles. We have adopted special techniques to mitigate this problem, for example, resampling for re-weighting, changing decision thresholds, and using asymmetric loss functions. We also use balanced accuracy, the average of specificity and sensitivity, as my model evaluation metric.

The results of the research are satisfactory. We have met the main research goal of finding a high-performing model. My best-performing model, the RF with CV-tuned decision thresholds (RF-CV), achieves a balanced accuracy of 93.0%. It can reliably detect equity bubbles and predict market crashes. For variable importance inference, we find that long-term market returns and market fundamental indicators are the most important factors, whereas short-term market returns and macroeconomic data have weak predictive power. On the technical question of the efficiencies of empirical methods, the research confirms that machine learning effectively detects equity bubbles and predicts market crashes. RF models perform best among the machine learning models, followed by logit models. This is expected since RF models can capture the non-linear relationships that logit models cannot detect. RNN models have unsatisfactory results, possibly due to the insufficient size of training data. This implies that deep learning techniques should be used cautiously since they do not always guarantee high performance. Finally, the best-performing model, the RF-CV, shows that we are likely in a stock bubble as of April 2021, and market crashes are likely in the next six months.

The research has two main contributions. First, we use a new methodology that uses the power of machine learning algorithms and feature-selection insights from economics theories. This methodology avoids the theory dependence problem, such as the joint hypothesis problem, and can boost the model performance by including more features. Second, we build a high-performing predictor (RF-CV) for detecting equity bubbles and predicting market crashes. To my knowledge, it is the highest-performing machine learning model for predicting market crashes. Believe this can be used as a useful tool for financial market practitioners.

The rest of the paper is organized as follows. The background section provides a literature review of previous research and my motivation for this research. The data section lists all the input variables, their theoretical justification of relevance, and my definition of the output variable, the bubble. The model section introduces the machine learning models we selected and the reasons for choosing them. The model section also introduces the methods to mitigate the imbalanced data set problem. The result section presents the model performance and the inference of variable importance. The discussion section provides a model diagnosis, conducts a robustness analysis, and discusses the limitations of the research. Finally, the conclusion summarizes this research's key findings and contributions.

2 Background

Much previous research has been done in the study of bubble detection and financial crisis prediction. However, we find previous research is limited in two ways: too many theoretical assumptions and insufficient feature inclusion. In this section, we will give a literature review of the previous research and explain my motivation for this research.

There is plenty of theoretical research on asset bubbles. At the foundation level, it is still debated whether bubbles exist. Eugene Fama, the proposer of the Efficient Market Hypothesis (EMH), is generally skeptical of the notion of the bubble [10]. He argues that people should not use the word "bubble" without careful definition and empirical validation. For those who believe in the existence of bubbles, their views on the cause of bubbles diverge. As described by [23], bubble models can be categorized into rational and behavioral bubble models, i.e., irrational bubble models.

The family of rational bubble models relies on the rational decision-making assumptions of the agents. They typically explain bubbles by asymmetrical information and the herding behaviors of investors. For example, [1] argues that if participants' beliefs are private, an asset bubble is possible when all participants rationally believe they can sell the stock to a new buyer "before its true value is publicly revealed." On the other hand, behavioral bubble models do not presume the rationality of agents. Irrational behaviors are central to the behavioral bubble models. For instance, it has been found that people are generally biased as they rely on representativeness for decision-making [22]. This effect can cause people to overact on market news in the financial markets and lead to biased asset pricing [8].

The interest in bubbles is not limited to theoretical research. On the empirical side, the phenomenon of bubbles is of great importance to policymakers and investors since an expected bubble could lead to huge losses. Researchers have tried various methods to detect bubbles accurately.

Some researchers believe that the high volatility of the stock price signals underlying bubbles. Since the stock's fundamental value is unlikely to change rapidly, they argue the high volatility must be due to irrational, speculative behaviors. Following this line of thought, [18] proposes the variance-bound test for bubble detection. If the variance of the stock is higher than a given bound, we can reject the null hypothesis of no bubble. Later research aims to test whether the time series variable of the bubble is stationary. Similar to the reasonings for volatility, researchers hold that the variables related to the asset should be stationary if there is no bubble. Craine [7] employs the Augmented Dicky-Fuller (ADF) test to check whether the price-dividend ratio is stationary. He argues that a rational bubble is likely if the ADF test rejects the null hypothesis of stationary.

However, these econometric empirical tests are not without criticism. A survey conducted by [12] finds that the conclusion of the econometric tests of bubbles is uncertain. He finds that models that fit data equally well can have different conclusions about whether a bubble exists. Even if the results of the econometric test are conclusive, he argues, it is not sufficient to demonstrate the existence of bubbles. It could be that the assumptions of the models are questionable. As noted by [23], many of the econometric tests suffer from the joint hypothesis testing problem: since the test depends on models, the test is simultaneously testing against the assumption of no bubble and the validity of the model assumptions. It could be the failure of any other model assumptions that lead to rejecting the null hypothesis. Also, [9] finds that the stationary tests for the bubble are flawed in that they cannot detect "an important class of rational bubbles."

Contrary to the econometric tests approach of bubble detection, some researchers choose to directly model the practical consequences of the bubble: market crashes. While econometric tests depend on their model assumptions, which are often unobservable and untestable, market crashes are directly observable. Jacobsson [13] proposes a method called Log-Periodic Power Law to detect speculative bubbles. More recently, researchers have used advanced machine learning methods to study the pattern of market crashes. Machine learning methods are powerful in capturing complex patterns in the data. Chatzis et al. [5] use models including Support Vector Machines, Random Forests, and Deep Neural Networks for predicting crisis on the 20-day horizon and 1-day horizon. Bashchenko and Marchal [2] and Moser [15] use Recurrent Neural Networks (RNNs) with Long short-term memory (LSTM).

However, these studies typically include a few types of features. Bashchenko and Marchal [2] and Moser [15] use almost exclusively daily prices of the stocks or market indexes. The features included in [5] are stock variables and bond market data.

Given the academic context, we aim to solve the limitations of the previous research. In light of the model-reliance problem of econometric tests, we choose the model-free machine learning approach for this research. On the other hand, we improve upon previous machine learning studies for market crashes by including more variables as a feature. The features are suggested by economic theories as relevant to the financial market performance. An introduction of the data is presented in the next section.

3 Data

This section will give an overview of the data used in this research. First, we will introduce the input variables, a.k.a. features, and my reasons for selecting them. The features include macroeconomic data, market return trends, and fundamental indicators. Then, we will explain my definitions of equity bubbles and financial crashes. The range of the data is from January 1960 to December 2020 in the US. Unless otherwise specified, the frequency of the variable is monthly. The raw data are retrieved from Federal Reserve Economic Data (FRED) and *multpl.com*.

3.1 Input Variable

GDP growth rate The real Gross Domestic Product growth rate measures the productivity growth of a nation. It is one of the most important macroeconomic metrics. Given its importance, sophisticated machine learning models might gain insights from the growth pattern for bubble detection. The frequency of the GDP data is quarterly rather than monthly. We replicate the quarterly GDP growth rate to all the corresponding months to make the data frequency consistent.

Inflation The Consumer Price Index's growth rate measures a nation's price inflation. Similar to the GDP growth rate, it is one of the most important macroeconomic indicators. In financial markets, previous research finds an inverse relationship between inflation and lower share prices [11]. It might be the case that a high level of inflation is associated with a higher possibility of market crashes.

T-bill yield The 10-year Treasury constant maturity rate is a key benchmark in the debt market. It captures the long-term expected yield of a risk-free investment. From the perspective of an investor, the debt investment and the equity investment are substitutions. It has been found that bond returns tend to be higher than stock returns when the market uncertainty increases [6]. There is reason to believe that inflation could help detect equity bubbles.

S&P 500 returns The Standard & Poor is one of the most widely used stock market indexes. It measures the stock returns of the largest 500 companies listed in the US. As discussed in the subsection, the output variables–equity bubbles and market crashes are constructed from S&P 500 returns. As a common practice in time-series forecasting, using historical returns to forecast future data is natural. To include the price trends in different periods, we include 1-, 3-, 6-, 12-, and 60-month returns of S&P 500 as features.

Consumer confidence As noted by [8], the psychological factors of investors have a significant influence on investment decisions. We use the data of consumer opinion surveys as a proxy for the public sentiment of the general business environment. While the capital market sentiment does not always align with consumer confidence, this is the most easily accessible data. Also, unlike other newly invented stock market sentiment indexes, the consumer confidence data is complete between 1960 and 2020.

Shiller P/E ratio Invented by [19], the Shiller P/E is also known as the cyclically adjusted price-to-earnings ratio. It measures stock prices as a ratio to the average inflation-adjusted earnings [16]. If the Shiller P/E ratio is too high, it suggests that the stock prices could be overvalued based on the corporate earnings.

Market capitalization-to-GDP ratio Warren Buffet believes that "the market value of all publicly traded securities as a percentage of the country's business" is a useful metric for investment decisions [4]. Given the investment success of Warren Buffet, the effectiveness of this metric seems worthwhile to explore. Since collecting data for all publicly traded securities is cumbersome, we calculate the ratio between the S&P 500 index and the nominal GDP as a proxy. Similar to the real GDP growth rate, the frequency of the nominal GDP is quarterly. We replicate the GDP data to match the monthly frequency.

Table 7 presents a summary of input variables. *Min* is the minimum value of the variable; q_1 is first quantile of the variable; $\tilde{x}$ is the median value of the variable; q_3 is the third quantile of the variable; *Max* is the Maximum value of the variable; $\bar{x}$ is the mean value of the variable; s is the sample standard deviation of the variable.

An interpretation of the median data in the table is the following: the real GDP growth rate of the current quarter is 3.0% annually; the inflation rate is 3.2% annually; the 10-year Treasury Bond yield is 5.7% annually; the Shiller P/E ratio of the month is calculated to be 20.5; the consumer confidence is 100.5, higher than the baseline value of 100; the market capitalization-to-GDP ratio, measured by the ratio between the S&P 500 index and the nominal GDP in the current quarter, is 0.1; the S&P 500 returns in the past 1, 3, 6, 12, and 60 months are 1, 2.4, 4.5, 9.7, and 45.4% respectively.

3.2 Output Variable

We follow the definition of [5] and define a market crash as an extreme market downturn. An extreme market downturn is a one-percentile market return. Given this definition, the bubble is a pre-stage of a market crash. The bubble is a binary variable: at a given month, it can take the value 1, which denotes that there is a bubble, or take the value 0, which denotes that there is no bubble. As an operational definition, the bubble variable takes the value 1 if and only if: (1) there will be a market crash in the next six months; (2) the current period does not see a market crash.

The benefit of this bubble definition is that it is entirely based on observable data. It does not involve a theoretical definition of the fundamental value of the stock nor prescribe the behaviors of investors. As discussed in the background section, these theoretical discussions could make the model depend on the assumptions of theories. They would limit the power of bubble detection models. Of course, the

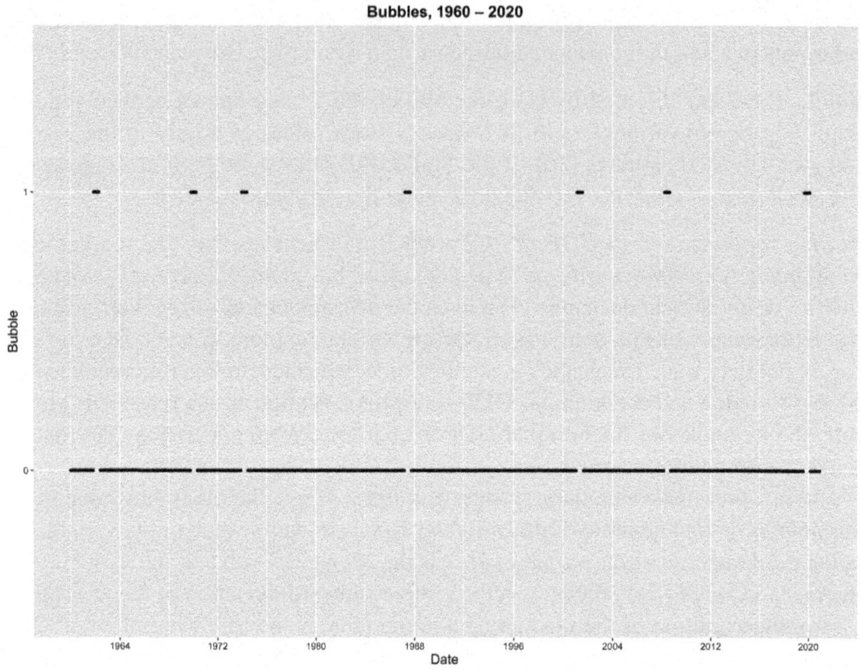

Fig. 1 The definition of bubbles

bubble definition in this research is not suited for a theoretical debate of bubbles. It only aims to offer empirical value for financial market practitioners.

In the above definitions, "one percentile" and "six-month" may seem to be arbitrary choices. Indeed, there is no obvious reason why we should not use "two percentile" and "five months" for the definitions. In the discussion section, we will alter the percentile and period length for the robustness analysis of the model.

Figure 1 shows the bubble categories through time. The graph shows that our bubble definition correctly identifies most pre-stages of historical market crashes, including the dot-com crash around 2001 and the US housing market crash around 2008. This confirms that our bubble definition is in line with reality.

Table 1 shows the distribution of output data. We see that non-bubble cases are much more than bubble cases. This is not unexpected since bubbles are abnormal market phenomena. However, as we will see in the next section, this imbalanced nature of the output variable would create a problem for modeling.

Finally, as a typical practice for machine learning problems, we randomly split the full data into training and testing data. The training data is 80% of the full data, while the testing data is the remaining 20%.

Table 1 Bubble distribution

Variable	Levels	n	%	$\sum$ %
Bubble	0	690	94.3	94.3
	1	42	5.7	100.0
	all	732	100.0	

4 Model

With the data set up in the previous section, the central question of the research becomes a supervised binary classification problem: we aim to predict the output variable given input variables through machine learning models. In this section, we will give an introduction to the models that we used in this research. Also, we will explain a special modeling challenge in this research—the problem of imbalanced data and the ways to mitigate the problem.

4.1 Model Selection

In this research, we fit three models to the data: logistic regression (logit), Random Forests (RF), and Recurrent Neural Networks (RNN).

Logistic Regression Logistic regression is one of the most widely used binary classifiers. It is easy to implement, and the interpretation of its results is straightforward. Logistic regression predicts the probability of the binary output variable through the logistic transformation. Equation 1 shows the logistic regression formula, where X is the design matrix of the input variables and Y is the output variable, the bubble. The formula shows that the logistic transformation of the probability of bubbles is just the log odds of bubbles. Therefore, we can also interpret the logistic regression as modeling the log odds of the bubble using linear models.

$$\text{logit}(P(Y)) = \log \frac{P(Y)}{1 - P(Y)} = X\beta \tag{1}$$

While logistic regression is easy to implement, it is limited because it cannot capture the non-linear pattern in the data. The relationship between the input and dependent variables (after the logistic transformation) is restricted to a linear form. However, bubbles and financial markets are complicated, and it is reasonable to assume that the true relationship is more than linearity. This is why we employ more complex models, such as Random Forests and Recurrent Neural Networks, to learn the non-linear pattern in the data.

Random Forests The RF is an ensemble learning method based on decision trees. It fits many decision trees to the data through bootstrap and randomly excludes some features. Then, it aggregates the predictive results of a bag of decision trees to yield the final result. Thanks to the cursive splitting algorithm of the decision tree and the bagging method, the RF is good at detecting non-linear patterns in the data. Besides being a non-linear classifier, the main advantage of the RF is that the bagging mitigates the overfitting problem of the simple decision trees algorithm. Also, the RF is a promising method for finance research. According to a study conducted by [14], soft classification techniques, including ensemble learning methods, "appear to be the direction for future research" in financial crisis prediction. The nature of financial crisis studies is similar to the study of market crashes. It is reasonable to assume that the techniques that work well for financial crisis studies could also yield good results for this study. As a strong ensemble learning classifier, the RF method is a candidate for the best-performing model in this research.

Besides the predictive power, the RF is also good for inference. Unlike many advanced machine learning methods, the RF enables us to analyze variable importance. Since the RF comprises a bag of decision trees, it can calculate the importance of a feature by comparing the decision tree performance with and without the input variable. If the inclusion of an input variable increases decision tree performance significantly, it means the input variable is important for making the prediction. This inference power of the RF is valuable for this research because one of our research questions is identifying factors important to prediction.

Recurrent Neural Networks Deep learning models, or artificial neural networks with multiple hidden layers, are becoming popular in finance research. They are shown to have superior results to traditional machine learning methods. The RNN is a family of deep learning methods good at predicting sequential data, including time-series data. In this research, we choose a specific type of RNNs—RNNs with bidirectional long short-term memory networks (RNN-BiLSTM). The long short-term memory (LSTM) is a neural network unit that helps the model to learn from the long-term pattern in the data. The bidirectional component enables the model to detect patterns in the sequential data in two ways. Readers interested in knowing more about how the RNN models work can refer to [17] and the appendix of [2].

In forecasting financial time series, previous research has shown that RNNs with BiLSTM generally perform better than traditional time series prediction methods like ARIMA [20]. A recent asset bubbles detection study uses this method and gets satisfactory results [2]. Based on previous studies and the nature of my study, select this method as a candidate for the best classifier.

However, a drawback of the RNN is that the result would be hard to interpret. The behaviors of deep learning models are known to be "black box", which means the model's results are often inexplicable except for prediction. Unlike the RF, the model-fitting process of deep neural networks does not shed light on the relative importance of input variables.

4.2 The Challenge of Imbalanced Data

Apart from model selection, this research also faces the special challenge of a data imbalance. As shown in Table 1, the distribution of the output variable is highly imbalanced. There are way more zeros ("non-bubble") than ones ("bubble"). This imbalanced distribution is expected since bubbles and market crashes are rare, yet it could cause trouble for the modeling.

The default behavior of a typical machine learning fitting algorithm is to minimize misclassification cases in all training data. Since the non-bubble case represents the over-majority of the data without any special treatments, the model fitters are expected to focus more on the non-bubble cases than the bubble cases. This behavior is undesirable because the research is interested in predicting bubbles. As an extreme case, even a naïve classifier that predicts zero in all cases would have a high accuracy rate. Its prediction is very likely to be correct since it is true that the market does not have bubbles most of the time. However, this kind of naïve classifier is useless and offers no practical value. We gain no insights into forecast market crashes using it. To make the research meaningful, we want to make our models less prone to suffer the same problem as naïve predictors.

Besides, from a practical perspective, the loss caused by the failure to predict a market crash is larger than that caused by a false alarm. The latter may lead to over-conservative investments or policy decisions, but an unexpected market crash can cause huge financial losses and incur big social costs. This practical concern also suggests that the default model-fitting prioritizes predicting non-bubbles and offers little practical value. We should focus on correctly predicting the bubbles.

We have tried several ways to tackle this challenge, including changing decision thresholds, re-sampling for re-weighting and adopting an asymmetric loss function. These methods make the models focus more on the bubble cases than the non-bubble cases. No more than one of these methods should be used concurrently; otherwise, there will be an overcorrection. Also, to reflect our preference for the prediction result, we use the balanced accuracy rate rather than the overall accuracy rate as my model evaluation metrics.

Changing the Decision Thresholds For all the binary classification methods we selected, we would first predict the probability of the bubble before making a categorical decision. The default decision threshold, as suggested by BaynesâŁ™ theorem of classification, is 0.5. For example, given a specific set of input data, the model may predict that the probability of bubbles is 0.4. Since 0.4 is smaller than 0.5, the model would make a categorical prediction of non-bubble.

We can lower the decision threshold to make the model more likely to make a "bubble" decision. Under a decreased decision threshold, a lower bubble probability is enough to trigger an alarm of the bubble. If we changed the decision threshold to 0.3, the model would give a "bubble" prediction in the above example since 0.4 is greater than 0.3. In this way, we make the model treat the bubble probability with higher weights. It can offset the over-representation problem of non-bubble cases.

How do we decide the new decision threshold? Setting the threshold to 0.3 seems arbitrary. In this research, we use two methods to determine the new threshold. First, the model can use the prevalence of positive cases, i.e., the frequency of bubbles in the training data, as the threshold. This would offset the imbalanced data problem completely. Also, the model can treat the threshold as a hyperparameter and use cross-validation (CV) to tune it. The CV method is particularly suitable for the RF model because the out-of-bag samples of the RF can be easily used for the CV.

Re-Sampling for Re-Weighting We can also directly change the re-weight in the data through resampling. For each bubble case, we sample the data more than once, while in the non-bubble case, we only sample it once. Essentially, the method replicates the bubble data. We replicate until the ratio between the number of bubble cases and the number of non-bubble cases is close to one. Using this method, the distribution of the output variable is perfectly rebalanced.

Asymmetric Loss Function A loss function for classification problems is a function that characterizes the misclassification. When fitting a classifier, we minimize the loss function on the training data to make a better prediction. The binary cross-entropy function is the default loss function for most binary classification algorithms. As shown in Eq. 2, the cross-entropy loss function does not distinguish losses coming from for the bubble misclassification and non-bubble misclassification.

$$CE(p) = -\log(p) \tag{2}$$

If we want to make the model focus more on bubble prediction, we can make the model fitter consider bubble misclassification as a bigger loss than non-bubble misclassification. To achieve this, we can adopt an asymmetrical loss function. In this research, we use the focal loss function, which proposes citefocal. The function takes the form as Eq. 3. α and γ are hyperparameters that determine how asymmetric the loss should be and the customized treatments for misclassification losses. we set them as the same as the original paper. This method is easiest to implement in the RNN model since the TensorFlow package, a widely used deep learning library, enables us to use customized loss functions for model fitting.

$$FL(p) = -\alpha(1-p)^{\gamma}\log(p) \tag{3}$$

Model Evaluation Use balanced accuracy as the model evaluation metric. It is defined as the average value of sensitivity and specificity. Sensitivity measures the proportion of correctly predicted positive (bubble) cases, while specificity measures the proportion of correctly predicted negative (non-bubble) cases. A formula form of the balanced accuracy is shown in Eq. 4.

$$\text{Balanced Accuracy} = \frac{\text{True Negative}}{\text{Predicted Negative}} + \frac{\text{True Positive}}{\text{Predicted Positive}} \tag{4}$$

This metric is preferred compared to a simple, overall accuracy rate. It measures the model performance separately on positive (bubble) and negative (non-bubble) cases. Under this metric, a naïve classifier that predicts all cases as non-bubble can only achieve 50% since it has 0 sensitivity.

5 Results

5.1 Prediction

Combining the models and the special consideration to the imbalanced data, we fit five sets of models to the training data: logistic regression, using prevalence as the decision threshold (logit-P); logistic regression, using re-weighted data (logit-RW); RF, using CV-tuned decision threshold (RF-CV); RF, using re-weighted data (RF-RW); RNN + BiLSTM, with a focal loss function (RNN-BiLSTM-focal). Their performances are shown in Table 2. Figure 2 presents a visualization of model performances by sensitivity and specificity. The performances are measured on the testing data, which are 20% randomly selected samples from the full data set. The testing data do not overlap with the training data.

The RF model with a CV-tuned decision threshold performs best, with a balanced accuracy of 93%. The sensitivity and specificity show that the model correctly identifies 100% of the bubble cases and 86% of the non-bubble cases. Most other models perform decently well, with balanced accuracy rates around 70%. The RNN model is a disappointment. It performs no better than a naïve model. We will attempt to explain the difference in the predictive power of the models in the discussion section.

Figures 3 and 4 provide more information about the RF-CV model. Figure 3 shows the decision threshold process. Cross-validation chooses the decision threshold of 0.08 since it yields the lowest balanced error in the CV data. Figure 4 shows the prediction results (red area) of the RF-CV model when imposed on the complete data set. As we can see, the RF-CV model correctly classifies almost all cases.

When feeding with the most recent data (April 2021, when this paper was written), the RF-CV model predicts *bubble*. It suggests that we are likely in a bubble, and financial crashes are likely in the next six months. This result is not financial advice,

Table 2 The performance of models

Model	Sensitivity	Specificity	Balanced accuracy
1 Logit-P	0.727	0.721	0.724
2 Logit-RW	0.727	0.735	0.731
3 RF-CV	1	0.86	0.93
4 RF-RW	0.455	1	0.727
5 RNN-BiLSTM-focal	0	1	0.5

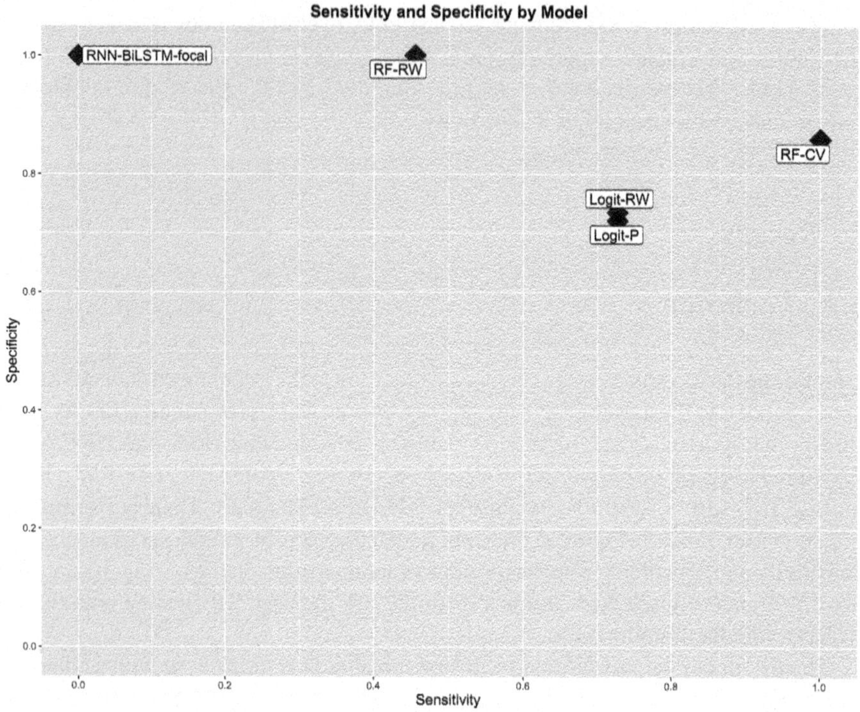

Fig. 2 Sensitivity and specificity by model

and we do not recommend the reader make risky financial decisions based on this result. If the reader still wants to take this result as a reference, we highly recommend the reader read the discussion section for the limitations of this research before making any decisions.

5.2 *Inference*

We can draw further insights from the research besides the prediction results. We can identify important factors for prediction. We use two of the best-performing models for inference: the RF-CV model and the logit-RW model.

Figure 5 shows the variable importance in the RF-CV model. We see that consumer confidence, long-term investment returns, e.g., 5-year S&P returns, and fundamental indicators, e.g., market capitalization-to-GDP ratio, is the most important. On the other hand, the short-term S&P 500 return and short-term macroeconomic data are among the least important features.

Table 3 shows the regression table for the logit-RW model. It mostly agrees with the results of the RF-CV variable importance plot. The fundamental indicators

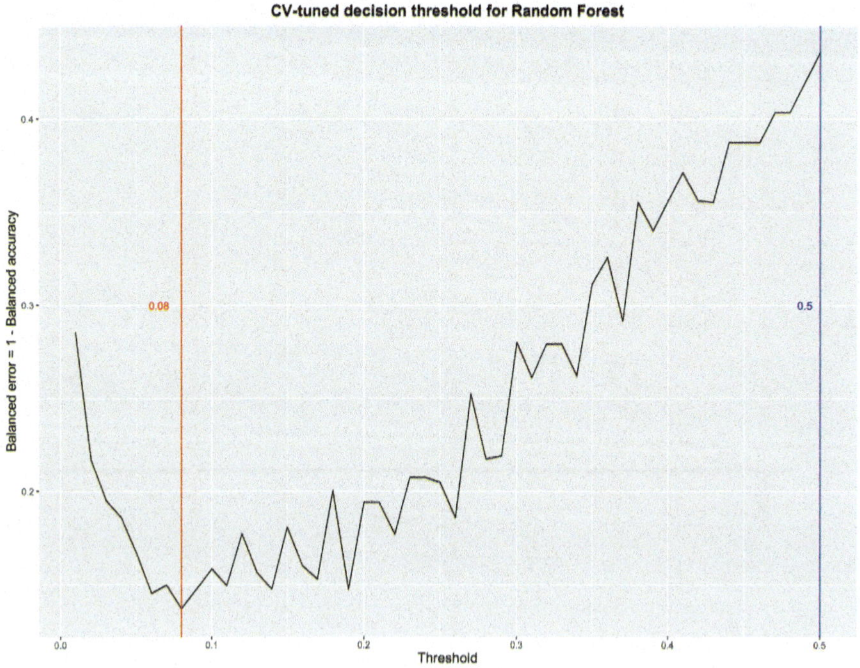

Fig. 3 The decision threshold tuned by CV

and long-term investment returns are significant at 99.9% confidence. The 3- and 6- month market returns are not significant. However, contrary to the result of the RF-CV model, the logit-RW model finds inflation and one-month market return significant for prediction.

Based on the above inference, it seems that our model favors the investment philosophy held by value investors in the context of avoiding market crashes. If the stock prices are too high for corporate earnings and national production, it would be wise to heed the risks of market crashes. Also, these inference results explain why the machine learning models from previous studies do not perform well. As mentioned in the background section, most previous machine learning studies for predicting financial crashes only include short-term market return data as features. Those variables have the lowest predictive power among all the input variables selected.

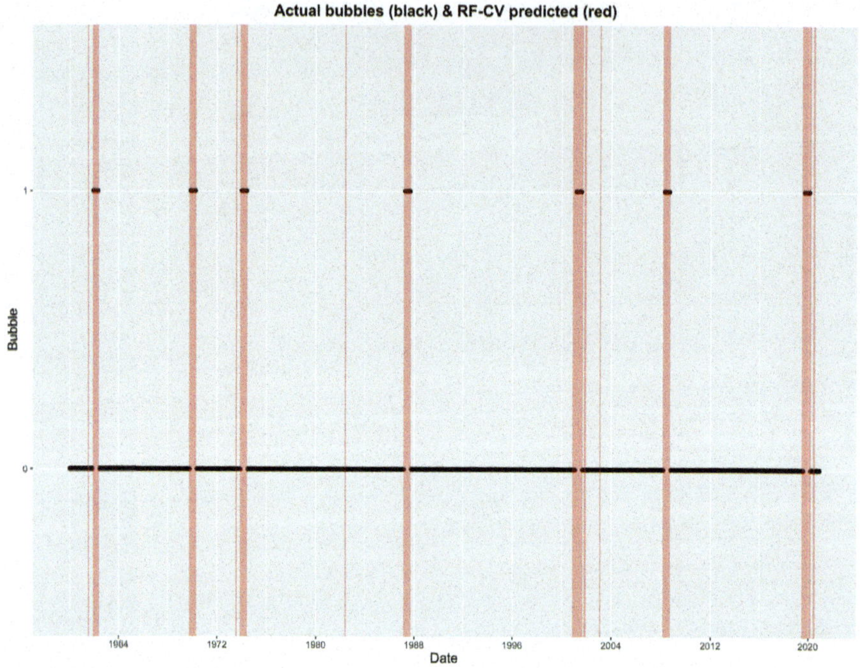

Fig. 4 RF-CV prediction of bubbles

Fig. 5 The variable importance in the RF-CV model

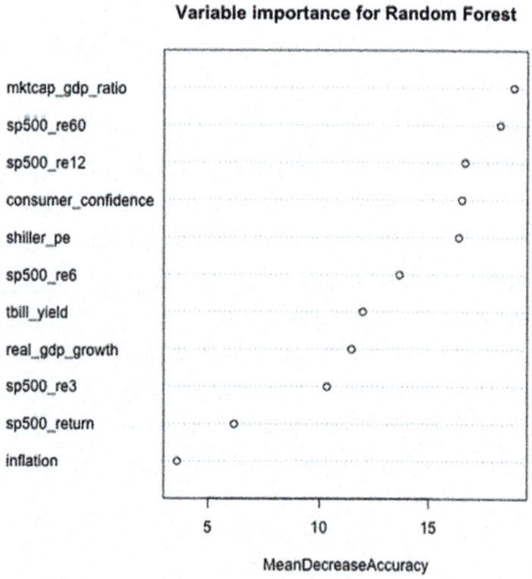

Table 3 Regression table for Logit - RW model using re-weighted data

	Dependent variable:
	Bubble
real_gdp_growth	−0.007
	(0.035)
inflation	0.099***
	(0.028)
tbill_yield	0.237***
	(0.073)
shiller_pe	−0.658***
	(0.058)
consumer_confidence	−1.182***
	(0.147)
mktcap_gdp_ratio	216.592***
	(18.361)
sp500_return	−0.105***
	(0.033)
sp500_re3	0.018
	(0.022)
sp500_re6	−0.021
	(0.020)
sp500_re12	−0.143***
	(0.016)
sp500_re60	0.075***
	(0.006)
Constant	106.314***
	(13.641)
Observations	1,112
Log Likelihood	−410.346
Akaike Inf. Crit.	844.693

*p<0.1; **p<0.05; ***p<0.01

6 Discussion

6.1 Model Diagnosis

Having seen the model performance in Table 2, it is natural to ask why certain models perform better than others. While it is unsurprising that the logistic regression models perform less well than more complex models, it is worthwhile to explore why the RF-RW and the RNN-BiLSTM-focal models do not perform as well as the RF-CV models.

Table 4 The Balanced Accuracy of RF-RW and RF-CV on the training data and the testing data

	Training data (%)	Testing data (%)
RF-CV	89.5	93.0
RF-RW	99.3	72.7

RF-RW After investigating the model performance on the training data, we find an overfitting problem for the RF-RW model. The model performs worse on the testing data than on the training data. As shown in Table 4, the RF-RW model has a 99.3% balanced accuracy rate on the training data but only 72.7% on the testing data. This implies that the model fits too much on the patterns in the training data. On the other hand, the RF-CV model does not have this problem.

Why would the RF-RW overfit but not the RF-CV? It might be that the re-weighting method makes the model over-trained for the bubble cases in the training data. The re-weighting method replicates the bubble cases, which could cause the model to fit too much on peculiar bubble patterns in the training data. The patterns are not universal for all bubble cases and are likely not present in the testing data. This is perhaps why the RF-RW model performs worse on the testing data. On the other hand, the RF-CV alters the decision threshold. This method makes the model focus on the bubble cases generally but not restricted to cases seen in the training data. Without the over-training on the peculiar bubble patterns in the training data, the RF-CV model avoids the overfitting problem.

RNN-BiLSTM-Focal The RNN-BiLSTM-focal model's prediction results are no better than that of a naïve predictor. This is quite a disappointment, given the relevant research that endorses the model. It is natural to suspect that the poor result is due to a particular model configuration, and a calibration of the model can improve the performance. Following this line of reasoning, we have tried various techniques, including,

1. Adjusting hyperparameters: the number of neurons, the number of layers, the learning rate, the batch size
2. Changing the type of layers: BiLSTM, LSTM, dense layers
3. Changing the loss function: adjusting the parameters of the focal loss function
4. Re-assigning the training and testing data: measure the model performance on a new randomly selected testing data
5. Combinations of the above

However, none of these techniques works. The performance of the model does not see any significant improvement. While the above techniques cannot fully dismiss the suspicion of insufficient model calibration, at least simple model re-configuration does not help.

examined the probability output of the model to analyze the problem. The distribution of the probability output is shown in Fig. 6. It shows that the variance of probability outputs of the model is small: almost all outputs are clustered around 0.37.

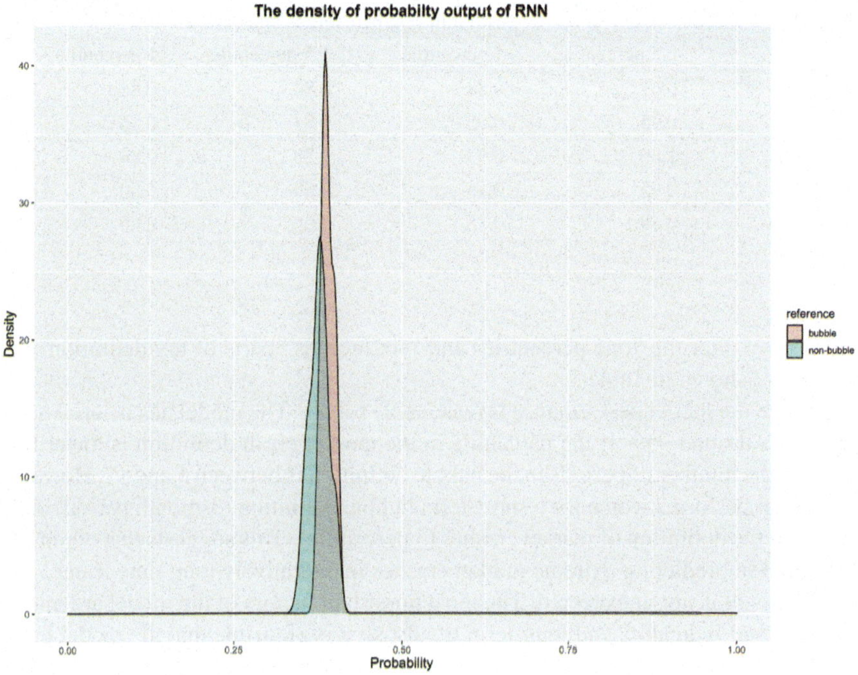

Fig. 6 The probability output of RNN-BiLSTM-focal

Also, the outputs of bubbles and non-bubbles overlap with each other. The model cannot distinguish the two based on the output. Based on this probability output, we suspect that the model is under-trained. This is likely because the model is too complex for our data size. A study on the impacts of data size on LSTM efficiency finds that three-year daily data is required for training an effective model [3]. Three-year daily data is approximately 1000 data points. This study only has 556 data points to train the RNN. It could be that the training data is insufficient to train the RNN model.

It is worth noting that the above model diagnoses for the RF-RW model and the RNN-BiLSTM-focal model are largely speculative. Due to the complexities of the models, there does not seem to be an easy way to check my explanations.

6.2 *Robustness Analysis*

We did the robustness analysis for the best-performing model, the RF-CV model. As discussed in the data section, the quantitative definitions of market crashes and bubbles are rather arbitrary. To check the model performance under different defi-

Table 5 Balanced accuracy of RF-CV with different percentiles and periods

	0.5 percentile	1 percentile	1.5 percentile	5 percentile
3 months	0.67	0.88	0.84	0.85
4 months	0.95	0.95	0.83	0.86
6 months	1.00	0.93	0.88	0.84
7 months	0.95	0.95	0.89	0.80
12 months	0.89	0.85	0.96	0.86

nitions, we alter the "one percentile" and "six months" parts of my definition. The results are shown in Table 5.

We see the model performance is reasonably robust. The model has balance accuracy rates around 95% if the percentile in the market crash definition is lower than one and the number of periods in the bubble definition is between 5 and 12. However, the model performs worse for a short-term bubble definition (3-month period) and a less extreme definition of market crashes (5 percentile). This suggests that our model is limited in predicting extreme market crashes in a relatively long time frame.

The result is not unexpected. The most important factors of the model are market fundamental indicators and long-term trends, so it is plausible that the model based on these factors has less predictive power in the short term. Besides, changing the decision threshold is deployed specifically for the imbalanced data problem. This method will not be effective if the data is less imbalanced due to a less extreme market crash definition.

6.3 *Limitations*

Besides the limitation that the model is unsuitable for short-term and less extreme market crash prediction, as discussed in the robustness analysis subsection, two other limitations are worth mentioning.

The first limitation is regarding the interpretation of the results. A balanced accuracy rate should not be confused with a precision rate. A balanced accuracy rate stands for the balanced percentage of correctly identified cases. On the other hand, a precision rate means, among all cases predicted as bubbles, the percentage of cases that are actual bubbles. While the RF-CV model has a highly balanced accuracy rate, the precision rate is much lower, at 36.7%. This means that if the model gives a *bubble* prediction, there is only a 36.7% possibility that a market crash would occur in the next six months. From a practical perspective, the model user should heed the possibility of false alarms when the model predicts bubbles.

Why would the two metrics differ? It can be best illustrated by the confusion matrix of the model prediction. As shown in Table 6, the model correctly identifies most cases in both categories, but precision (11/30) is still modest. Although the

Table 6 Confusion matrix of the RF-CV model, reference(columns) and prediction(rows)

	Bubble	Non-bubble
Bubble	11	19
Non-bubble	0	117

proportion of misclassified non-bubble cases is low (19/177), the number is non-negligible. Non-bubble cases are much more than bubble cases, so a small portion of misclassified non-bubbles (19) is more than the correctly identified bubbles (11). This drives the precision much lower than the accuracy.

The second limitation is that the training and testing data are not independent. While they are split randomly from the full data, they are both from a similar period. For example, the data from May 2008 is in the training set, whereas the data from June 2008 could be in the testing set. The two periods are positively correlated due to their proximity in time. At the training stage, the model fitter learns that May 2008 is in a situation bubble and incorporates its pattern into the model. When faced with the data of June 2008 in the testing set, it is easy for the model to correctly classify it as a bubble case since it has seen a similar instance in the training data. However, for real-world applications, the model would encounter cases with much less similarity to the training data. When asked to check if there is a bubble as of April 2021, the model does not have any information about whether March 2021 or May 2021 has a bubble. This is a harder problem for the model than the question of June 2008. It is dubious if the model would perform equally well. In short, the dependence between the training data and the testing data will likely make the models' testing performance overstated.

7 Conclusion

we start this research to detect equity bubbles and predict financial crashes. Previous empirical studies are limited by their theoretical assumptions or feature selection techniques. To overcome these limitations, we choose the model-free machine learning approach, and the features are selected based on theories but do not rely on theories. Based on the nature of the research problem and relevant studies, we use logistic regression, Random Forests, and Recurrent Neural Networks as my models. To tackle the challenge of the imbalanced output variable, we change the decision thresholds, re-sample the data for weighting, and adopt an asymmetric loss function.

The key findings of the research are as follows. The primary research goal is realized - we find a high-performance predictor. The Random Forest with CV-tuned performs the best, with a balanced accuracy rate of 93.0%. The model's high performance shows that machine learning is a viable method for equity bubble detection. In particular, it confirms the findings of [14] that ensemble learning methods are promising in financial crisis prediction. Also, the unsatisfactory result of the RNN model suggests that deep learning methods may not always outperform traditional methods. Especially in the case of small and medium data sizes, the complexity of the RNN model could lead to under-training. In the research, we also identified important factors for the prediction. Behind the predictive models, it is found that long-term market returns and market fundamental indicators have the highest predictive power, whereas short-term market returns and macroeconomic data are not as important. Finally, it is worth mentioning that the best-performing model predicts that market crashes are likely in the next few months (starting from April 2021).

The research has two main contributions. First, we use a new research methodology. we improve the prediction accuracy of machine learning methods by utilizing variable importance insights from economics theories. The result of this research encourages future machine learning studies to select more features based on theories. Second, we build a high-performing predictor (RF-CV) for detecting equity bubbles and predicting market crashes. While the caveat is that the model only has a modest precision rate, we believe the model can be useful for investment practitioners. For example, policymakers can use it as an early warning system, and investors can use it as a risk management tool.

The prediction horizon and the type of market downturn limit the research. The model cannot be generalized to short-term, e.g., 1-month and mild market crash predictions. From a practical perspective, the model does not have a high precision rate, so it is prone to sending false alarms. Besides, it is suspected that a positive correlation between the training data and the testing data makes the model overstate its testing performance.

Future research is needed to analyze deeper and overcome the correlation problem. Also, the extension of this research can include more features in the predictive models. Lag values can be included to capture the time series pattern of the data better. Besides, through feature engineering, we can use the result of econometric tests as input variables. For example, for the test proposed by [7], we can include a dummy variable of whether the non-stationary null hypothesis is rejected. Using the same methodology as this research, feature engineering is not subjected to theoretical challenges such as the joint hypothesis problem since it only includes the test result as a feature. In this way, we can include potentially important factors and test the empirical performance of the econometric tests. Future research can also consider including more machine learning algorithms, such as gradient boosting and support vector machines. Hopefully, using these methods, future research can improve on the findings of this research and make the results more robust.

Overall, the novelty of the paper lies in its innovative approach to combining economic theory with advanced machine learning techniques, addressing imbalanced data issues, and providing valuable insights into the predictors of financial market crashes.

8 Appendix*

See (Table 7).

Table 7 Summary of features

Variable	Min	q_1	$\tilde{x}$	q_3	Max	$\bar{x}$	s
Annualized GDP growth rate (%)	−31.4	1.4	3.0	4.7	33.4	3.0	4.5
Annualized inflation (%)	−19.3	1.5	3.2	5.4	24.0	3.7	3.9
10-year T-bill yield (%)	0.6	3.9	5.7	7.7	15.3	6.0	2.9
Shiller P/E ratio	6.6	15.0	20.5	25.7	44.2	20.6	8.0
Consumer confidence index	95.7	98.9	100.5	101.1	103.1	100.0	1.6
Market capitalization-to-GDP ratio	0.0	0.1	0.1	0.1	0.2	0.1	0.0
1-Month S&P 500 return (%)	−20.4	−1.2	1.0	2.8	12.0	0.6	3.6
3-Month S&P 500 return (%)	−31.1	−1.7	2.4	6.2	25.9	2.0	7.0
6-Month S&P 500 return (%)	−37.8	−2.4	4.5	9.9	38.0	4.0	10.6
12-Month S&P 500 return (%)	−42.5	−0.8	9.7	18.3	52.7	8.0	15.3
60-Month S&P 500 return (%)	−32.6	9.4	45.4	70.4	213.9	46.1	46.5

References

1. Allen F, Morris S, Postlewaite A (1993) Finite bubbles with short sale constraints and asymmetric information. J Econ Theory 61(2):206–229
2. Bashchenko O, Marchal A (2020) Deep learning for asset bubbles detection. Capital Markets: Market Efficiency eJournal
3. Boulmaiz T, Guermoui M, Boutaghane H (2020) Impact of training data size on the lstm performances for rainfall-runoff modeling. Modeling Earth Systems and Environment 6:2153–2164
4. Buffett W, Loomis C (2001) Warren Buffett on the stock market. Fortune Investor's Guide, pp 80–94
5. Chatzis SP, Siakoulis V, Petropoulos A, Stavroulakis E, Vlachogiannakis N (2018) Forecasting stock market crisis events using deep and statistical machine learning techniques. Exp Syst Appl 112:353–371. https://doi.org/10.1016/j.eswa.2018.06.032, https://www.sciencedirect.com/science/article/pii/S0957417418303798
6. Connolly R, Stivers C, Sun L (2005) Stock market uncertainty and the stock-bond return relation. J Financ Quant Anal 161–194
7. Craine R (1993) Rational bubbles: a test. J Econ Dyn Control 17(5–6):829–846
8. Daniel K, Hirshleifer D, Subrahmanyam A (2005) Investor psychology and security market under-and overreaction. Adv Behav Financ II:460–501
9. Evans GW (1991) Pitfalls in testing for explosive bubbles in asset prices. Am Econ Rev 81(4):922–930. http://www.jstor.org/stable/2006651
10. Fama EF (2014) Two pillars of asset pricing. Am Econ Rev 104(6):1467–85
11. Feldstein M (1978) Inflation and the stock market. Working Paper 276, National Bureau of Economic Research (08 1978). https://doi.org/10.3386/w0276, http://www.nber.org/papers/w0276

12. Gürkaynak RS (2008) Econometric tests of asset price bubbles: Taking stock*. J Econ Surv 22(1):166–186. https://doi.org/10.1111/j.1467-6419.2007.00530.x, https://onlinelibrary.wiley.com/doi/abs/10.1111/j.1467-6419.2007.00530.x
13. Jacobsson E (2009) How to predict crashes in financial markets with the log-periodic power law. Master dissertation. Department of Mathematical Statistics, Stockholm University
14. Lin W, Hu Y, Tsai C (2012) Machine learning in financial crisis prediction: a survey. IEEE Trans Syst Man Cybern Part C (Appl Rev) 42(4):421–436 (2012). https://doi.org/10.1109/TSMCC.2011.2170420
15. Moser R (2019) Predicting stock market crashes. An attempt with statistical machine learning techniques and neural networks. https://towardsdatascience.com/predicting-stock-market-crashes-with-statistical-machine-learning-techniques-and-neural-networks-bb66bc3e3ccd (01 2019). Accessed 23 Apr 2021
16. multpl: Shiller pe ratio. https://www.multpl.com/shiller-pe. Accessed 17 Apr 2021
17. Olah C (2015) Understanding lstm networks. http://colah.github.io/posts/2015-08-Understanding-LSTMs/
18. Shiller R (1980) Do stock prices move too much to be justified by subsequent changes in dividends?
19. Shiller R (2015) Irrational exuberance, 3 edn. Princeton University Press. https://EconPapers.repec.org/RePEc:pup:pbooks:10421
20. Siami-Namini S, Tavakoli N, Namin AS (2019) A comparative analysis of forecasting financial time series using arima, lstm, and bilstm. arXiv:abs/1911.09512
21. Taleb NN (2007) The black swan: the impact of the highly improbable, vol 2. Random House
22. Tversky A, Kahneman D (1973) Availability: a heuristic for judging frequency and probability. Cogn Psychol 5(2):207–232. https://doi.org/10.1016/0010-0285(73)90033-9, https://www.sciencedirect.com/science/article/pii/0010028573900339
23. Wöckl I (2019) Bubble detection in financial markets-a survey of theoretical bubble models and empirical bubble detection tests. Behav Exper Financ eJournal

Dynamic Approach for Pre-processing of Microarray Gene Expression Data

Hemant B. Mahajan and K. T. V. Reddy

Abstract The rapid growth of microarray gene expression (MAGE) databases allows computer models for disease diagnosis, prognosis, prediction, etc. Soft computing, data mining, machine learning, and pattern recognition have helped researchers develop computationally effective models to identify new disease classes gain insight into disease mechanisms, and develop diagnostic or therapeutic targets through molecular classification. Molecular classification is a supervised pattern recognition problem that challenges the development of novel classification and gene selection models by addressing high dataset dimensionality, overfitting due to imbalanced data elements in different classes, and computational complexity. This paper focused on the high dimensionality problem by proposing the dynamic pre-processing algorithm. The purpose of this approach is to use a dynamically determined threshold value to eliminate redundant and noisy gene data. The complex MAGE data is converted into the simple using the proposed pre-processing algorithm. The core part of this model is converting the input MAGE data using Discrete Cosine Transform (DCT), applying dynamic thresholding, and then performing inverse DCT. Using publicly accessible datasets, the suggested algorithm's effectiveness is assessed. The simulation results revealed the proposed MAGE data reduction algorithm discards the redundant genes by 13.79% more and minimizes the processing time by 6.04% compared to existing methods.

Keywords Classifiers · Dimensionality reduction · Discrete cosine transform · Data processing · Dynamic thresholding · Gene data filtering · Microarray gene expression

H. B. Mahajan (✉) · K. T. V. Reddy
Datta Meghe Institute of Higher Education and Research, Wardha, India
e-mail: mahhemant@gmail.com

K. T. V. Reddy
e-mail: ktvreddy.feat@dmimsu.edu.in

A. C. Frery et al. (eds.), *Computational Intelligence and Data Analytics*, Lecture Notes on Data Engineering and Communications Technologies 236,
https://doi.org/10.1007/978-981-96-0451-7_7

1 Introduction

The rapid change in gene expression databases expands the scope of the possibilities for the development of computer models for disease diagnosis, prognosis, disease prediction, and other related purposes [1]. The advancements in computer diagnosis have helped researchers overcome the challenges they faced in developing computationally effective models to identify the new class of disease. These advancements have also assisted in gaining insight into disease mechanisms and developing diagnostic or therapeutic targets, which is referred to as molecular classification [2]. Molecular classification, or simply classification [3, 4], is a supervised pattern recognition problem that poses challenges for the development of novel classification and gene selection models by addressing issues related to the high dimensionality of the datasets, overfitting of the datasets due to imbalanced data elements in different classes, and the computational complexity issues [5]. These challenges can be broken down into three categories: overfitting of the datasets due to imbalanced data elements in different classes.

Several other classifiers have previously been constructed and received recognition for their use in the categorization of gene expression datasets [6, 7]. The scholars have also conducted a great deal of comparative study as well. When it comes to gene expression datasets, there are certain variations of genes that are simple to categorize as either detrimental or innocuous to a human person, whereas, the importance of other variants is yet unknown [8]. Therefore, analyzing the data is an essential part of the diagnostic process in medicine. The curse of dimensionality, which refers to the high number of dimensions that input data has, makes it difficult to extract relevant genes from gene expression data. As a result, the identification of vital genes continues to be a difficulty for both academics and other stakeholders involved in the study of gene expression data. The solution to this issue is called gene selection, and it involves writing specific requirements and limitations into an algorithm to choose a small number of genes that are nevertheless among the most informative [9]. These genes may then be utilized by data mining tools such as classifiers to accurately categorize the dataset. The most significant problem is that achieving a level of classification accuracy of one hundred percent is extremely difficult and is dependent on the particular selection methods used. Therefore, to achieve a higher level of accuracy, it is necessary to provide the classifier with the most informative genes possible. The approaches of gene selection need to reduce the amount of computational instability as much as possible while simultaneously improving the classification methods' level of precision. It is necessary to determine the relevance of the particular group of genes that was chosen. The difficulty in the computing element of an algorithm causes a delay in the performance of the learning algorithms. This is because the dataset has a large dimensionality, which is one of the qualities that causes the delay [10]. Therefore, dimension reduction has a certain benefit, and to solve any classification issue, it is necessary to choose the most effective and informative gene subset. This subset should have the fewest dimensions possible of the most vital traits, and it should be able to solve the problem effectively [11]. Because, initially, such things are unknown, it is an extremely difficult task to attain such a relevant and meaningful subset of features. Because of this, dimension reduction methods are utilized to learn about the undiscovered subset of characteristics (genes) [12].

We design a novel method for MAGE pre-processing to reduce dimensionality by discarding noisy, irrelevant, and redundant gene profiles from the raw input MAGE data. We designed a pre-processing algorithm that consists of a series of steps including the dynamic thresholding-based DCT approach. The proposed algorithm is experimentally evaluated using the publicly available research dataset. The novelty of the proposed model is highlighted via below list of contributions below:

- Initially, the low expressed genes, low entropy genes, low variance genes, and relevance indexed-based filtering are applied to identify the redundant genes.
- To reduce the noisy genes, we applied the DCT on gene profiles and then reduced the noisy gene profiles using the dynamically discovered threshold value.
- The threshold value for each gene profile data is dynamically discovered by a novel approach of dividing the data into foreground and background gene data.
- The proposed model is evaluated using breast and lung cancer gene profiles dataset.

In the second section, a discussion of several relevant works is presented. In Sect. 3, we will discuss the methodology that has been suggested. The findings of the simulation are presented in Sect. 4. In the fifth section, the conclusion is presented.

The first paragraphs that follow a table, figure, equation, etc., do not have an indent, either.

2 Related Works

In this section, we have reviewed different gene pre-processing and selection techniques those are recently proposed.

2.1 State-of-the-Arts

In this investigation, the confusion factor and risk factor were utilized to identify physiologically significant genes for cancer prediction [13]. Group cancer genes were identified using the weighted general group lasso (WGGL) framework [14]. The investigation followed these steps. A system that weighs genes and groups based on their mutual information was created to determine their relative value. The ISVM-RFE model was presented for GFS and classification [15]. The intrinsic, linear, and non-linear properties were examined. The study [16] used a unique deep fuzzy neural network to manage gene data ambiguity and diagnose cancer. The SDSPCA method was proposed for gene selection and cancer classification [17].

An unsupervised two-stage Gene Feature Selection (GFS) technique for cancer sample classification uses a gene encoder [18]. In the first step, we used spectral-based feature selection filters, PCA, and a correlation filter. The GA used autoencoder-based clustering to examine the chromosome. In [19], authors developed an improved linear regression-based embedded general equilibrium forecasting system. The ensemble feature technique was created via boosting. It boosts linear regression-based GFS performance. A tri-level GFS method using transformation and wavelets classifies prostate cancer [20]. A method is given in [20]. The [21] multilayer recursive feature elimination (MGRFE) method uses an embedded integer-coded GA. It sought the gene combination that required the fewest observations while delivering the most information about the events under study. SVM-mRMRe is a hybrid gene selection model [22]. It selects the most useful and important genes from high-dimensional microarray data using ensemble, filter-based, and embedding techniques. Two GFS algorithms were utilized to compare and intersect six distinct approaches in [23]. The authors [24] suggest two distinct ways to rank multivariate traits: pairwise consistency and pairwise correlation. In [25], writers compared feature selection and classification algorithms to discover which had the best prediction accuracy. An empirical research using three cancer-related microarray datasets assessed each combination's efficacy.

At-tribute Selection and PCA were used to reduce gene expression datasets [26]. The pre-processing stage applied both reduction procedures before experimental testing. A unique feature selection method that selects meaningful, relevant, and non-redundant properties from gene data was proposed [27]. The suggested method found non-redundant features by maximizing patient pair similarity using an approximative k-cover strategy. Molecular feature selection tool, a novel ensemble-based feature selection method for '-omics' biomarker discovery [28]. The molecular feature selection method found immune-related genes that may be biomarkers for three prostate cancer subtypes. Petinrin et al. [29] described a cancer detection strategy that included feature selection, oversampling, and classification models. The authors of [30] reviewed recently published MAGE data reduction and classification algorithms to identify cancers and evaluate their effectiveness.

2.2 Research Gaps

We have reviewed the different techniques [13–30] of gene data pre-processing, but these methods suffer from the challenges. Microarray gene expressions are complicated by high dimensionality, redundancy, and data noise. Recently, gene profile analysis has paid less attention to these issues. Existing redundancy identification and removal methods [13–30] were static. Such methods failed to identify redundant genes and left numerous gene profiles. It causes poor gene profile analysis and classification. These problems motivate us to propose a novel approach for gene data redundancy removal according to the contributions mentioned earlier.

3 Proposed Methodology

The proposed approach of pre-processing the high-dimensional and redundant gene profile data is designed in this section. Figure 1 shows the steps of the proposed algorithm which consists of a combination of different filtering techniques and a dynamic thresholding-based noise reduction method. Let D is input gene profile data to the proposed model. After acquiring the input data D, we first removed the gene profiles that contain the empty or NaN values. In most of the existing studies, we have not found such steps. The presence of empty or NaN gene profiles misleads the classification. After the removal of NaN or empty gene profiles, we have further applied the Relevance Index-based Filtering (RIF) technique to remove the gene profiles having small variance values. After that next step is to remove genes that have very low absolute expression values using the *genelowvalfilter*(.) function. The low entropy genes also cause redundancy in the gene data processing; therefore, we have removed such genes using the *geneentropyfilter*(.) function. Thus, three different filtering techniques were applied that focused on removing the genes whose having low variance, low entropy, and low expression values from the input data D.

The steps are expressed in the below Equations. Equation (1) shows the removal of low variance genes removal using the RIF technique. Equation (2) shows the removal of genes with low expression values. And Eq. (3) shows the removal of genes with low entropy. Let D be the output data after removing empty and NaN genes.

$$f1 = genevarfilter(D) \tag{1}$$

$$f2 = genelowvalfilter(f1) \tag{2}$$

$$f3 = geneentropyfilter(f2) \tag{3}$$

After that, we applied the 1D DCT on the $f3$ data for the removal of noisy gene profiles using the dynamically computed threshold value T as shown in Fig. 1. Equation (4) applies the DCT to input data and produces the transformed gene profile data $f4$.

$$f4 = dct(f3) \tag{4}$$

After that, on each gene profile d in $f4$, we have compared its value with the dynamically computed threshold value T. The processing of computing threshold value is given below.

Step 1: The mean operation is used to compute the first threshold value in Eq. (5).

$$t1 = mean(mean(f4)) \tag{5}$$

Step 2: Using the first threshold $t1$, divide the gene data into two sections.

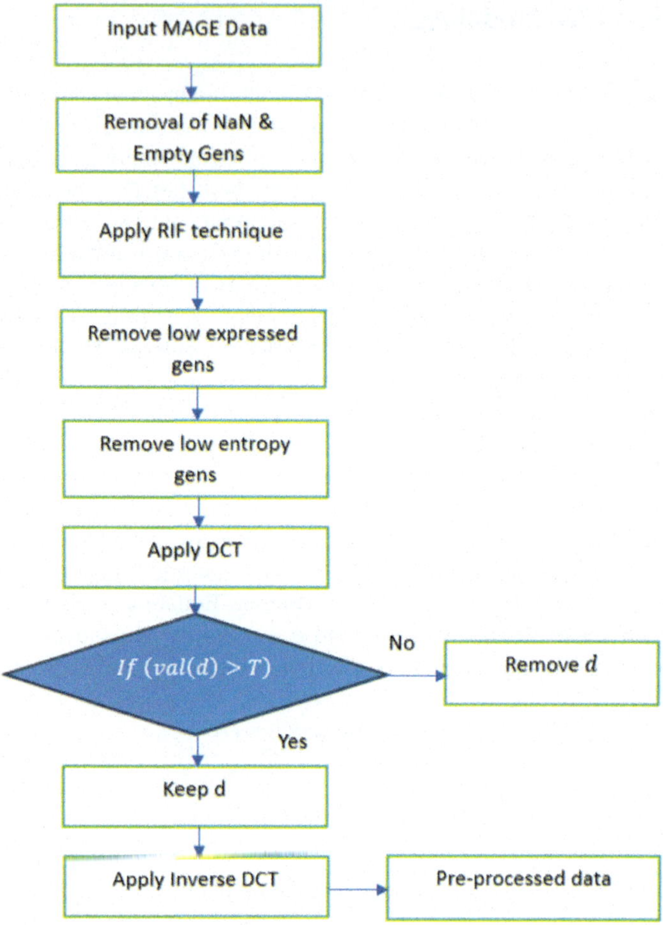

Fig. 1 Proposed approach for MAGE data pre-processing

$$p^1 = \begin{cases} f4(i), & f4(i) \le t1 \\ 0, & Else \end{cases} \tag{6}$$

$$p^2 = \begin{cases} f4(i,j), & f4(i,j) > t1 \\ 0, & Else \end{cases} \tag{7}$$

Step 3: Average mean of p^1 and p^2.

$$t2 = mean(mean(p^1)) \tag{8}$$

$$t3 = mean(mean(p^2)) \tag{9}$$

Step 4: Finally, compute the threshold value T as:

$$T = (t2 + t3)/2 \tag{10}$$

As shown in Fig. 1, the mean value of each gene data d is compared with dynamically computed threshold value T. Accordingly, the gene data d is discarded or kept in the final pre-processed data P using Eq. (11).

$$P = \begin{cases} f4(i), & mean(d) > T \\ remove, & Else \end{cases} \tag{11}$$

where, $d \in f4(i)$.

4 Simulation Results

We used the MATLAB simulation program to develop the proposed model. The machine must have a Windows 10 operating system, 8 GB of RAM, and a minimum hard drive size of 40 GB. The MATLAB 2020a version with the signal processing toolbox was used. The proposed method was tested using The Cancer Genome Atlas Lung Adenocarcinoma (TCGA-LUAD) Dataset [31]. The TCGA-LUAD dataset's attributes are listed below.

- Platform: cDNA microarrays
- Diagnostic classes:
- Normal: 59 examples
- Tumor: 517 examples
- Gene count: 20,429
- Sample count: 576

Yet another dataset was used to investigate the proposed approach. Breast Cancer Gene Expression (BCGE) [32] Dataset was used.

- Platform: cDNA microarrays
- Number of Classes: 6
- Number of genes: 54,676
- Number of samples: 151.

The results of the suggested pre-processing algorithm are shown in step-by-step form in the following figures (Figs. 2, 3 and 4). The results of this process are depicted in Fig. 2, which may be accessed by clicking here. Because the supplied data do not contain any entries of that kind, the removal has not appeared in the results. In addition to that, we made use of the RIF, low expression gene removal, and low entropy gene removal approaches.

Experimental Screens	Significance	Reduced Data
Name Value dbclass 576x1 double genes 20429x1 cell genexp 576x20429 double nnclass 576x2 double	Original acquired Lung DB	None
Name Value dbclass 576x1 double genes 20429x1 cell genexp 576x20429 double nnclass 576x2 double	Removal of NaN and Empty Gene profiles	N/A

Fig. 2 Removal of NaN and empty genes using TCGA-LUAD dataset

Experimental Screens	Significance	Reduced Data
Name Value dbclass 518x1 double genes 518x1 cell genexp 518x20429 double nnclass 518x1 double	Relevance Index-based Filtering	58
Name Value dbclass 518x1 double genes 518x1 cell genexp 518x20429 double nnclass 518x1 double	Remove low-expressing genes	N/A
Name Value dbclass 440x1 double genes 440x1 cell genexp 440x20429 double nnclass 440x1 double	Remove low-entropy genes	136

Fig. 3 Removal of low variance, low expressed, and low entropy gene profiles using TCGA-LUAD

Experimental Screens	Significance	Reduced Data
dbclass 400x1 double genes 400x1 cell genexp 400x20429 double nnclass 400x1 double	DCT-Dynamic threshold-based removal	40

Fig. 4 DCT-based filtering of noisy gene data using TCGA-LUAD

It reveals that a total of 194 gene profiles were narrowed down by utilizing these procedures. Figure 4 displays, as a final point of interest, the further reduction of redundant genes that was achieved by employing dynamic threshold-based gene profile reduction. The profiles of a total of forty genes were trimmed down. As a result, 176 gene profiles were eliminated from a total of 576 gene profiles since the pre-processing method that was proposed deemed them to be unnecessary, noisy, or redundant. To demonstrate that the model that has been proposed is effective, we have compared its performance in Tables 1 and 2 with that of newly proposed data reduction approaches found in [27–30] using both lung and breast cancer datasets. We have evaluated the results in terms of the overall number of decreased genes as well as the processing time for each. In comparison to the methods that are already in use, the suggested approach is far more effective in reducing the large number of noisy genes. Additionally, the reduction of a greater number of noisy gene profiles is accomplished in a shorter amount of time. The simplification of gene profiles is an additional step towards ensuring secure transmission [33] to ensure the security against different cyber threats [34] and privacy preservation [35].

The dynamically discovered threshold value, in comparison to the predetermined or static threshold approaches that are currently in use, is the primary cause for the performance gain that was achieved by employing the proposed method of dimensionality reduction. Not only does it preserve some genetic redundancy, but it also requires more time. The gene profiles with low variance, low expression, and low entropy were eliminated before the DCT analysis began, which also contributed to better overall results using the proposed approach.

Table 1 Data reduction techniques analysis using TCGA-LUAD dataset

Methods	Total genes	Reduced genes	Pre-processing time (s)
[27]	576	486	1.77
[28]	576	463	1.54
[29]	576	445	2.11
[30]	576	462	1.89
Proposed	576	400	1.71

Table 2 Data reduction techniques analysis using BCGE dataset

Methods	Total genes	Reduced genes	Pre-processing time (s)
[27]	151	132	0.72
[28]	151	125	0.63
[29]	151	121	0.98
[30]	151	124	0.78
Proposed	151	108	0.67

5 Conclusion

MAGE data pre-processing is the act of discovering and correcting (or eliminating) corrupt or erroneous records from a dataset. Alternatively, it may refer to the identification of wrong, incomplete, or irrelevant elements of the data, followed by the replacement, modification, or deletion of the coarse or noisy data. In this paper, we have designed an efficient approach to perform the MAGE data filtering using the combination of different filtering techniques with a dynamic threshold-driven DCT approach for the removal of irrelevant, noisy, and redundant gene profiles. We have evaluated the proposed approach using MATLAB and publicly available breast and lung gene profile datasets. Simulation results showed that the proposed MAGE data reduction approach discards 13.79% more redundant genes and reduces processing time by 6.04% compared to the state-of-the-art. Applying feature selection and classification methods for disease prediction will be the possible future direction for this work.

References

1. Mayer CD, Glasbey CA (2005) Statistical methods in microarray gene expression data analysis. In: Husmeier D, Dybowski R, Roberts S (eds) Probabilistic modeling in bioinformatics and medical informatics. Advanced information and knowledge processing. Springer, London. https://doi.org/10.1007/1-84628-119-9_7
2. Tarca AL, Romero R, Draghici S (2006) Analysis of microarray experiments of gene expression profiling. Am J Obstet Gynecol 195(2):373–388. https://doi.org/10.1016/j.ajog.2006.07.001
3. Kaushik R, Sivaselvan B, Kamakoti V (2023) Clinical thermography for breast cancer screening: a systematic review on image acquisition, segmentation, and classification. IETE Tech Rev. https://doi.org/10.1080/02564602.2023.2238683
4. Alves S, Lanna C, Figueiredo K, Boroni M, Vellasco M (2021) Multi-omic data integration applied to molecular tumor classification, pp 2171–2178. https://doi.org/10.1109/BIBM52615.2021.9669609
5. Koumakis L, Potamias G, Tsiknakis M, Zervakis M, Moustakis V (2015) Integrating microarray data and GRNs. In: Guzzi P (eds) Microarray data analysis. Methods in molecular biology, vol 1375. Humana Press, New York, NY. https://doi.org/10.1007/7651_2015_252
6. Hamed BA, Ibrahim OAS, Abd El-Hafeez T (2023) Optimizing classification efficiency with machine learning techniques for pattern matching. J Big Data 10:124. https://doi.org/10.1186/s40537-023-00804-6
7. Cheung LWK (2012) Classification approaches for microarray gene expression data analysis. In: Wang J, Tan A, Tian T (eds) Next generation microarray bioinformatics. Methods in molecular biology, vol 802. Humana Press. https://doi.org/10.1007/978-1-61779-400-1_5
8. Khaire U, Dhanalakshmi R (2020) Stability investigation of improved whale optimization algorithm in the process of feature selection. IETE Tech Rev 39:1–15. https://doi.org/10.1080/02564602.2020.1843554
9. Richard AC, Lyons PA, Peters JE et al (2014) Comparison of gene expression microarray data with count-based RNA measurements informs microarray interpretation. BMC Genomics 15:649. https://doi.org/10.1186/1471-2164-15-649
10. Aziz R, Verma CK, Srivastava N (2017) Dimension reduction methods for microarray data: a review. AIMS Bioeng 4:179–197. https://doi.org/10.3934/bioeng.2017.1.179

11. Dai J, Lieu L, Rocke D (2006) Dimension reduction for classification with gene expression microarray data. In: Statistical applications in genetics and molecular biology, vol 5. Article 6. https://doi.org/10.2202/1544-6115.1147
12. Wang H, van der Laan MJ (2011) Dimension reduction with gene expression data using targeted variable importance measurement. BMC Bioinform 12:312. https://doi.org/10.1186/1471-2105-12-312
13. Paul A, Sil J (2018) Identification of differentially expressed genes to establish new biomarker for cancer prediction. IEEE/ACM Trans Comput Biol Bioinform 1–1. https://doi.org/10.1109/TCBB.2018.2837095
14. Wang Y, Li X, Ruiz R (2018) Weighted general group lasso for gene selection in cancer classification. IEEE Trans Cybern 1–14. https://doi.org/10.1109/TCYB.2018.2829811
15. Luo K, Wang G, Li Q, Tao J (2019) An improved SVM-RFE based on F-statistic and mPDC for gene selection in cancer classification. IEEE Access 7:1–1. https://doi.org/10.1109/ACCESS.2019.2946653
16. Bamunu Mudiyanselage T, Xiao X, Zhang Y, Pan Y (2019) Deep fuzzy neural networks for biomarker selection for accurate cancer detection. IEEE Trans Fuzzy Syst 1–1. https://doi.org/10.1109/TFUZZ.2019.2958295
17. Feng C-M, Xu Y, Liu J-X, Gao Y-L, Zheng C-H (2019) Supervised discriminative sparse PCA for com-characteristic gene selection and tumor classification on multiview biological data. IEEE Trans Neural Netw Learn Syst 1–12. https://doi.org/10.1109/TNNLS.2019.2893190
18. Khan U, Al-Obeidat F, Tubaishat A, Shah B, Halim Z (2022) Gene encoder: a feature selection technique through unsupervised deep learning-based clustering for large gene expression data. Neural Comput Appl 34. https://doi.org/10.1007/s00521-020-05101-4
19. Seo H, Cho D-H (2020) Cancer-related gene signature selection based on boosted regression for multilayer perceptron. IEEE Access 1. https://doi.org/10.1109/ACCESS.2020.2985414
20. Prabhakar S, Lee S-W (2020) Transformation based tri-level feature selection approach using wavelets and swarm computing for prostate cancer classification. IEEE Access 1. https://doi.org/10.1109/ACCESS.2020.3006197
21. Peng C, Wu X, Yuan W, Zhang X, Li Y (2019) MGRFE: multilayer recursive feature elimination based on an embedded genetic algorithm for cancer classification. IEEE/ACM Trans Comput Biol Bioinform. https://doi.org/10.1109/TCBB.2019.2921961
22. Elkafrawy P, Fathi H, Qaraad M, Kelany A, Chen X (2021) An efficient SVM-based feature selection model for cancer classification using high-dimensional microarray data. IEEE Access 1. https://doi.org/10.1109/ACCESS.2021.3123090
23. Houssein E, Hassan H, Al-Sayed M, Nabil E (2021) Gene selection for microarray cancer classification based on manta rays foraging optimization and support vector machines. Arab J Sci Eng. https://doi.org/10.1007/s13369-021-06102-8
24. Jiménez F, Sánchez G, Palma J, Miralles Pechuan L, Botía J (2022) Multivariate feature ranking with high-dimensional data for classification tasks. IEEE Access 10:1. https://doi.org/10.1109/ACCESS.2022.3180773
25. Zanella L, Facco P, Bezzo F, Cimetta E (2022) Feature selection and molecular classification of cancer phenotypes: a comparative study. Int J Mol Sci 23:9087. https://doi.org/10.3390/ijms23169087
26. Souza J, Francisco A, Macedo D (2019) Dimensionality reduction in gene expression data sets. IEEE Access 7:61136–61144. https://doi.org/10.1109/ACCESS.2019.2915519
27. Rana P, Thai P, Dinh T, Ghosh P (2021) Relevant and non-redundant feature selection for cancer classification and subtype detection. Cancers 13(17):4297. https://doi.org/10.3390/cancers13174297
28. Gerolami J, Wong JJM, Zhang R, Chen T, Imtiaz T, Smith M, Jamaspishvili T et al (2022) A computational approach to identification of candidate biomarkers in high-dimensional molecular data. Diagnostics 12(8):1997. https://doi.org/10.3390/diagnostics12081997
29. Petinrin OO, Saeed F, Salim N, Toseef M, Liu Z, Muyide IO (2023) Dimension reduction and classifier-based feature selection for oversampled gene expression data and cancer classification. Processes 11(7):1940. https://doi.org/10.3390/pr11071940

30. Osama S, Shaban H, Ali A (2022) Gene reduction and machine learning algorithms for cancer classification based on microarray gene expression data: a comprehensive review. Expert Syst Appl 213:118946. https://doi.org/10.1016/j.eswa.2022.118946
31. https://lce.biohpc.swmed.edu/lungcancer/datasetsearch.php?datasetid=60
32. https://www.kaggle.com/datasets/brunogrisci/breast-cancer-gene-expression-cumida
33. Mahajan H, Reddy KTV (2023) Secure gene profile data processing using lightweight cryptography and blockchain. Cluster Comput. https://doi.org/10.1007/s10586-023-04123-6
34. Kadam MV, Mahajan HB, Uke NJ, Futane PR (2023) Cybersecurity threats mitigation in internet of vehicles communication system using reliable clustering and routing. Microprocess Microsyst 102:104926. https://doi.org/10.1016/j.micpro.2023.104926
35. Uke NJ, Lokhande SA, Kale P, Pawar SD, Junnarkar AA, Yadav S, Bhavsar S, Mahajan H (2024) Distributed privacy preservation for online social network using flexible clustering and whale optimization algorithm. Clust Comput. https://doi.org/10.1007/s10586-024-04295-9

FPGA-Accelerated Deep Learning for Soil Organic Matter Prediction: An Innovative Machine Learning Approach

Vaishali S. Patil, Santosh A. Shinde, and Nandkishor M. Dhawale

Abstract This paper presents the results of the prediction of soil organic matter (SOM) content in agricultural soil samples by training and testing of various deep learning (DL) models in Python, followed by their implementation and testing on a Field Programmable Gate Arrays (FPGA). This paper investigates the transformative potential of cutting-edge technologies, with special attention to Image Processing and Image Analysis using DL models while implementing them on FPGAs to predict SOM content in soil samples. First, Deep Convolutional Neural Network (DCNN) models were trained using 180 soil sample images categorized into three classes, alongside predictor variables, with datasets partitioned into training, validation, and testing subsets. By leveraging TensorFlow and Keras Deep Learning frameworks, pre-trained weights, and data augmentation techniques, the DCNN model demonstrates exceptional performance, minimizing prediction errors. Moreover, the integration of Artificial Intelligence (AI) such as DCNN in image processing proves instrumental in precisely detecting and categorizing different soil types. This study incorporates popular DCNN models, including VGG-16, MobileNetV2, and InceptionV3, achieving notable accuracies. Notably, the proposed DCNN model surpasses the accuracy and outperforms the results reported in the literature. This showcases its efficacy in advancing SOM prediction through innovative DL approaches. Next, the integration of model optimization techniques, such as quantization and TensorFlow Lite, adds to the robustness of the findings, emphasizing the significance of these advancements in sustainable precision agriculture and environmental preservation.

V. S. Patil · S. A. Shinde (✉)
Department of Electronics, Shivaji University, Kolhapur, Maharashtra, India
e-mail: sas_eln@unishivaji.ac.in

V. S. Patil
e-mail: vsp.rs.eln@unishivaji.ac.in

N. M. Dhawale
Department of Instrumentation Engineering, P.V.P. Institute of Technology Budhgaon, Sangli, Maharashtra 416304, India

A. C. Frery et al. (eds.), *Computational Intelligence and Data Analytics*, Lecture Notes on Data Engineering and Communications Technologies 236,
https://doi.org/10.1007/978-981-96-0451-7_8

At last, but not least the resultant embedded FPGA-based system is not only user-friendly but also energy-efficient, representing a significant leap forward in enabling comprehensive soil analysis and fostering sustainable, precision farming practices.

Keywords Artificial intelligence · Deep learning · FPGA · Image classification · Convolutional neural networks · Precision agriculture · Soil organic matter

1 Introduction

Soil Organic Matter (SOM) is a vital carbon-containing component in soil, originating from decomposed plants, animals, and microorganisms. It plays a crucial role in soil health by acting as a nutrient reservoir, improving water retention, influencing soil structure, supporting microbial activity, and affecting the Cation Exchange Capacity (CEC). SOM is essential for nutrient cycling, water retention, and fostering the growth and function of beneficial soil microorganisms, contributing significantly to overall soil health. Various factors influence SOM content in the soil, including climate, vegetation, soil management practices, land use changes, microbial activity, soil texture, waterlogged conditions, and parent material. Climate affects decomposition rates, with warm and humid environments accelerating the process. Different vegetation types contribute varying amounts of organic matter, impacting SOM content. Soil management practices, land use changes, microbial activity, soil texture, and parent material also play roles in determining SOM levels.

Image processing, particularly leveraging AI and ML performances, presents a valuable tool for accurately identifying and categorizing soil types based on soil images. This study utilizes TensorFlow and Keras Deep Learning frameworks, employing CNN model, to categorize soil images. The models achieve good validation accuracies, with the CNN model surpassing DCNN models and previous literature works. These image processing techniques provide an efficient and accurate means for soil analysis, enhancing agricultural research and environmental monitoring. Advanced image processing techniques for assessing SOM include diverse computational models like DenseNet, ANN, Support Vector Regression (SVR), and random forests. Deep learning models, including multilayer perceptron and CNNs, offer viable approaches for analyzing soil data. FPGA-based systems demonstrate high accuracy and real-time processing capabilities for monitoring SOM content, promising efficient and accurate soil assessment techniques. Edge computing is increasingly essential for applications requiring real-time processing and low latency, such as IoT and autonomous systems. FPGAs offer a promising solution due to their reconfigurability and parallel processing capabilities. This paper details the process of training machine learning models in Python, their performance metrics, and the subsequent re-implementation of FPGA hardware for edge deployment. The study focuses on soil classification using DL frameworks, specifically TensorFlow and Keras, highlighting the advantages of image processing for quick and accurate estimation of SOM. The advantages of image processing for SOM estimation include high accuracy, efficiency, real-time processing capabilities, and customization for specific tasks. Traditional methods for SOM assessment can be costly, labor-intensive, and time-consuming. Soil color, captured through digital cameras, proves to be a valuable indicator of soil composition, enabling rapid and precise measurements of soil organic matter distribution. The primary goal of the study is to accurately classify soil images using pre-trained DL models, emphasizing the comparison and evaluation of different neural network architectures. The proposed CNN model outperforms DCNN models, showcasing the effectiveness of image processing in soil analysis.

The paper follows a structured organization to delve into various facets of the research. Section 2 initiates a comprehensive literature review, comparing different deep learning models to establish a foundation. Moving forward, Sect. 3 delineates the data acquisition process, encompassing image collection and the setup of the web camera, while also addressing image preprocessing and dataset preparation. Section 4 offers a visual representation through the presentation of the system's block diagram. Section 5 delves into the core of the research, focusing on image processing for Soil Organic Matter (SOM) analysis employing Convolutional Neural Network (CNN) overlay. Transitioning seamlessly, Sect. 6 explores transfer learning using pre-trained models, including VGG16, Inception-v3, and MobileNet. The subsequent sections, Sects. 7 and 8, delve into the performance and results of these pre-trained models, followed by a comparative analysis. Finally, Sect. 9 draws the paper to a close with a succinct conclusion, summarizing the findings and proposing avenues for future research.

2 Literature Review

Modern methods for assessing SOM include deep learning for predicting content when correlated with soil spectral reflectance [1]. Various models, including DenseNet, ANN, SVR, and random forests, have been adapted for non-linear relationships between spectral data and soil features [2–4]. Researchers have also used soil color data from low-cost digital cameras and hyperspectral imaging to derive strong correlations between color attributes and SOC [5]. Various models, including PLSR, BPN, SVM, LR, and RBF-based neural networks, have proven effective in predicting SOM content [6] Deep learning models like multilayer perceptron and CNNs offer viable approaches for analyzing soil data [7]. Novel FPGA-based systems, proposed by [8, 9], have demonstrated high accuracy and real-time processing capabilities for monitoring SOM content, promising efficient and accurate soil assessment techniques [10, 11]. The study focuses on soil organic matter classification and its impact on agricultural productivity and ecosystem health. TensorFlow and Keras are DL frameworks used for image processing and classification, enabling accurate categorization of soil types based on images. These frameworks automate the identification and classification of soil types, enhancing accuracy and time efficiency [12]. The study aims to develop a CNN classification model for image processing to evaluate soil organic matter quickly. The study employed two deep learning models, CNN and ANN, and a subset of specific spectral data to assess their accuracy. The most effective model was used to fit the spectral reflectance data, and the study aimed to determine if certain data preprocessing steps could be omitted for streamlining the process.

SOM is a crucial component in soil analysis, influencing properties like fertility, pH, water retention, structure, and health [13–15]. It aids in sustainable land management, agriculture, and environmental preservation. Traditional methods for SOM assessment include Loss on Ignition, Walkley–Black method, Dry Combustion Analyzer, Soil Respiration, and Differential Scanning Calorimetry. Digital cameras offer advantages such as ease of sampling, avoidance of chemicals and hazardous materials, and speed, cost-effectiveness, and precision [16]. This study aims to accurately classify soil images using pre-trained Deep Learning (DL) models and compare the performance of different neural network architectures. The central objective is to identify the most suitable DL model for this task by thoroughly assessing and comparing their performances. The projected CNN model excelled both DCNN models, demonstrating the advantages of using image processing for SOM estimation.

3 Data Acquisition and Image Collection

Soil samples were collected manually from various locations in the Sangli district, particularly Miraj Tehsils, covering 1,640 km^2. A total of 180 samples were obtained from Kanchanpur in the Miraj Tehsil/Taluka of Sangli district, Maharashtra, India [17]. Each sample was carefully collected from 30 × 30 × 30 cm fields, sealed, and labeled before transport to the laboratory in Kanchanpur. Traditional soil organic carbon determination methods involving potassium dichromate and sulfuric acid titration were used, but they are time-consuming and hazardous. To address this, an image acquisition system was developed for efficient soil analysis, reducing light interference, and utilizing data for modeling and algorithm development. Figure 1 displays soil sample images captured during data collection, resized to meet classification model input requirements. Resizing adjusts image dimensions while maintaining the aspect ratio, ensuring standardized inputs for subsequent processing. This step reduces computational complexity, ensures consistent model input, and optimizes memory usage. Resized images facilitate meaningful feature extraction, contributing to classification model success. Commonly done using image processing libraries like OpenCV or deep learning frameworks such as TensorFlow or PyTorch.

The setup featured a 3MP Logitech C922 Pro Stream Webcam placed on a custom-built box with a compartment to reduce light reflection. A rectangular tray holder held the soil sample petri dish, with LED strips providing even illumination. The camera, positioned 20 cm from the sample, captured JPG photographs (1000 × 1000) under controlled lighting. A Python script cropped the images for analysis. Users may benefit from clear instructions on operating the camera for high-quality image capture and basic training on interpreting analysis results (Fig. 2).

Fig. 1 Soil samples image data set (representative ones)

Fig. 2 Illustration of FPGA kit assembled with webcam for image acquisition

3.1 Image Preprocessing and Preparation of the Dataset

The dataset was resized from 1000 × 1000 to 256 × 256 pixels using random cropping for machine learning model training. Data augmentation techniques such as reversing and scaling increased the dataset to 6480 images, enhancing model efficiency. Soil Organic Matter (SOM) classification into low, medium, and high classes was based on expert opinion. For instance, soil with SOM less than 0.865 was classified as low, while SOM between 0.87 and 1.29 was categorized as medium, and SOM above 1.29 was considered high illustrated in Fig. 3 [18]. Fine-tuning involved freezing the initial layers and adjusting the remaining layers for image

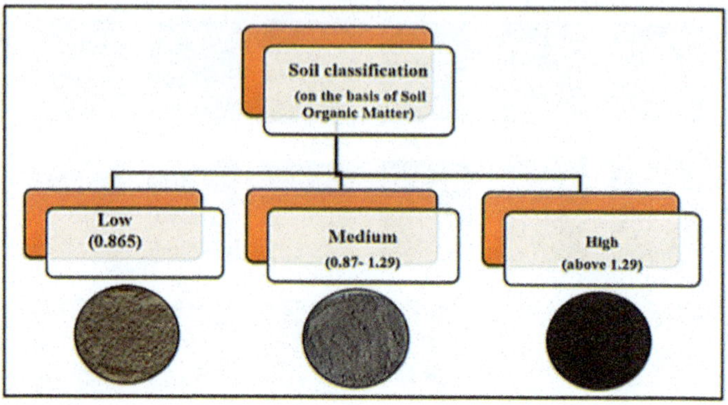

Fig. 3 Soil classification based on organic matter

qualities. Optimization included setting appropriate learning rates and regularization for model training, with an emphasis on larger datasets for improved deep learning performance and generalization. Preprocessing techniques involved rescaling images to normalize pixel values and applying various data augmentation methods such as random rotations, shifts, shearing, zooming, and flips to increase dataset diversity. These steps significantly improved the diversity and quality of the training dataset, facilitating robust model training and enhanced generalization. Additionally, models were optimized using techniques such as learning rate scheduling, regularization, data augmentation, early stopping, model architecture, batch size, and optimization algorithms. These techniques help improve model robustness, reduce overfitting, and enhance training efficiency. Learning rate scheduling adjusts the learning rate during training to avoid local minima, while regularization prevents overfitting by penalizing complex models. Data augmentation increases dataset diversity, reducing overfitting and improving generalization. Early stopping monitors the model's performance on a validation dataset. By fine-tuning hyperparameters, regularization strategies, and data augmentation methods, researchers can create more robust CNN models with higher accuracy and performance.

4 Block Diagram of System

The complete drawing representation of the current system, illustrated in Fig. 4, is fixed on the KRIA KV260 FPGA device [19, 20]. The developed system is strategically engineered to proficiently monitor a spectrum of vital farm parameters, encompassing SOM. The integration strategy employs intuitive techniques for sensing, detection, and prediction of these pivotal parameters, ensuring a seamless and accurate analysis.

The developed system is developed for the efficient monitoring of soil SOM. To predict SOM, the system integrates a deep learning model. Various pre-trained

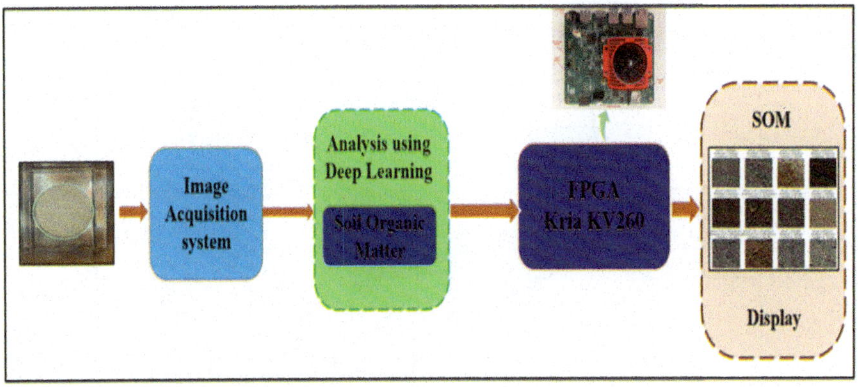

Fig. 4 Block diagram of the complete system

models such as VGG16, MobileNetV2, and InceptionV3 undergo testing and training, with a focus on factors like training time, model size, and accuracy, to determine the most suitable model for compatibility with the Kria KV260 Vision AI Starter FPGA board. Despite challenges in accuracy faced by the basic CNN model on soil image datasets, the adoption of VGG16 with fine-tuning algorithms in Python yields remarkable results. This approach exhibits excellent accuracy on both validation and testing datasets, particularly for new soil images, marking a significant advancement in the system's performance.

5 Image Processing for SOM Analysis

5.1 Convolutional Neural Network Architecture

The CNN architecture shown in Fig. 5, widely used in various applications including soil prediction, comprises essential elements such as input layer, convolutional layers, activation function, pooling layers, fully connected layers, and dropout layer. The input layer receives data, convolutional layers extract features, activation function introduces non-linearity, pooling layers reduce dimensionality, fully connected layers generate predictions, and dropout layer prevents overfitting [21–25].

The CNN architecture includes layers such as Conv2D, MaxPooling2D, Dense, and Flattened, with specific configurations like 32 filters, kernel size (3, 3), ReLU activation, pool size (2, 2), and 64 units. The final Dense layer serves as the output layer for classification into three classes.

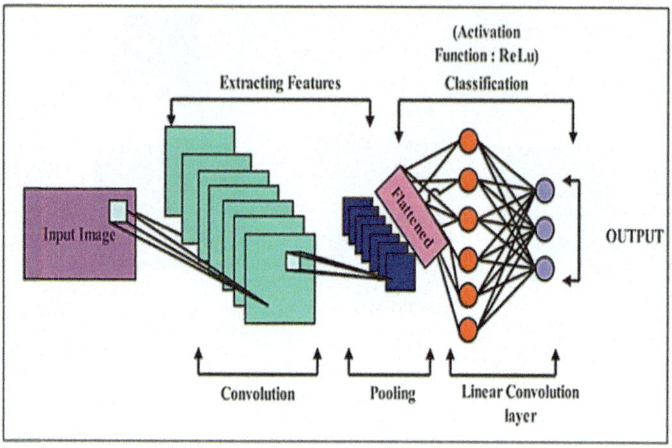

Fig. 5 Conventional neural network (CNN) architecture

6 Transfer Learning Using Pre-trained Models

Transfer learning with pre-trained models involves taking a deep learning model already trained on a large dataset, removing its top layers, adding new task-specific layers, and fine-tuning it on a new, smaller dataset. This process leverages the pre-existing model's knowledge to expedite learning and enhance performance on a related task with limited data. Deep learning is supported by various frameworks like Caffe, Python, MATLAB, Theano, TensorFlow, and Torch. To analyze images, annotated data is given for training. We use Python, and TensorFlow for algorithms. Pre-trained models are used as feature extractors, while semi-supervised or unsupervised learning methods can be employed if unlabeled data is available [26, 27].

6.1 Models of Deep Convolutional Neural Networks (DCNNs)

This article evaluates pre-trained CNN models (VGG16, InceptionV3, and MobileNet) using the ImageNet dataset. These models are suitable for large-scale tasks like object identification and image categorization, reducing the need for extensive labeled data and computing power. Transfer learning accelerates convergence and improves accuracy by leveraging information from pre-trained models for new tasks. However, fine-tuning may be necessary to adapt the model to specific problem domains. Keras Applications offers pre-trained deep learning models for tasks like prediction and feature extraction, adaptable to various image data formats. The VGG16 model is designed for 224 × 224 × 3 color images and features pre-trained weights from ImageNet. Google's MobileNetV2 is designed for mobile and embedded vision applications. The Inception V3 model for SOM prediction was tested on a dataset of 6480 and 180 images, achieving good number accuracy without straining computational budget [28–30].

7 Results

The soils are dark in color and have a texture analogous to clay. These have a clay-like texture, are granular to blocky in structure, and range in color from reddish brown to medium black. Laboratory analysis results of soil organic matter add a table on SOM for soil samples shown in Table 1.

7.1 Evaluation of Convolutional Neural Networks for Soil Organic Matter Prediction

The study evaluated various networks for classifying SOM in agricultural land, focusing on rural farmers' benefits. The soil has a clay-like texture and brown color, ranging from reddish brown to medium black. A dataset of 180 soil images was used to categorize the images into three distinct soil SOM classes. The primary objective was to develop a CNN for SOM classification with high accuracy, helping farmers understand soil properties and predict organic matter content through image-based classification, providing valuable insights for agricultural practices.

Figure 6 shows the training, validation accuracy, and loss of a machine learning model on a dataset of 6480 images across 3 classes. The model was trained using a sequential architecture with convolutional and pooling layers followed by dense layers.

The result in Fig. 7 is a 4×4 grid of subplots, each showing a test image along with its actual class, predicted class, and confidence level. This visualization helps evaluate how well the model is performing on the test dataset and identify any misclassifications.

Various machine learning and deep learning models were initially developed and tested using Python. The models include MobileNet V2, VGG16, and InceptionV3. The performance metrics were recorded for both validation and test datasets to evaluate their accuracy, precision, recall, and F1-score. The study reveals that VGG16 and InceptionV3 outperform MobileNet V2 in terms of overall performance and generalization. InceptionV3 shows the highest test performance, indicating strong generalization but not as high as VGG16's, indicating potential overfitting. VGG16 maintains a good balance between validation and test performance, indicating robust generalization without significant overfitting. MobileNet V2 shows low precision and recall, suggesting the need for further tuning or a different architecture. VGG16

Table 1 Laboratory analysis SOM value for soil samples

Sample ID	SOM level	Sample ID	SOM level	Sample ID	SOM level	Sample ID	SOM level
1	0.90	31	1.60	61	1.80	91	2.50
2	0.90	32	1.60	62	1.80	92	0.50
3	0.10	33	1.70	63	1.90	93	3.10
4	0.20	34	1.70	64	1.90	94	0.50
5	0.10	35	1.70	65	1.90	95	1.80
6	0.20	36	1.70	66	1.10	96	0.60
7	0.03	37	1.80	67	1.15	97	0.70
8	0.03	38	1.80	68	1.20	98	0.80
9	0.03	39	1.80	69	1.22	99	0.90
10	0.30	40	1.80	70	0.40	100	0.90
11	0.40	41	1.90	71	1.30	101	0.90
12	0.50	42	1.90	72	1.40	102	0.90
13	0.50	43	1.90	73	1.40	103	0.22
14	0.60	44	1.20	74	1.41	104	0.22
15	0.70	45	1.30	75	1.42	105	0.81
16	0.80	46	1.40	76	1.49	106	1.00
17	0.90	47	1.40	77	1.50	107	1.00
18	0.90	48	1.49	78	1.53	108	1.00
19	0.90	49	1.53	79	1.64	109	1.10
20	0.90	50	1.50	80	1.65	110	1.50
21	0.22	51	1.50	81	0.50	111	1.60
22	0.22	52	1.06	82	1.68	112	1.80
23	0.81	53	1.60	83	1.84	113	1.80
24	1.00	54	1.60	84	1.90	114	1.80
25	1.00	55	1.60	85	1.68	115	0.10
26	1.00	56	1.70	86	2.00	116	1.80
27	1.10	57	1.70	87	2.10	117	1.80
28	1.50	58	1.70	88	2.10	118	1.90
29	1.60	59	0.03	89	2.30	119	1.90
30	1.60	60	1.80	90	2.40	120	1.90

shows strong and balanced performance on both validation and test sets, while InceptionV3 demonstrates excellent generalization but needs closer scrutiny for potential overfitting.

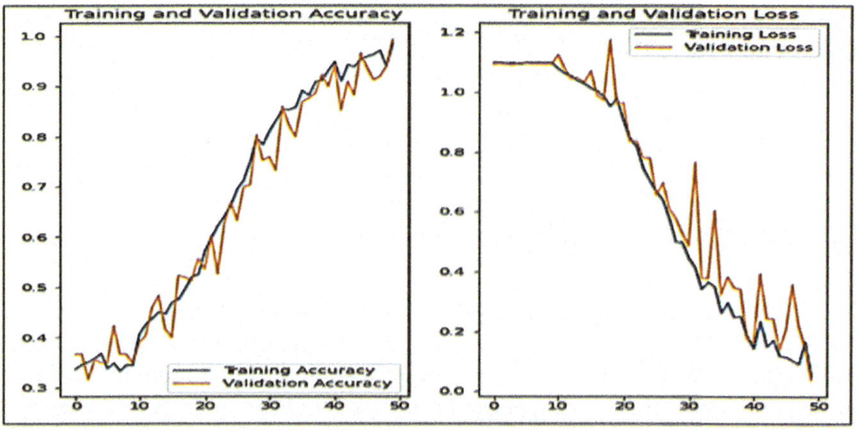

Fig. 6 Accuracy of training and validation for CNN models, together with training and validation loss

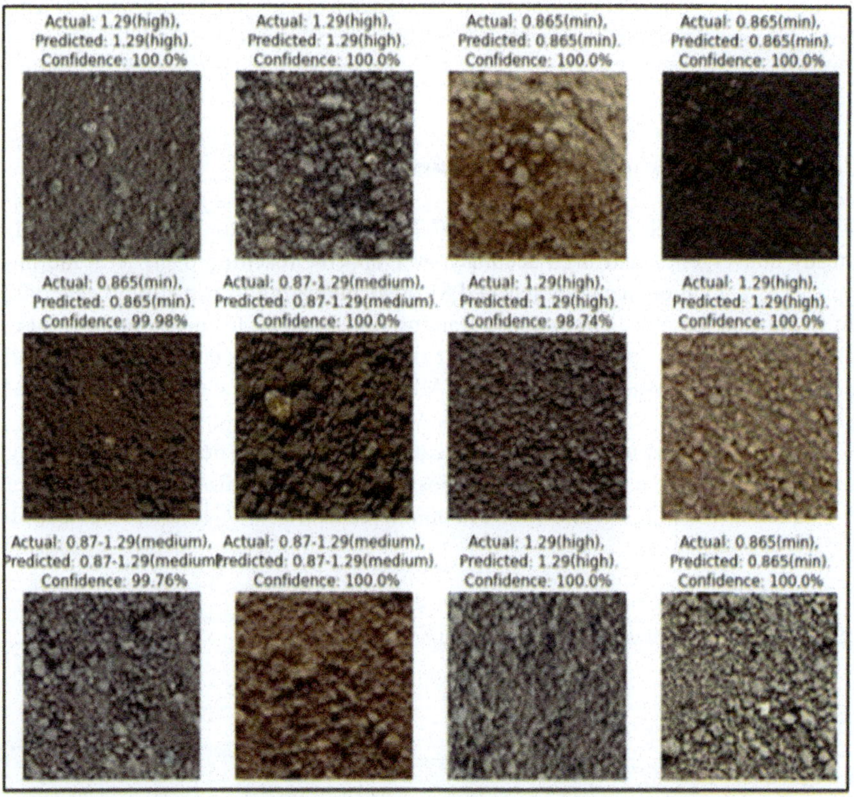

Fig. 7 Result prediction of SOM class of test images with its actual class, predicted class, and confidence level

Table 2 Precision, recall, and F1-score of pre-trained models

Model	Precision	Recall	F1 score
MobileNet V2	0.32732	0.32778	0.32698
VGG16	0.66000	0.65000	0.68000
InceptionV3	0.74483	0.74583	0.76442

7.2 Results of Pre-trained Models

The study analyzes a dataset of 6480 soil photographs divided into three groups to assess the performance of three pre-trained deep learning models: VGG16, MobileNetV2, and InceptionV3. The dataset was divided into training (80%) and testing (20%) sets after it was gathered from several places in the Sangli area. Five epochs were used to train the models, and each one's performance was evaluated. It highlights the adaptability of deep learning in precision agriculture, emphasizing the effectiveness of pre-trained weights in these models. The comparative analysis of the pre-trained model evaluation metrics, such as precision, recall, and F1-score, considering these metrics provide more information about the model is shown in Table 2.

7.2.1 VGG16 Performance and Analysis

The VGG16 model, with a size of 167 MB compressed to 56.3 MB, showed a steady convergence and high accuracy over epochs, reaching 69.76% at the final epoch. Its validation accuracy reached 81.66%, indicating its capacity to generalize fit to unobserved data. The model's training times ranged from 778 to 1637 s (1 h and 20 min) across epochs. The training times varied across epochs, indicating the computational demands of each epoch. All the performance and analysis results are shown in Table 3.

The model's learning and generalization ability were demonstrated through metrics such as a decrease in training loss from 113 to 68.37%, and a fluctuating validation loss. In Figs. 8 and 9 the evaluation results for VGG16 are presented. The

Table 3 Represents performance of VGG16 model

Model	Size .h5	.tflite file	Epochs	Duration (s)	Training loss (%)	Training accuracy (%)	Validation loss (%)	Validation accuracy (%)
VGG16	167 MB	56.3 MB	1	782	113	38.42	116	37.22
			2	778	101	46.89	87.77	52.77
			3	1637	84.68	61	68.73	72.77
			4	1094	75.06	66	82.71	62.22
			5	792	68.37	69.76	47.20	81.66

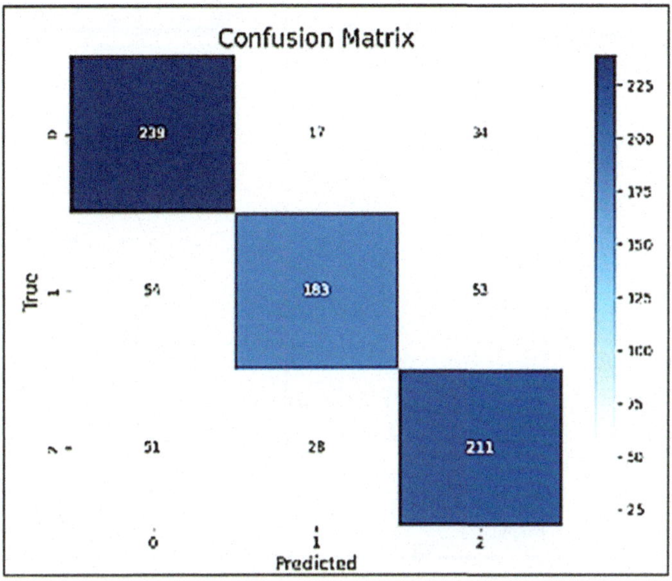

Fig. 8 Confusion matrix using VGG16 model

confusion matrix, training, validation loss, and accuracy are displayed. The model made 239, 183, and 211 correct predictions for classes 0, 1, and 2, respectively. However, there were also incorrect predictions, leading to a validation accuracy of 81.66%.

7.2.2 MobileNetV2 Performance and Analysis

The training of a model on a dataset of 6480 images with 3 classes was conducted for 5 epochs. The model showed improvement in accuracy during training and performance on the test set. The resulting training accuracy is 81.50%. The final training loss was 49.35%. The model also achieved a test accuracy of 78.33%. The model's implementation on the test set was reasonably good. All the performance and analysis results are shown in Table 4.

The evaluation of the MobileNetV2 model shows a validation accuracy of 78.33%. Figures 10 and 11 display the confusion matrix and training/validation loss, respectively. The model made 219, 251, and 231 correct predictions for classes 0, 1, and 2, but there were also incorrect predictions. Further analysis is needed to understand the nature of misclassifications and improve the model.

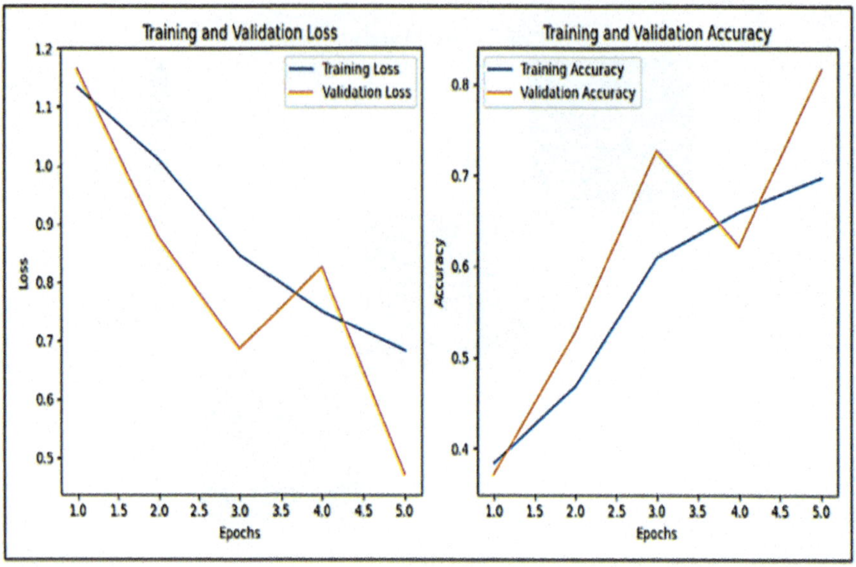

Fig. 9 Training and validation loss and accuracy using VGG16

Table 4 Represents performance of MobileNetV2 model

Model	Size .h5	.tflite file	Epoch	Duration (s)	Training loss (%)	Training accuracy (%)	Validation loss (%)	Validation accuracy (%)
MobilenetV2	16.4 MB	10.9 MB	1	227	88.63	58.72	86.26	62.22
			2	234	69.74	71.39	67.16	72.22
			3	225	61.14	75.93	60.01	77.22
			4	206	53.07	80.34	50.91	78.89
			5	209	49.35	81.50	51.75	78.33

7.2.3 InceptionV3 Performance and Analysis

The inception V3 model was trained on a dataset of 6480 images from three classes, 180 images from another three classes, and 870 images from the test dataset. The training process took 5435 s and involved 5 epochs. The model was saved in HDF5 format. The training metrics showed a training loss of 47.23%, an exactness of 82.24%. The model also got a validation accuracy of 85.56%. The model's size is 97.7 MB, but it showed improvement in accuracy during training and reasonable performance on the test set. All the performance and analysis result are shown in Table 5.

In Figs. 12 and 13, the evaluation results for InceptionV3 are presented. The confusion matrix, training, validation loss, and accuracy are displayed. The model made 236, 199, and 236 correct predictions for classes 0, 1, and 2, respectively.

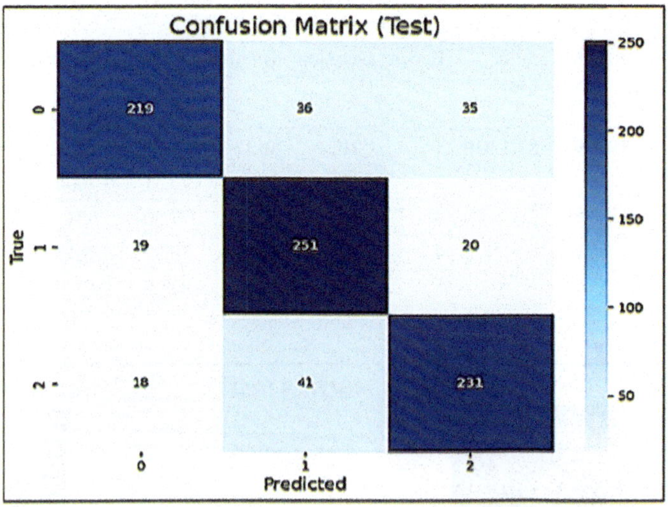

Fig. 10 Confusion matrix for MobileNetV2

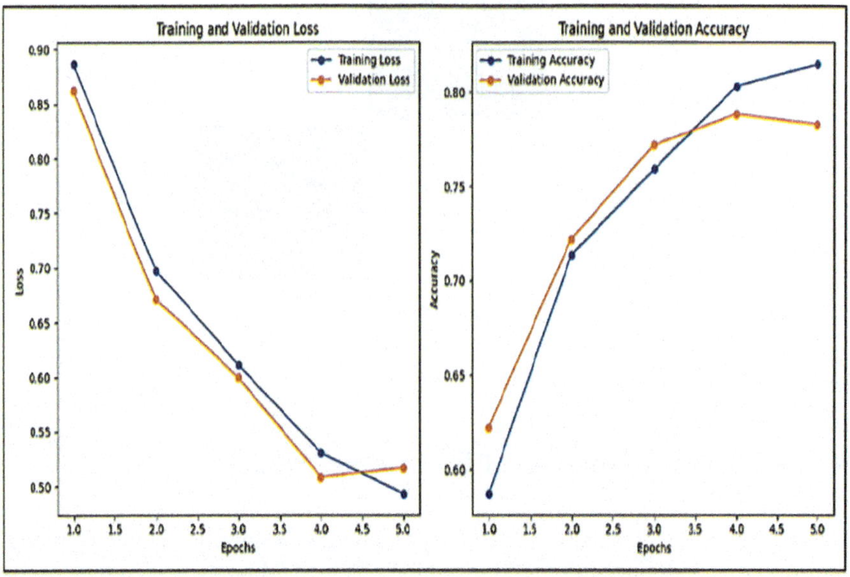

Fig. 11 Training and validation loss and accuracy using MobileNetV2 model

However, there were also incorrect predictions, leading to a validation accuracy of 85.55%.

Table 5 Represents performance of Inception V3 model

Model	Size .h5	.tflite file	Epoch	Duration (s)	Training loss (%)	Training accuracy (%)	Validation loss (%)	Validation accuracy (%)
InceptionV3	95.7 MB	87.1 MB	1	917	86.13	61.25	78.82	60.56
			2	871	6845	72.31	7692	65.56
			3	1404	5797	77.50	5543	83.33
			4	1167	5246	79.92	6618	74.44
			5	1106	47.22	82.23	47.35	85.55

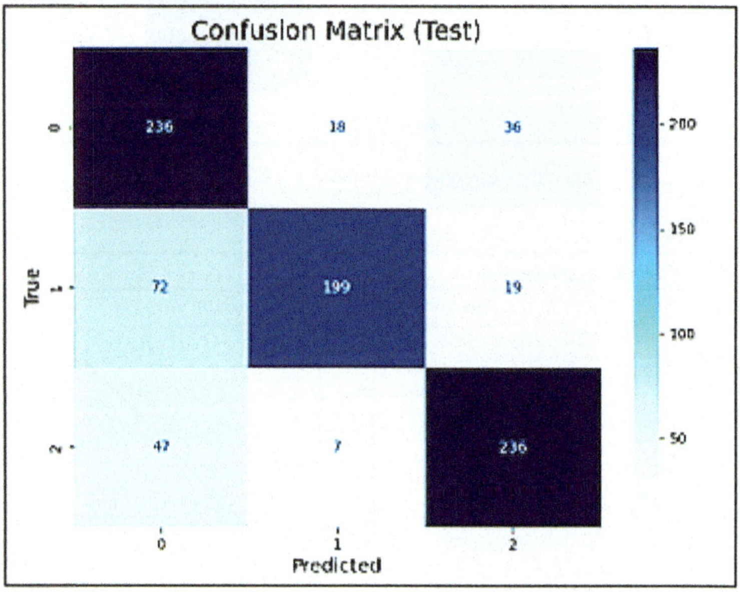

Fig. 12 Confusion matrix for InceptionV3

8 Comparative Analysis of Pre-trained Models

The study conducted a detailed evaluation of three pre-trained deep learning models like as VGG16, MobileNetV2, and InceptionV3 for SOM prediction using a dataset of 6480 soil images. The models demonstrated varying implementation in terms of precision, training time, and model size.

- **VGG16**

Despite a larger model size of 167 MB (.h5) and 56.3 MB (.tflite), VGG16 showed steady convergence, achieving a final training accuracy of 69.77% and a validation

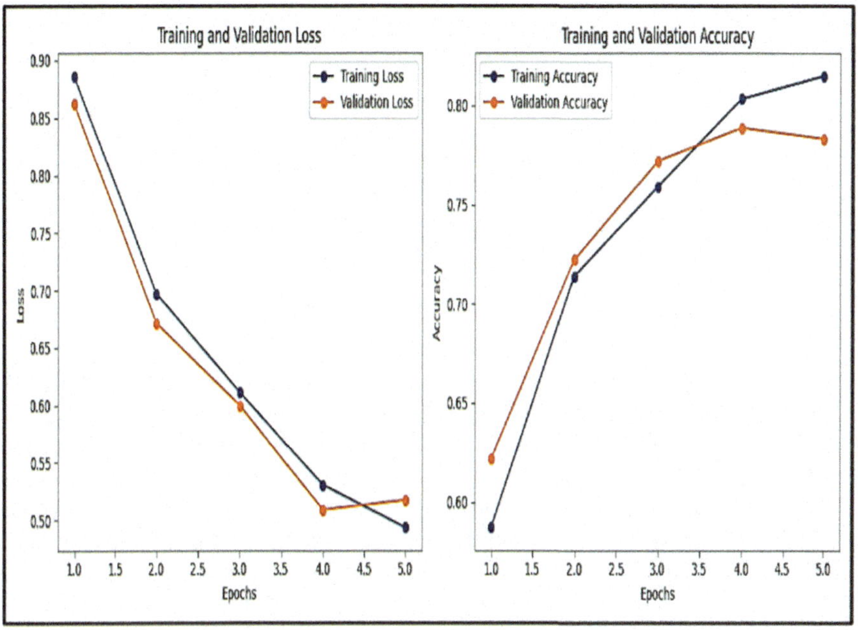

Fig. 13 Training and validation loss and accuracy using Inception V3 model

accuracy of 81.67%. The model exhibited good generalization to unseen data, with varying computational demands across epochs.

- **MobileNetV2**

The MobileNetV2 model, with a size of 16.4 MB (.h5) and 10.9 MB (.tflite), demonstrated an accuracy of 78.33% on the training data. The model showed efficiency in accurate classification, with a relatively shorter total training time of 18 min and a final loss of 49.35%, showcasing efficiency during training.

- **InceptionV3**

The InceptionV3 model, with a size of 95.7 MB (.h5) and 87.1 MB (.tflite), achieved a high training accuracy of 82.23% at the final epoch and a validation accuracy of 85.55%, with low losses. The total training duration was 91.08 min. In terms of model size, MobileNetV2 has the smallest size, making it more suitable for deployment on resource-constrained devices. Inception V3 achieved the highest accuracy during training and validation.

MobileNetV2 showed efficiency in terms of accuracy and training time, making it a good choice for scenarios where computational resources are limited. Training times vary, with VGG16 being the slowest, followed by InceptionV3 and MobileNetV2. The choice of the best model is determined by the specific requirements of the application, considering factors such as model size, training time, and desired accuracy. The comparative performance discussions are in Table 6.

Table 6 Comparative analysis of proposed model

Model	Size .h5	.tflite file	Epochs	Time (min)	Loss	Accuracy (%)	Validation loss	Validation accuracy (%)
VGG16	167 MB	56.3 MB	5	72	0.68	69.77	0.47	81.67
MobileNetV2	16.4 MB	10.9 MB	5	18.35	0.4935	81.50	0.5175	78.33
Inception V3	95.7 MB	87.1 MB	5	91.08	0.4722	82.23	0.4735	85.55

9 Conclusion

This paper presented the results on the prediction of soil organic matter (SOM) content in agricultural soil samples by training and testing of various deep learning (DL) models in Python, followed by their implementation and testing on a Field Programmable Gate Arrays (FPGA). As per our knowledge, till date no other researchers have reported such kind of work and we may be the first ones. Transfer learning has gained popularity in image classification and natural language processing, particularly in image classification and natural language processing. This research aimed to address the problem of measuring SOM using pre-trained models, specifically VGG16, MobileNetV2, and InceptionV3, obtained from Google's TensorFlow HUB. The pre-trained models delivered high accuracy within only 2–3 epochs, saving substantial computational power. Transfer learning proved to be a cost-effective and time-efficient solution, especially in the context of deep learning training, which is known for its computational demands. It not only saves computational time but also simplifies the training of highly complex models, making it a pragmatic choice for addressing sophisticated research problems. The VGG16 model demonstrated high convergence, accuracy, and reasonable training times, reaching 81.67%. The MobileNetV2 and Inception V3 models also showed improvement in accuracy, achieving 78.33% and 85.55% during training and reasonable test performance. However, warnings were issued about saving in HDF5 format, suggesting the use of the native Keras format for better compatibility. The proposed technique aims to provide a quick and cost-effective solution for farmers by predicting SOM through machine learning. The proposed technique incorporates two stages: identifying soil samples using CNN, with 80% of the images utilized for training and 20% for testing, and forecasting soil organic matter using VGG 16, MobileNetV2, and InceptionV3 architectures with pre-trained weights using transfer learning. This finding holds significant promise, particularly for underdeveloped countries facing resource constraints. Conventional soil testing methods would require substantial initial expenses for specialized equipment, skilled personnel, and maintenance costs, leading to higher costs for chemical soil analyses. By using machine learning models like CNN, VGG16, MobileNetV2, and InceptionV3, a more accessible, cost-effective, and efficient solution for estimating soil organic matter content can be offered, benefiting precision agricultural practices in regions with limited resources, reducing financial burdens on farmers, and facilitating more sustainable and productive farming practices. The research outlines the development and implementation of a FPGA-based system for SOM analysis using Python, which captures soil images and displays parameters on a display. The system uses the Kria KV260 Vision AI Starter Kit for parameter integration and uses a transferred learning classification model for soil organic matter content prediction.

Acknowledgements We are deeply thankful to Chhatrapati Shahu Maharaj Research, Training, and Human Development Institute (SARTHI), Pune, for the invaluable support provided through the CSMNRF 2020 scholarship. Our heartfelt gratitude to them for being a catalyst in my academic and research journey. With great appreciation, the authors would like to thank Dr. Dinkar Patil, Senior Scientist and Head of Krushi Vigyan Kendra (Agriculture Science Center) in Kanchanpur, Sangli, as well as his hardworking staff, Shri. Shailesh Patil, for their vital advice and assistance throughout this research. We also appreciate Dr. Dilip Kathmale, who oversees the Agriculture Research Center (ARC) in Kasabe Digraj, Sangli, and his staff for sharing their expertise. This initiative was made possible in large part by the cooperation and support of these specialists.

References

1. Patil VS, Shinde SA, Dhawale NM (2021, December) A review on the determination of soil organic matter and soil moisture content using conventional methods and image processing techniques. In: 2021 IEEE Pune section international conference (PuneCon). IEEE, pp 1–6
2. Dhawale NM, Adamchuk VI, Prasher SO, Viscarra Rossel RA, Ismail AA (2022) Evaluation of two portable hyperspectral-sensor-based instruments to predict key soil properties in Canadian soils. Sensors 22(7):2556. https://doi.org/10.3390/s22072556
3. Dhawale NM, Adamchuk VI, Prasher SO, Viscarra Rossel RA (2021) Evaluating the precision and accuracy of proximal soil vis–NIR sensors for estimating soil organic matter and texture. Soil Syst 5(3):48. https://doi.org/10.3390/soilsystems5030048
4. Dhawale NM, Adamchuk VI, Prasher SO, Viscarra Rossel RA, Ismail AA, Kaur J (2015) Proximal soil sensing of soil texture and organic matter with a prototype portable mid-infrared spectrometer. Eur J Soil Sci 66:661–669. https://doi.org/10.1111/ejss.12265
5. Gholizadeh A, Saberioon M, Rossel RV, Klemen A, Boruvka L (2019, January) Estimation of soil organic carbon using colour features derived from digital camera. In: Geophysical research abstracts, vol 21
6. Xie W, Zhao XM, Guo X, Ye YC, Sun XX, Kuang LH (2018) Spectrum-based estimation of the content of soil organic matters in mountain red soil using RBF combination model. Scientia Silvae Sinicae 54(6):16–23
7. Wang C, Wu XH, Li LQ, Wang YS, Li ZW (2018) Convolutional neural network application in prediction of soil moisture content. Spectrosc Spect Anal 38(1):24–47

8. Ma L, Zhao L, Cao L, Li D, Chen G, Han Y (2022) Inversion of soil organic matter content based on improved convolutional neural network. Sensors 22(20):7777. https://doi.org/10.3390/s22207777
9. Yan Y, Yang J, Li B, Qin C, Ji W, Xu Y, Huang Y (2023) High-resolution mapping of soil organic matter at the field scale using UAV hyperspectral images with a small calibration dataset. Remote Sens 15(5):1433. https://doi.org/10.3390/rs15051433
10. Hinton GE, Salakhutdinov RR (2006) Reducing the dimensionality of data with neural networks. Science 313(5786):504–507
11. Patil VS, Shinde SA, Dhawale NM (2021) Use of machine learning for soil analysis: a scoping review. J Shivaji Univ: Sci Technol 47(1):49–56. ISSN: 0250-5347
12. Gu J, Wang Z, Kuen J, Ma L, Shahroudy A, Shuai B, Chen T et al (2018) Recent advances in convolutional neural networks. Pattern Recogn 77:354–377
13. Wright AL, Wang Y, Reddy KR (2008) Loss-on-ignition method to assess soil organic carbon in calcareous Everglades wetlands. Commun Soil Sci Plant Anal 39(19–20):3074–3083
14. Walkley A, Black IA (1934) An examination of the Degtjareff method for determining soil organic matter, and a proposed modification of the chromic acid titration method. Soil Sci 37(1):29–38
15. Ji W, Adamchuk VI, Biswas A, Dhawale NM, Sudarsan B, Zhang Y, Viscarra Rossel RA, Shi Z (2016) Assessment of soil properties in situ using a prototype portable MIR spectrometer in two agricultural fields. Biosyst Eng 152:14–27. https://doi.org/10.1016/j.biosystemseng.2016.06.005
16. Patil VS, Shinde SA, Dhawale NM (2021, May) Smart phone camera-based weighing scale for kitchens in household applications. J Phys: Conf Ser 1921(1):012025. IOP Publishing
17. Central Ground Water Board (CGWB) (2023) District profile—Sangli. http://cgwb.gov.in/District_Profile/Maharashtra/Sangli.pdf. Accessed 23 Feb 2023
18. Acción Á, Argüello F, Heras DB (2020) Dual-window superpixel data augmentation for hyperspectral image classification. Appl Sci 10(24):8833. https://doi.org/10.3390/app10248833
19. Foundries.io. Kria KV260 vision AI starter kit reference manual. Foundries.io documentation. https://docs.foundries.io/latest/reference-manual/boards/kv260.html
20. Mouser Electronics. Kria KV260 vision AI starter kit datasheet. https://www.mouser.com/datasheet/2/903/ds986_kv260_starter_kit-2301120.pdf
21. Krizhevsky A, Sutskever I, Hinton GE (2012) ImageNet classification with deep convolutional neural networks. In: Adv Neural Inf Process Syst
22. Azizi A, Gilandeh YA, Mesri-Gundoshmian T, Saleh-Bigdeli AA, Moghaddam HA (2020) Classification of soil aggregates: a novel approach based on deep learning. Soil Tillage Res 199:104586
23. Hegazi EH, Yang L, Huang J (2021) A convolutional neural network algorithm for soil moisture prediction from Sentinel-1 SAR images. Remote Sens 13(24):4964
24. Ng W, Minasny B, de Sousa Mendes W, Dematte JAM (2019) Estimation of effective calibration sample size using visible near-infrared spectroscopy: deep learning vs machine learning. Soil Discuss 1–21
25. Patil VS, Shinde SA, Sarwade MP, Dhawale NM (2023) Smart robotic soil pH measurement system. In: 2023 National conference on smart agriculture technologies (NCSAT). Indian Soc Tech Educ J 46(Special Issue):22–30
26. Simonyan K, Zisserman A (2014) Very deep convolutional networks for large-scale image recognition. arXiv preprint arXiv:1409.1556
27. Zhang W, Zhong J, Yang S, Gao Z, Hu J, Chen Y, Yi Z (2019) Automated identification and grading system of diabetic retinopathy using deep neural networks. Knowl-Based Syst 175:12–25
28. Mascarenhas S, Agarwal M (2021) A comparison between VGG16, VGG19 and ResNet50 architecture frameworks for image classification. In: 2021 International conference on disruptive technologies for multi-disciplinary research and applications (CENTCON), vol 1, pp 96–99

29. Sandler M, Howard A, Zhu M, Zhmoginov A, Chen L-C (2018) Mobilenetv2: inverted residuals and linear bottlenecks. In: Proceedings of the IEEE conference on computer vision and pattern recognition, pp 4510–4520
30. Szegedy C, Vanhoucke V, Ioffe S, Shlens J, Wojna Z (2016) Rethinking the inception architecture for computer vision. In: Proceedings of the IEEE conference on computer vision and pattern recognition, pp 2818–2826

Node Classification Using Graph Neural Network for Low Resource Malayalam Language

Sincy V. Thambi and P. C. Reghu Raj

Abstract Graph Neural Networks (GNN) is a type of graph embedding approach designed to convert the nodes and edges in a graph into low-dimensional vector representations while preserving important structural and relational information. Node classification is the process in which the nodes of a graph are assigned labels according to the characteristics that are associated with those nodes. The objective of this paper is to perform node classification for Malayalam, an official language in India using various GNN approaches. Among all the GNN techniques used for node classification in Malayalam, the Relational Graph Convolutional Network (RGCNConv) model demonstrated the highest average accuracy of 70.87%. While this work may not be on par with the English research, our innovative outcome in Malayalam using GNN represents the first attempt in this area. Given the resource constraints associated with the Malayalam language, we have developed and made publicly available a node classification dataset that will serve as a foundation for future node classification research in Malayalam.

Keywords Graph embedding · Graph Neural Network · Graph Convolution Network · Malayalam node classification

Supported by Centre for Engineering Research and Development(CERD) of APJ Abdul Kalam Technological University, Kerala.

S. V. Thambi (✉)
Government Engineering College, Palakkad, India
e-mail: sincyvt@gmail.com

P. C. Reghu Raj
Government Engineering College, Kozhikode, India

S. V. Thambi · P. C. Reghu Raj
APJ Abdul Kalam Technological University, Thiruvananthapuram, Kerala, India

A. C. Frery et al. (eds.), *Computational Intelligence and Data Analytics*, Lecture Notes on Data Engineering and Communications Technologies 236,
https://doi.org/10.1007/978-981-96-0451-7_9

1 Introduction

Graphs can be used wherever there is an interconnection in the underlying data. These days, graphs are widely utilized among other domains, including speech recognition, natural language processing (NLP), social networks, recommendation systems, and search engines. By maintaining the meaning and the context, embedding is a viable method for turning text into a low-dimensional vector representation. The idea is that similar words will be mapped into similar embedding representations. The existing word embedding methods include Word2vec, Glove, BERT [1], and so on. GNNs are embedding approaches that have been specifically designed to execute directly on graphs, which enables them to recognize complex relationships and dependencies among nodes. This renders them highly suitable for tasks in which it is important to understand the context of every node in the graph, such as node classification.

Node classification is to predict the category of every node in the graph by analyzing its characteristics and the graph's structure. A GNN model for node classification is trained to make predictions about unknown labels of nodes by comparing their similarity or dissimilarity to the nodes with known labels. The task of node classification holds significant importance across multiple areas, encompassing social network research, recommendation systems, knowledge graph-based question answering systems (KGQA), bioinformatics, citation networks, and other related fields. In the area of KGQA, node classification plays a vital role in finding the expected answer type of a question. For a knowledge graph $G = (V, E)$, V represents the nodes that are real-world entities, and E is the edge that represents the relation connecting those nodes/entities. The classes are the type/category of the answers to which the nodes are expected to belong like *person, number, location*, etc. Thus node classification helps to improve the answer precision in KGQA systems.

In response to the growth of smart devices and cross-linguistic research, low resource languages are gaining prominence. Low-resource languages are increasingly adopting machine translation, searching, and question-answering systems for practical purposes. The Dravidian language family includes Malayalam, the official language of Kerala in India. Malayalam has unrestricted word order. Malayalam is challenging to comprehend due to its agglutinative character. About 20 million people prefer Malayalam material on social media and the internet [2]. Our effort aims to perform node classification using various GNN approaches for Malayalam.

The rest of the paper is structured as follows: Section 2 deals with the current approaches to node classification. Section 3 discusses the various GNN approaches. Section 4 discusses the implementation of the proposed node classification procedures for Malayalam. The sections that follow go over the evaluation criteria, results, and conclusion. Now a survey of related work is given in the next section.

2 Related Work

When comparing the advancements in node classification research in English and Malayalam, it becomes evident that the latter is still in its early stages. This can be attributed to the absence of fundamental NLP techniques and the limited availability of data. We searched for comparable studies in the English language and found a plethora of such research.

Graph representation is a universal language for expressing systems with an underlying interconnection of information items. The work by Gori et al. established neural networks operating on graphs [3]. In order to handle data representation using graphs, Scarsellil et al. developed a novel neural network model known as the GNN model, which extends the conventional neural networks [4]. The limitation of the model proposed by Scarsellil et al. is reduced by Li et al. by extending GNN to sequential outputs and incorporating contemporary techniques for recurrent neural network training into the original GNN architecture [5]. The works by Shi et al., Zeng et al., demonstrate how adaptable GNNs are in solving applications such as recommendation systems, drug discovery, social network analysis, traffic flow prediction, citation network analysis, cybersecurity, robotics, and autonomous systems that use graph-based modeling of the underlying data [6, 7].

Graph Convolutional Networks (GCNs) execute convolutions directly on graphs to develop node representations and are good for semi-supervised node classification [8]. Graph Attention Networks (GATs) use attention mechanisms to learn weighted combinations of neighbor node attributes, improving node classification performance [9]. In GNN, neighbor sampling approaches are introduced by GraphSAGE [10] and FastGCN [11]. Recent works increase the depth of GNN models [12]. However, deeper models are susceptible to over-smoothing, which occurs when numerous GNN layers are stacked and the features of the nodes become indistinguishable, resulting in performance degradation of the model. KGQA needs to categorize nodes in the KG into specific categories which will be beneficial in producing accurate results [13].

When compared to the unaffected low resource Malayalam language, the review of literature relevant to node classification demonstrates that English has made enormous progress in establishing a node classification system using GNN approaches. We are doing node classification for Malayalam with various GNN-based techniques here. To the best of our knowledge, our work is the first attempt at node classification; this will be an initiative in Malayalam KG-based research.

3 Graph Neural Networks

In order to extract node features while preserving their structure during graph analysis, we must find embeddings (Z_v) of a node v, which maps node data to a low-dimensional embedding space. This mapping is done in such a way that the distance in the embedding space corresponding to node positions in the original graph is preserved. GNNs rely on a message-passing framework. At each iteration, every node aggregates information from its local neighborhood. Every node's embedding now contains information from its nearest graph neighbors, or 1-hop neighborhood, after the first iteration. Following the second iteration, each node embedding has data from its 2-hop neighborhood-i.e., the nodes in the graph that can be accessed by a path of length 2, and so on. As these iterations continue, each node embedding learns an increasing amount of information from deeper regions of the graph [14].

GNNs show promise in a range of NLP tasks and offer an effective way to model graph-structured data [15, 16]. The fundamental idea of GNN is illustrated in (1). The embeddings of node v in layer l, $H_v^{(l)}$ is updated by aggregating the embeddings $H_u^{(l-1)}$ from its local neighbours $u \in (N(v))$ from the previous layer $l-1$ using non-linear activation functions [17].

$$H_v^{(l)} = \sigma(W^{(l)} \sum_{u \in N(v)} \frac{H_u^{(l-1)}}{|N(u)|} + B^{(l)} H_v^{(l-1)}), \forall l \in \{0, ..., L-1\} \tag{1}$$

where $H_v^{(l-1)}$ is the previous layer embedding of node v, $\sum_{v \in N(u)} \frac{H_u^{(l-1)}}{|N(u)|}$ is the average of the neighbour's previous layer embedding, $W^{(l)}$ and $B^{(l)}$ are the trainable matrices. Initial node features x_v are considered as the layer 0 embedding H_v^0 of the node v.

The embeddings for every node Z_v are obtained by running L iterations of the GNN, as shown in Eq. (2). By stacking L GCN layers, the network may construct node representations based on each node's L-hop neighborhood [18].

$$Z_v = H_v^{(L)} \tag{2}$$

GNN's fundamental architecture is depicted in Fig. 1. Here, adjacent nodes B, C, and F that are directly connected and are at a distance of one hop are selected for a target node A in the corresponding graph. Embeddings from those neighboring nodes are aggregated and combined with the embeddings of the target node. Then a

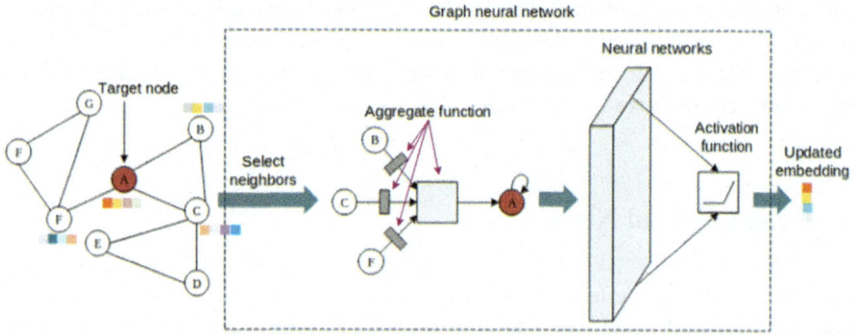

Fig. 1 Basic Architecture of GNN [7]

non-linear activation function is applied to generate the updated node embeddings of the target node.

As a knowledge graph, our system requires both node and edge features to function properly. We experimented with many GNN models and convolutional layers, including Graph Convolutional Network (GCNConv), GCN with Edge Convolution, TransformerConv, GENeralized Graph Convolution (GENCONV), Graph Attention Network (GATConv), Relationall GATConv (RGATConv), and Relational GCN-Conv (RGCNConv). They accept multidimensional edge and node features. After training, a single node in the graph will contain the information about that node along with its adjacent nodes and relations. If our model is having 2 layers, then a node will contain information about those nodes which are up to a distance of 2 nodes and 2 edges away from the present node. Thus nodes can aggregate features of both nodes and edges from their neighbours.

3.1 Graph Convolutional Network(GCN)

GCN employs a straightforward graph convolution procedure to aggregate data from neighbouring nodes by calculating the average or weighted sum of their feature vectors [8, 18]. GCN replaces self-loops with multi-layer perceptrons, configurable aggregations, integrated edge feature support, and skip connections. Graph convolutions in GCNs are based on message forwarding [8]. GCN layers update node features by transmitting information from neighbouring nodes. To compute a new node feature in message forwarding, we first combine information from the node and its neighbors using an aggregation function. We then update the node feature by running the aggregated value through a differentiable function. GCN is frequently regarded as more scalable for large graphs due to its simpler aggregation mechanism. With GCN, the updated node features $H_v^{(l+1)}$ of a node v can be derived using the Eq. (3).

$$H_v^{(l+1)} = \sigma\left(W^{(l+1)} \sum_{v \in N(u) \cup v} \frac{H_u^{(l)}}{\sqrt{|N(v)||N(u)|}}\right) \tag{3}$$

where $\sum_{v \in N(u)} \frac{H_u^{(l-1)}}{\sqrt{|N(v)||N(u)|}}$ is the average of the neighbour's previous layer embedding and $W^{(l+1)}$ is the trainable matrix. The basic GCN layers are shown in Fig. 2.

3.2 Edge Convolution (EdgeConv)

EdgeConv operates on graphs in each layer of the network to update node features with the associated edge features [19]. As illustrated in Fig. 3, the EdgeConv output

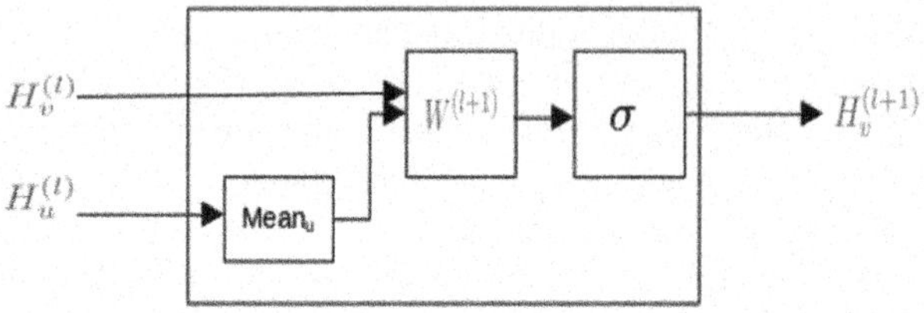

Fig. 2 A GCN layer

Fig. 3 Node updation using EdgeConv

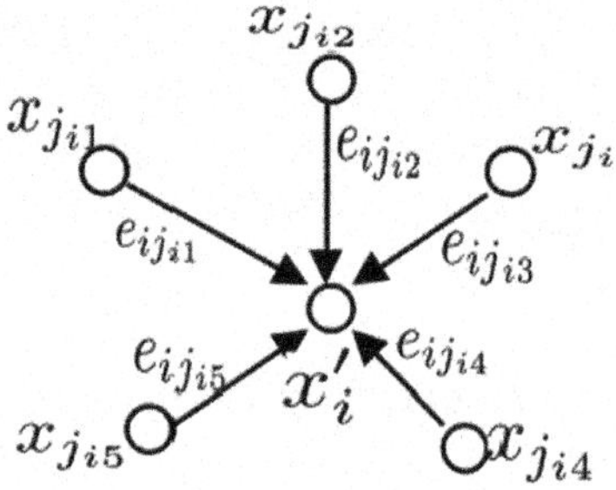

is computed by aggregating the edge features associated with all of the edges coming from each connected vertex. In addition to including local neighborhood data, EdgeConv can be stacked to learn global shape characteristics and capture semantic features in the initial embedding over potentially large distances. For a given graph $G = (V, E)$, where $V = \{x_1, ..., x_n\}$ and $E \subseteq V \times V$ represents the vertices and edges, respectively. Node features $x_i \in \mathbb{R}^F$ and edge features can be defined as $e_{ij} = h_\theta(x_i, x_j)$, where $h_\theta : \mathbb{R}^F \times \mathbb{R}^F \rightarrow \mathbb{R}^{F\prime}$ is a nonlinear function with a set of learnable parameters θ.

EdgeConv involves applying a channel-wise symmetric aggregating operation (e.g.,$\sum$ or max) on the edge features from each vertex [19]. Equation 4 describes the output of EdgeConv at the vertex i.

$$x_i' = \sum_{j:(i,j)\in E} h_\theta(x_i, x_j) \tag{4}$$

The EdgeConv layer is differentiable, may be integrated into the existing GNN architectures [19].

3.3 *TransformerConv*

TransformerConv presents a masked label prediction technique that predicts the occurrence of label leakage by randomly masking the label of some training examples [6]. The feature matrix $X \in \mathbb{R}^{n\times m}$, which typically consists of dense vectors with m dimensions, describes the nodes. In the same space as node features, we suggest embedding the partially observed labels $\hat{Y} \in \mathbb{R}^{n\times m} \rightarrow \hat{Y}_d \in \mathbb{R}^{n\times m}$ which consist of the label embedding vector for labeled nodes and zeros vectors for the unlabeled. By simply combining the node characteristics and label vectors as propagation information, we can integrate the label propagation into Graph Transformer, $H^0 = X + \hat{Y}_d \in \mathbb{R}^{n\times m}$. As seen in Eq. (5), updated node embeddings of node v at layer $l+1$ are generated, where β is the gated function and A^* is the normalized adjacency matrix.

$$H_v^{(l+1)} = \sigma(((1-\beta)A^* + \beta I)H^{(l)}W^{(l)}) \tag{5}$$

3.4 GENeralized Graph Convolution (GENConv)

When we go deeper, we see that the GCNs have problems with over-smoothing, overfitting, and vanishing gradient. These difficulties restrict GCNs' ability to depict large-scale graphs. The GENeralized Aggregation function was proposed to improve GCN performance [20]. In addition to GCN, skip-connections rather than self-loops, customized aggregations, multi-layer perceptrons (MLP), and integrated edge feature support are all used in Generalized Aggregation Networks (GEN). Message passing, generalised message aggregators, and message normalization layers constitute GEN [21]. Equation 6 depicts the message generation process for constructing individual messages $m_{vu}^{(l)}$ for each neighbor $u \in N(v)$. Equation 7 uses generalized message aggregators to generate an aggregated message $m_v^{(l)}$, which is then normalised to obtain updated node characteristics $m_v^{(l)}$ and final message normalization for getting updated node features $H_v^{(l+1)}$ in Eq. (8).

$$m_{vu}^{(l)} = ReLU(H_u^{(l)} + H_{e_{vu}}^{(l)}) + \epsilon, u \in N(v) \tag{6}$$

$$m_v^{(l)} = \zeta^{(l)}(m_{vu}^{(l)}/u \in N(v)) \tag{7}$$

$$H_v^{(l+1)} = MLP\left(H_v^{(l)} + s.\|H_v^{(l)}\|_2 \frac{m_v^{(l)}}{\|m_v^{(l)}\|_2}\right) \tag{8}$$

where, $H_u^{(l)}$: Features of node u
$H_{e_{vu}}^{(l)}$: Edge features between node v and the neighboring node $u \in N(v)$
ϵ : Small positive constant(10^{-7})
$\zeta^{(l)}$: Message aggregation function either power mean/softmax [21]
MLP : Multilayer perceptron
s : Learnable scaling factor initialized to 1 was updated using $s = \frac{\|m_v^{(l)}\|_2}{\|H_v^{(l)}\|_2}$ i.e., $m_v^{(l)}$ is first normalized by their $l2$ norm and then scaled by the $l2$ norm of $H_v^{(l)}$ by a factor of s.

3.5 Graph Attention Network Convolution (GATConv)

Attention mechanisms in deep learning are used to selectively focus on the most relevant parts of the input when making a prediction [22]. The relevance of the nodes in a given context is determined by the attention coefficient α_{uv}. Apart from GCN, an attention coefficient is calculated for every neighboring node $N(u)$ of node v in GAT [9]. This yields an expressive and interpretable model. A masked self-attention method that determines an attention score for every pair of nodes is used to construct these coefficients. Next, a SoftMax function is used to normalize the scores over the neighborhood of each node. If F is the number of node features, $H_u \in \mathbb{R}^F$, $H_v \in \mathbb{R}^F$ indicate the node features for nodes u and v, respectively. The weight matrix $\mathbf{W} \in \mathbb{R}^{F' \times F}$ represents the number of features F' in the final node feature. Equation (9) shows how a shared attentional mechanism, $a : \mathbb{R}^{F'} \times \mathbb{R}^{F'} \rightarrow \mathbb{R}$ can be used to compute the attention coefficient e_{vu}, which indicates the relevance of node u's attributes to node v. LeakyReLU uses a tiny negative slope for negative function inputs, allowing neural networks to maintain improved information flow during and after training [23].

$$e_{vu} = a(\mathbf{W}H_v, \mathbf{W}H_u) = LeakyReLU(\tilde{w}_a^T[\mathbf{W}H_v || \mathbf{W}H_u]) \tag{9}$$

$$\begin{aligned} \alpha_{vu} &= Softmax_v(e_{vu}) \\ &= \frac{exp(LeakyReLU(\tilde{w}_a^T[\mathbf{W}H_v || \mathbf{W}H_u]))}{\sum_{k \in N(u)} exp(LeakyReLU(\tilde{w}_a^T[\mathbf{W}H_v || \mathbf{W}H_k]))} \end{aligned} \tag{10}$$

where $\tilde{w}_a^T$ is a learnable weight vector. The softmax function is then used to convert the attention coefficient to normalized attention coefficient α_{vu} as indicated in Eq. (10). The output feature H_v of node v is obtained by applying nonlinearity σ to a linear combination of input features and normalized attention coefficients, as illustrated in Eq. (11).

$$H_v' = \sigma(\sum_{u \in N(v)} \alpha_{vu} \mathbf{W} H_u) \tag{11}$$

3.6 RGCNConv

Relational Graph Convolutional Networks (RGCNs) are an extension of GCNs for relational graphs [16]. Multigraphs, specifically knowledge graphs, represented as $G = (V, E, R)$, with node $v \in V$, labeled edges $v, r, u \in E$ where $r \in R$ denotes the relation type, are the target application for RGCNs [24]. Equation (12) is used to get updated node embeddings for node v.

$$H_v^{(l+1)} = \sigma\left(\sum_{r\in R}\sum_{u\in N_v^r}\frac{1}{c_{v,r}}W_r^{(l)}H_u^{(l)} + W^{(l)}H_v^{(l)}\right) \tag{12}$$

The collection of neighbor indices of node v under relation $r \in R$ is denoted by N_v^r. The normalization constant $c_{i,r} = |N_v^r|$ is a problem-specific normalization constant that can either be learned or chosen in advance. Depending on the kind and the direction of the edge, a relation-specific change occurs here. Here, the node's embedding information is collected along with activation from every neighboring node for both incoming and outgoing edges. After that, the output is combined and sent through a non-linear activation function (ReLU).

3.7 RGATConv

A variation on GNNs, the graph attention network(GAT) uses masked self-attention layers to represent graph structure [25]. Dependency label characteristics are not included in the vanilla GAT since it uses an adjacent matrix as structural information instead. To provide more informative representations, RGAT integrates relational characteristics into the attention calculation and aggregation process. By acting on graph topologies, these models induce a masked self-attention that considers both node attributes and local relationship structure. This makes it possible to dynamically assign a priority for various nodes in the graph, taking into account both the nodes and their properties under particular relations [25]. In contrast to GAT, RGAT accounts for relational aspects. This can entail by considering the features that depict the relationship between nodes along with the node features. For each relation type, attention logits $a_{vu}^{(r)}$ which indicates the importance of node node u to that of node v under relation r. $a_{vu}^{(r)}$, composed of query and key kernels are computed, as indicated by Eq. (13) [25].

$$\mathbf{q}_v^{(r)} = \mathbf{W}_1^{(r)}\mathbf{x}_v \cdot \mathbf{Q}^{(r)} \quad \text{and} \quad \mathbf{k}_v^{(r)} = \mathbf{W}_1^{(r)}\mathbf{x}_v \cdot \mathbf{K}^{(r)} \tag{13}$$

Equation (14) illustrates the two methods that have been suggested to calculate attention logits $\mathbf{a}_{v,u}^{(r)}$ for each relation type r, i.e., additive attention and multiplicative attention.

$$\mathbf{a}_{v,u}^{(r)} = \text{LeakyReLU}(\mathbf{q}_v^{(r)} + \mathbf{k}_u^{(r)}) \quad \text{and} \quad \mathbf{a}_{v,u}^{(r)} = \mathbf{q}_v^{(r)} \cdot \mathbf{k}_u^{(r)} \tag{14}$$

After that, for every relation type r, the attention coefficients $\alpha_{v,u}^{(r)}$ are determined using one of the two distinct attention processes, $across-relation$ or $within-relation$ attention mechanisms as indicated by the Eqs. (15) and (16).

$$\alpha_{v,u}^{(r)} = \frac{\exp(\mathbf{a}_{v,u}^{(r)})}{\sum_{k\in\mathcal{N}_r(v)}\exp(\mathbf{a}_{v,k}^{(r)})} \tag{15}$$

$$\alpha_{v,u}^{(r)} = \frac{\exp(\mathbf{a}_{v,u}^{(r)})}{\sum_{r' \in \mathcal{R}} \sum_{k \in \mathcal{N}_{r'}(v)} \exp(\mathbf{a}_{v,k}^{(r')})} \tag{16}$$

where $\mathcal{R}$ denotes the set of relations, i.e. edge types. Each logit $E_{v,u}^{(r)} = a(H_v^r W, H_u^r W)$ is formed by combining the intermediate representations for nodes in its neighbourhood under each relation r with the intermediate representations for node v. For each type of relation, a softmax is taken over the logit matrix to create the attention coefficients $\alpha_{v,u}^{(r)}$. For every relation, these attention coefficients create a weighted sum over the neighbourhood nodes. A softmax is taken over each logit matrix for each relation type to form the attention coefficients $\alpha_{v,u}^{(r)}$. These are then aggregated and passed through a nonlinear function to produce the updated representation of the node v [25].

4 Node Classification for Malayalam

This section describes the dataset used to evaluate our model, along with an explanation of the experimental setup and findings.

4.1 Dataset

As part of this work, a node classification dataset for Malayalam comprising 2863 training samples and 716 test samples was generated. The dataset was a part of MetaQA KG,[1] which is presently accessible in the English language and is translated to Malayalam using Google Translator. The dataset comprises of 3579 nodes and 4670 edges which are film-related information that can be categorized into 8 distinct classes. This curated dataset is made publically available.[2] This Malayalam KG database is a comprehensive graph that encompasses all the data, depicted as edges and nodes. 80% of the data is taken as a training set and the remaining 20% as a testing set. The training set is split again into a training set and a validation set.

4.2 Experimental Setup

Training a GNN to predict the labels of nodes based on the nodes' characteristics and neighborhood structure constitutes node classification. The GNN model is conceptualized as a function $f(X, A)$ that is dependent on the adjacency matrix A and

[1] https://github.com/yuyuz/MetaQA.

[2] https://github.com/sincyvt/Malayalam-Node-Classification.git.

Table 1 Node classification results

GNN Model	Accuracy	F1-score
GCNConv	52.93	67.34
GCNConv with Edge Conv	31.31	40.32
Transformer Conv	54.89	57.87
GENConv	54.19	55.03
GATConv	46.22	47.93
RGATConv	26.20	28.58
RGCNConv	54.89	**70.87**

node features X. By aggregating the characteristics of n hops of neighbors, GNN generates node embeddings of a node. The softmax function and calculation of the cross-entropy error across all labeled training examples are then performed. Back-propagation of the gradients of loss occurs across the layers of the GNN. The model can be used to predict the identities of nodes in the test set once it has been trained.

In this graph, the nodes and the edges are encoded using the BERT version mBERT (multilingual BERT) [26]. mBERT embeddings of nodes and edges are considered as the initial feature vector X and passed to the GNN model. The initial feature vector of size 768 is further reduced to 128, by using principal component analysis (PCA). PCA reduces the computational complexity while retaining the semantic meaning in the feature vector. The reduced feature vector of each node v is then passed through multiple layers of the GNN model which aggregates neighbouring node/edge information and updates each node's feature vector H_v^l. Various GNN models tried out in this experiment are listed in Table 1. The final node embeddings Z_v as per Eq. 2 are obtained after passing through a single layer of the GNN model layer by aggregating neighbouring node/edge information and *ReLu* activation function. Z_v is then passed through a classifier (a dropout layer followed by a softmax function) to predict the class probabilities for each node. Thus each node will be classified into their corresponding classes. Training is performed using cross-entropy loss using the predicted class label $\hat{Y}$ and the actual class label Y. The model is optimized further using Adam optimizer. The GNN model is described in Fig. 4. After training the GNN model converged almost for 10 epochs to obtain the final GNN classifier model with updated weights. This classifier model can then be used for predicting the class of new nodes in our KG. Here we tried various GNN models such as GCNConv, GCNConv with Edge Conv, TransformerConv, GENConv, GATConv, RGATConv, and RGCNConv. The Node Classification Model Architecture is shown in Fig. 5.

Using GNNs for node classification, neighboring information is crucial because it takes use of the structure of the graph and captures contextual dependencies. Here node features along with edge features up to a distance of 1-hop is considered. Performance is usually enhanced by adding layers, but adding more layers might cause overfitting or the over-smoothing issue, in which node representations become

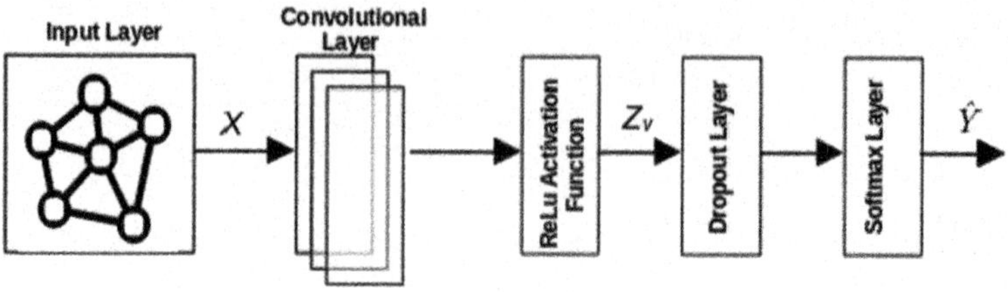

Fig. 4 GNN model

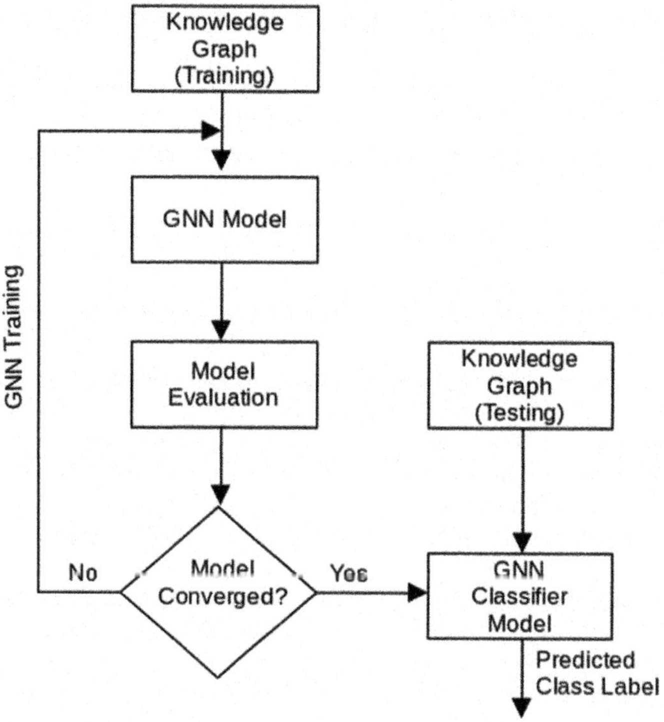

Fig. 5 Node classification model architecture

essentially identical. Results obtained by utilizing various GNN approaches on this dataset are detailed in the following section.

4.3 Results

The design and evaluation of the system were carried out using a dropout rate of 0.03 and an Adam optimizer to train the model for 10 epochs since the best results are obtained for this criterion by the trial and error process. Adam optimizer and binary cross-entropy loss are used to train the model. The model is implemented with PyG (Pytorch Geometric)[3] which is a library for building GNN models. In addition, NetworkX 3.0[4] is used for graph processing. The findings from the experiments are shown in Table 1.

Table 1 compares the performance of our GNN-based node classification models such as GCNConv, GCNConv with Edge Conv, TransformerConv, GENConv, GATConv, RGATConv, RGCNCONV. With Malayalam KG nodes as input, for embedding, we used the mBERT embedding model. This mBERT embeddings are used as initial input features of nodes and are passed through the above said various GNN models. We experimented with the models and obtained the performance in terms of evaluation measures such as F1 score and Accuracy. The accuracy and macro-averaged F1-score is presented, which computes the F1 score independently for each class, then takes the average and treats all classes equally.

We can see from the results that the RGCNConv model can reach cutting-edge performance with mBERT as the initial embedding feature. This is higher than the GCNConv model by 3.53% in the F1-score. This is because rich semantic information embedded in many different types of relations can be captured and modeled in a powerful way using RGCNConv. It is evident from the results that certain models exhibit better results compared to the others. Compared to GCNConv, RGCNConv can perform better in structured multi-relational KG, with a well-defined set of relationship types by effectively capturing the structure of the graph. Another reason for the enhancement in performance compared to GATConv and RGATConv is due to the less dynamic requirements where the importance of neighbors does not need to be dynamically adjusted. Certain models experience a marked decrease in model accuracy and F1-score when compared to others. This may be the result of the modeling capabilities of the models presently in use being insufficient. A bar graph of the result comparison of various GNN models is shown in Fig. 6.

5 Conclusion

To the best of our knowledge, this is the first attempt in Malayalam for node classification, which plays an important role in the Malayalam KGQA process since it helps distinguish the type of answer. Graph embedding is the procedure of transforming nodes, edges, or entire graphs into a continuous vector space. Several graph embedding models have been developed in recent years, and further study will persist in the future. In this paper, we covered various deep-learning approaches for graph embeddings. Also addressed GNNs which use the graph features to learn the complex patterns from the underlying graph and their various types. There exist multiple variations of GNN models, each characterized by distinct methods for gathering

[3] https://pytorch-geometric.readthedocs.io/en/latest/.

[4] https://networkx.org/.

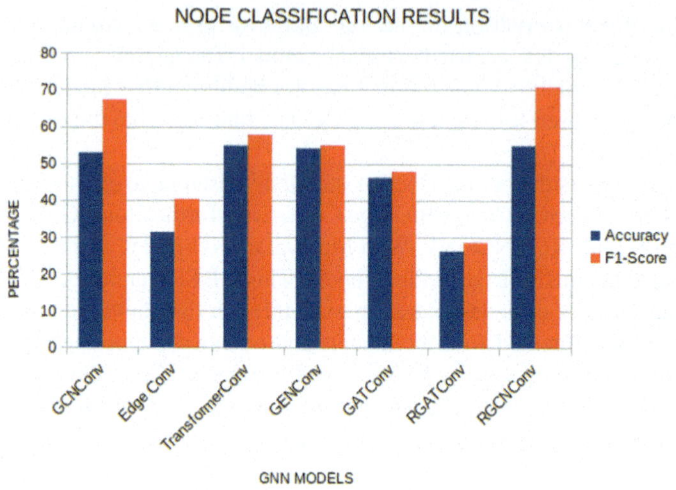

Fig. 6 Comparison of accuracy and precision of various GNN models

information and updating the hidden states of nodes. Due to the restricted resources available in Malayalam, we have used the MetaQA KG in English to generate an annotated corpus in the language for eight classes. Node classification for various existing GNN models is tried out. According to our findings, the RGCNConv model fared better at classifying nodes than other GNN approaches. To promote further research in the node classification, we have made the resources utilized in this work available to the general public.

A potential drawback in GNN models is that they may incur significant computational costs, particularly when applied to large graphs, which are frequent in real-world applications. Such complexities can be reduced with the help of introducing pruning mechanisms by only considering the necessary portion of the graphs during embedding. By providing weights, GNN's capability can be further enhanced. GNNs can operate more effectively with KG-based QA systems by including edge features in addition to node features.

Acknowledgements The authors would like to thank the CERD of APJ Abdul Kalam Technological University, Kerala.

References

1. Sun F et al (2019) Bert4rec: sequential recommendation with bidirectional encoder representations from transformer. In: Proceedings of the 28th ACM international conference on information and knowledge management, pp 1441–1450
2. Reghu Raj PC et al (2015) A memory based approach to Malayalam noun generation. In: International conference on control communication and computing India (ICCC). IEEE, pp 634–637

3. Gori M et al (2005) A new model for learning in graph domains. IEEE Int Joint Conf Neural Netw 2:729–734
4. Scarselli F et al (2009) The graph neural network model. IEEE Trans Neural Netw 20(1):61–80
5. Li Y et al (2015) Gated graph sequence neural networks. arXiv:1511.05493
6. Shi Y et al (2021) Masked label prediction: unified message passing model for semi-supervised classification, pp 1548–1554. arXiv:2009.03509
7. Zeng Y et al (2021) Rlc-gnn: an improved deep architecture for spatial-based graph neural network with application to fraud detection. Appl Sci 11:5656
8. Kipf TN et al (2017) Semi-supervised classification with graph convolutional networks. In: International Conference on Learning Representations
9. Velickovic P et al (2018) Graph attention networks. In: International conference on learning representations
10. Hamilton W et al (2017) Inductive representation learning on large graphs. In: NIPS, pp 1024–1034
11. Chen J et al (2018) FastGCN: fast learning with graph convolutional networks via importance sampling. In: ICLR
12. Li GM et al (2021) Training graph neural networks with 1000 layers. In: ICML
13. Thambi SV et al (2024) A novel technique using graph neural networks and relevance scoring to improve the performance of knowledge graph-based question answering systems. J Intell Inf Syst. https://doi.org/10.1007/s10844-023-00839-4
14. Gnn. https://blog.dataiku.com/graph-neural-networks-merging-deep-learning-with-graphs. Accessed 04 Jan 2024
15. Vashishth S et al (2020) Composition-based multi-relational graph convolutional networks. In: ICLR. OpenReview.net
16. Schlichtkrull M et al (2018) Modeling relational data with graph convolutional networks. In: The Semantic Web, Cham Springer International Publishing, pp 593–607
17. Hamilton W et al (2020) Graph representation learning. In: Synthesis lectures on artificial intelligence and machine learning, vol 14, pp 1–59
18. Dwivedi VP et al (2023) Benchmarking graph neural networks. J Mach Learn Res 24(43):1–48
19. Wang Y et al (2019) Dynamic graph cnn for learning on point clouds. ACM Trans Graph 38(5)
20. Li G et al (2020) Deepergcn: all you need to train deeper gcns. arXiv:2006.07739
21. Li G et al (2023) Deepergcn: training deeper gcns with generalized aggregation functions. IEEE Trans Pattern Anal Mach Intell
22. Vaswani A et al (2017) Attention is all you need. In: Advances in neural information processing systems, vol 30. Curran Associates, Inc
23. Leakyrelu. https://www.educative.io/answers/what-is-leaky-relu. Accessed 04 Jan 2024
24. Thanapalasingam T et al (2022) Relational graph convolutional networks: a closer look. PeerJ Comput Sci 8:e1073
25. Busbridge D et al (2018) Relational graph attention networks. arXiv:abs/1904.05811
26. Devlin J et al (2019) BERT: pre-training of deep bidirectional transformers for language understanding. In: Proceedings of the 2019 conference of the North American chapter of the ACL: human language technologies. Association for Computational Linguistics, pp 4171–4186

Regression Analysis for Longitudinal Aging Study in India Data

Maanya Saxena, Ashish Sharma, and S. Stephen Raj

Abstract This paper examines the Longitudinal Aging Study in India (LASI) and its role in providing valuable insights into the health, social, psychological, and economic well-being of the older Indian population. The paper examines the use of dependent independent variables in a multiple linear regression model, tests assumptions of linearity, and examines the significance of the overall model and the individual variables. There are 190 variables in the dataset being used. This paper presents the results of comparing the regression models obtained through basic, forward, and stepwise selection methods where the model obtained using the stepwise selection method, when all the linearity assumptions are satisfied, explains 86.51% of the variation in the dependent variable and the Adjusted R-squared of the model is 0.8374.

Keywords Regression · Longitudinal aging · Healthcare · Data analytics · Linearity assumptions · Selection methods · Statistics

1 Introduction

1.1 Longitudinal Aging Study in India (LASI)

The Longitudinal Aging Study in India (LASI) is a state wise dataset of aging and health. LASI is a national landmark, the largest in the world and the only of its kind in India. Its main goal is to provide valid, reliable, and continuous scientific data on

M. Saxena (✉) · A. Sharma · S. S. Raj
CHRIST (Deemed to be University), Ghaziabad 201003, India
e-mail: maanya.saxena@ems.christuniversity.in

A. Sharma
e-mail: ashish.sharma@christuniversity.in

S. S. Raj
e-mail: stephenraj.s@christuniversity.in

A. C. Frery et al. (eds.), *Computational Intelligence and Data Analytics*, Lecture Notes on Data Engineering and Communications Technologies 236,
https://doi.org/10.1007/978-981-96-0451-7_10

the health, social, mental, and economic well-being of India's older adult (aged 45 and above) population. These data will help to monitor existing programs and inform new national health and aging policies. LASI is undertaken as a globally harmonized study carried out in more than 44 developed and developing countries worldwide; and the survey schedules, protocols, and methodologies are harmonized across these countries. These are part of the Health and Retirement Study (HRS) family of longitudinal aging and retirement surveys. Therefore, LASI data will also serve as an important public resource for national policy-making and worldwide comparative research on aging. LASI is undertaken under the aegis of the Ministry of Health and Family Welfare (MoHFW), Government of India. The International Institute for Population Sciences (IIPS), Mumbai is the nodal institution for implementing LASI in collaboration with the Harvard T.H. Chan School of Public Health (HSPH) and the University of Southern California (USC) [1].

2 Literature Review

Shrestha [2] utilizes the correlation coefficients, variance inflation factor, and eigenvalue method used in detecting multicollinearity. Advanced regression procedures such as principal components regression, weighted regression, and ridge regression method can be used to determine the presence of multicollinearity. Yuan et al. [3] utilize the meta-regression analysis employed to examine the relationship between urbanization and the urban–rural income gap in China by integrating and analyzing data from a range of individual studies.

Moitra et al. [4] utilized a multi-level meta-regression approach to estimate the pooled relative risk (RR) of suicide associated with various mental disorders. The study also employed a series of covariates to quantify and adjust for sources of heterogeneity in the data, including demographic and methodological variables, as well as the Sociodemographic Index and the Health Access Quality Index. Rana et al. [5] explore the association between self-rated health (SRH) and various factors such as multi-morbidity, mental health, functional health, and social participation among middle-aged and older adults in India using data from LASI wave-1. With a sample size of 65,562 individuals aged 45 and above, the study reveals that multi-morbidity has the highest contribution in predicting poor SRH, followed by functional health, mental health, and social participation.

Mustafa [6] investigates factors influencing life satisfaction in Indian older adults. Utilizing data from the Longitudinal Aging Study in India, the study excludes single-participant households to apply a household fixed-effect model. Findings reveal significant associations between life satisfaction and age, education, functional health, food security, and self-rated health.

The LASI dataset (Longitudinal Aging Study in India) is a comprehensive dataset that contains extensive information on the health, economic, and social well-being of the elderly population in India. However, in the context of statistical modeling, particularly regression analysis, there appears to be a limitation. Specifically, there is no existing regression model within the current research or published literature that uses the death rate across all age groups (not just the elderly) as the dependent variable. This means that while the dataset might be used for various other analyses, there has yet to be an effort to create a model where the primary outcome or variable of interest being predicted or explained is the overall death rate, encompassing all ages.

3 Research Methodology

Figure 1 is designed by summarizing the steps involved in performing the regression analysis on the Longitudinal Aging Study in India (LASI)—Wave 1 dataset. The dataset consists of 190 indicators for the 28 Indian states and 8 Union Territories, also for India as a whole. Among all the variables we are considering the death rate (per thousand population) all ages, as a dependent Variable. Our aim is to find among other remaining variables how many are significantly affecting the dependent variable death rate (per thousand population) all ages.

3.1 Data Description

The study utilized the data from the Longitudinal Aging Study in India (LASI) Wave 1 Report (2021), Ministry of Health and Family Welfare, Government of India.

The dataset consists of 190 indicators for the 28 Indian states and 8 Union Territories, also for India as a whole. Among all the variables we are considering death rate (per thousand population) all ages, as a dependent Variable. Our aim is to find among other remaining variables how many are significantly affecting the dependent variable death rate (per thousand population) all ages.

3.2 Procedure

Multiple Linear Regression Model. So we will be constructing a Multiple Linear Regression Model using the method of least squares with "Death Rate (per thousand population) All ages" being as the dependent variable we have to find how other indicators are considered in LASI which will be the independent variables.

In order to do so we will be constructing a Multiple Linear Regression Model with "death rate (per thousand population) all ages" being as the dependent variable and we have to find a significant independent variable among other indicators considered in LASI. Since there are 190 indicators used in the dataset, first we will find the fourteen highly correlated independent variables with our dependent variable, and for

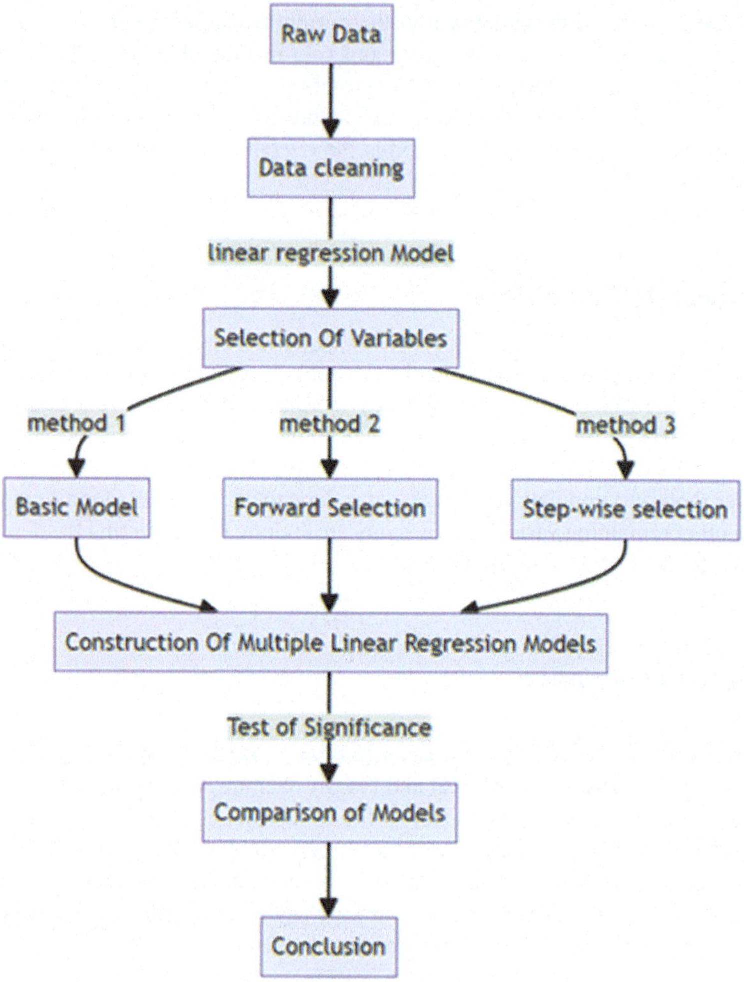

Fig. 1 Steps of regression analysis for LASI wave-1 dataset

the convenience of the model we will consider the following notations for dependent (Y) and independent variables (Xi; i = 1–14):

Y: "Death Rate (per thousand population) All ages"
X1: "Death Rate (per thousand population) 60 + population 2"
X2: "Mean income from government/public transfers by Source (in INR) 16"
X3: "Mean income from non-agricultural business or self-employed activities by Source (in INR)"
X4: "Sources of Finance for Health Care Services during the Last Hospitalization by Income (%) 96"
X5: "Agricultural and allied activities (%) 21"

X6: "Agricultural and allied activities 21"
X7: "70–79 Years of Household population (%)"
X8: "Mean income from agricultural and allied activities by Source (in INR) 13"
X9: "Walker/Walking stick using (%) 75"
X10: "Sought out-patient care in the past 12 months (%)"
X11: "Any Instrumental Activities of Daily Living (IADL) Limitations (%) 72"
X12: "75+ Years of Household population (%)"
X13: "60–74 Years of Household population (%)"
X14: "No schooling (%)".

4 Construction of Regression Models

4.1 Full Model

When we processed the data over software R then for the linear regression model the coefficient and p-value are mentioned in Table 1.

When constructing the linear regression model with all the 14 highly correlated independent variables with the dependent variable, we get the following fitted multiple linear regression model which is given by,

$$Y = -2.447e + 00 + (1.162e\text{-}01) * X1 + (6.804e\text{-}05) * X2$$

Table 1 Linear regression model coefficient and p-value

Full model		
	Estimate	p-Value
Intercept	−2.447e+00	0.1610
X1	1.162e-01	1.19e-05
X2	6.804e-05	0.7419
X3	5.252e-05	0.1121
X4	1.391e-02	0.2627
X5	1.565e-02	0.2658
X6	−8.931e-05	0.3759
X7	1.029e-01	0.8205
X8	2.321e-05	0.6711
X9	5.611e-02	0.6057
X10	2.810e-02	0.0577
X11	−2.410e-03	0.9146
X12	−3.028e-02	0.9158
X13	2.203e-01	0.2427
X14	−1.473e-02	0.4734

$$+ (5.252\text{e-}05) * X3 + (1.391\text{e-}02) * X4 + (1.565\text{e-}02) * X5$$
$$+ (-8.931\text{e-}05) * X6 + (1.029\text{e-}01) * X7 + (2.321\text{e-}05) * X8$$
$$+ (5.611\text{e-}02) * X9 + (2.810\text{e-}02) * X10 + (-2.410\text{e-}03) * X11$$
$$+ (-3.028\text{e-}02) * X12 + (2.203\text{e-}01) * X13 + (-1.473\text{e-}02) * X14$$

Significance of the Overall Model

Now we will check the significance of the model by applying the F test. For this, the null hypothesis is "there is no linear relationship between the dependent variable and the independent variables; that is, the regression model is not significant". Against the alternative hypothesis that "at least one of the regressors contributes significantly to the model, i.e., at least one regression coefficient is not zero". From the output of the model summary, the p-value is 6.602e-07, which is less than 0.05. Hence, we reject the null hypothesis. Therefore, the overall regression model is significant, and at least one of the regression coefficients is not zero.

Significance of Individual Variables

For the 5% level of significance, only X1 and X10 have a significant contribution to the regression model, since the p-value corresponding to the X1 variable is 1.19e-05 and the p-value corresponding to the X10 variable is 0.0577 which is less than 0.05 level of significance.

Interpretation

Hence, when constructing a full model with 14 highly correlated independent variables with the dependent variable, we conclude that the model is significant with X1: "Death Rate (per thousand population) 60 + population 2" and X10: "Sought outpatient care in the past 12 months (%)" contributing significantly to the regression model.

For the full regression model, the Multiple R-squared is 0.8756 which implies that 87.56% of the variation in the dependent variable is explained by the linear regression model and the Adjusted R-squared of the model is 0.7965.

4.2 *Forward Selection Method*

When we processed the data over software R then for the linear regression model by using the Forward Selection Method out of 14 variables, we got 7 variables for which AIC values were the least. The coefficient and p-value of these selected 7 variables are mentioned in Table 2.

The fitted multiple linear regression model is given by,

$$Y = -3.794\text{e} + 00 + (1.134\text{e-}01) * X1$$
$$+ (1.191\text{e-}01) * X7 + (2.019\text{e-}02) * X5$$

Table 2 Coefficient and the P-value for the variables X1, X7, X5, X10, X3, 13, and X4

Forward selection model		
	Estimate	P-value
Intercept	−3.794e+00	0.00221
X1	1.134e-01	7.75e-07
X7	1.191e-01	0.73667
X5	2.019e-02	0.00842
X10	2.438e-02	0.01676
X3	3.851e-05	0.01019
X13	2.593e-01	0.10088
X4	1.675e-02	0.11443

$$+ (2.438\text{e-}02) * X10 + (3.851\text{e-}05) * X3$$
$$+ (2.593\text{e-}01) * X13 + (1.675\text{e-}02) * X4$$

The Adjusted R-squared value of the new model taking only X1, X7, X5, X10, X3, 13, and X4 variables is 0.8325, which is greater than the Adjusted R-squared value of the entire model, therefore model obtained from the forward selection method is better than the full model. This shows that the multiple linear regression model with just X1, X7, X5, X10, X3, X13, and X4 variables explains the 86.51% variation in the values of dependent variable Y and there is no need of adding more variables in the multiple linear regression model. Since the AIC value of the forward selection model with the seven variables is −16.85 and on adding any extra variables in the model will not further decrease the AIC value, so adding any more variables in the model is not significant.

Significance of Forward Selection Model

Now we will check the significance of the model by applying the F test. For this, the null hypothesis is "there is no linear relationship between the dependent variable and the independent variables; that is, the regression model is not significant". Against the alternative hypothesis that "at least one of the regressors contributes significantly to the model, i.e., at least one regression coefficient is not zero". From the output of the model summary, the p-value is 5.563e-11, which is less than 0.05. Hence, we reject the null hypothesis. Therefore, the overall regression model is significant, and at least one of the regression coefficients is not zero.

Significance of Individual Variable

For the 5% level of significance, only intercept, X1, X5, X10, and X3 have a significant contribution to the regression model, since the p-value corresponding to the X1 variable is 7.75e-07, that of X5 variable is 0.00842, X10 variable is 0.01676, and the p-value corresponding to the X3 variable is 0.01019 which is less than 0.05 level of significance.

Interpretation

Hence, when constructing a multiple linear regression model will have X1, X7, X5, X10, X3, X13, and X4 variables, we conclude that the model is significant with X1: "Death Rate (per thousand population) 60 + population 2" and X10: "Sought outpatient care in the past 12 months (%)" contributing significantly to the regression model.

For the full regression model, the Multiple R-squared is 0.8651 which implies that 86.51% of the variation in the dependent variable is explained by the linear regression model and the Adjusted R-squared of the model is 0.8325. The Multiple R-squared of this model is slightly less than the full model but the Adjusted R-squared of this model is significantly greater than the full model which is 0.7965. Therefore, this model is better than the full model.

4.3 Assumptions of Linearity

Correlation

All the variables are correlated to the dependent variable, Y: "Death Rate (per thousand population) All ages" (Fig. 2).

Some predictor variables have strong positive correlations among themselves. For example, X6 and X13 have a strong positive correlation of 0.88, X7 and X13 have a strong positive correlation of 0.76, and X8 and X12 have a strong positive correlation of 0.76.

Multicollinearity. The strong positive correlations between "Death Rate (per thousand population) All ages" and various predictors suggest that these variables may be influential in explaining the variability in "Death Rate (per thousand population) All ages".

The presence of strong correlations between predictor variables (X6 and X13, X7 and X13, etc.) may indicate multicollinearity. This multicollinearity can complicate the interpretation of individual predictor effects in regression analysis. So we will remove this using R Programming.

Normality. Now we will test the assumption of normality for the model obtained using the forward selection method using the one-sample Kolmogorov–Smirnov test. For this, the null hypothesis is "the sample is drawn from the reference distribution (in this case, a uniform distribution)". Against the alternative hypothesis that "it tests whether the empirical distribution function of data differs significantly from the uniform distribution". From the output of the model summary, the p-value is 0.5734, which is less than 0.05. This suggests that, based on the Kolmogorov–Smirnov test, there is not enough evidence to conclude that the distribution of "Death Rate (per thousand population) All ages" significantly deviates from a uniform distribution. Therefore, the assumption of normality is satisfied.

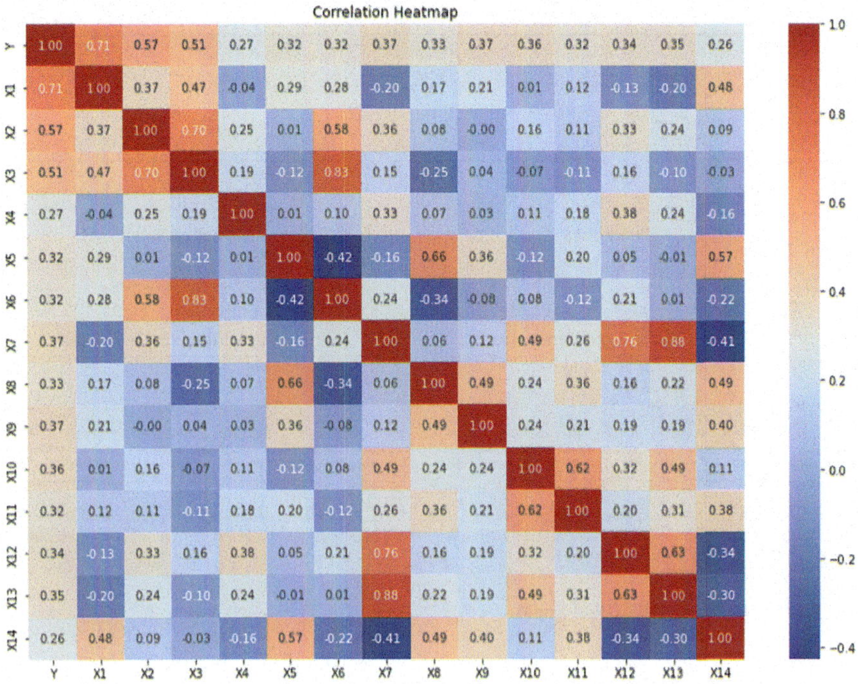

Fig. 2 Heat map showing the correlation between variables

Homoscedasticity

Interpretation: From the Residual vs Fitted graph, we conclude that for the fitted values the variation in the residuals is not exactly constant but lies in range so homoscedasticity occurs. From the Normal Q–Q graph, we conclude that the data almost exactly lies along a straight line, so the sample data appears to be normally distributed (Fig. 3).

Autocorrelation. Now we will test the assumptions of autocorrelation for the model obtained using the forward selection method using the Durbin–Watson test. For this, the null hypothesis is "there is no autocorrelation in the residuals of the regression model, that is, rho is zero". Against the alternative hypothesis that "there is a positive or negative autocorrelation in the residuals of the regression model that is, rho is not zero". Since D–W Statistic is 2.614404, which is greater than 2, also the p-value of the test is 0.076 which is greater than 0.05. Hence we fail to reject the null hypothesis and therefore there is no autocorrelation in the residuals of the regression model. The Durbin–Watson statistic is close to 2, which is the value indicating no autocorrelation. The p-value suggests that there is some evidence against the null hypothesis of no autocorrelation, but it's not strong evidence, and therefore there is no autocorrelation in the residuals of the regression model.

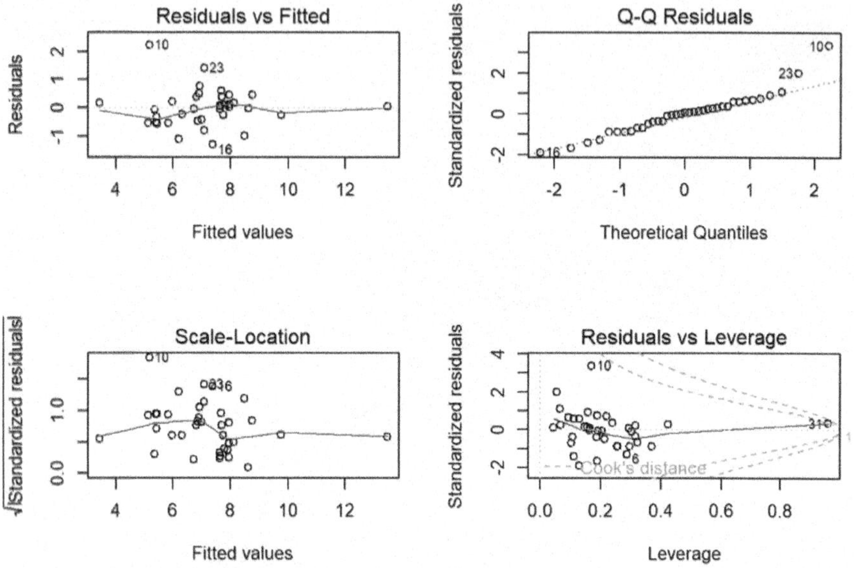

Fig. 3 The residual versus fitted graph

4.4 Step-Wise Selection Method

When we processed the data over software R then for the linear regression model by using the Stepwise Selection Method out of 14 variables, we got 6 variables for which AIC values were the least. The coefficient and p-value of these selected 7 variables are mentioned in Table 3.

Using the Stepwise Procedure, the best multiple linear regression model will have X1, X3, X4, X5, X10 and X13 independent variables. The fitted multiple linear regression model is given by,

$$Y = (-3.822e{+}00) + (1.115e{-}01) * X1 + (4.110e{-}05) * X3$$

Table 3 Coefficient and the P-value for the variables X1, X3, X4, X5, X10 and X13

Stepwise selection model		
	Estimate	P-value
Intercept	−3.822e+00	0.001717
X1	1.115e-01	2.86e-07
X3	4.110e-05	0.001288
X4	1.735e-02	0.092442
X5	1.985e-02	0.007831
X10	2.516e-02	0.010316
X13	3.053e-01	0.000143

$$+ (1.735\text{e-}02) * X4 + (1.985\text{e-}02) * X5$$
$$+ (2.516\text{e-}02) * X10 + (3.053\text{e-}01) * X13$$

The Adjusted R-squared value of the new model after satisfying the assumptions of linearity taking X1, X3, X4, X5, X10, and X13 variables is 0.8374a and the AIC value of the final model is −18.71, which is less among all the above models.

Significance of Overall Model

Now we will check the significance of the model by applying the F test. For this, the null hypothesis is "there is no linear relationship between the dependent variable and the independent variables; that is, the regression model is not significant". Against the alternative hypothesis that "at least one of the regressors contributes significantly to the model, i.e., at least one regression coefficient is not zero". From the output of the model summary, the p-value is 9.853e-12, which is less than 0.05. Hence, we reject the null hypothesis. Therefore, the overall regression model is significant, and at least one of the regression coefficients is not zero.

Significance of Individual Variable

For the 5% level of significance, all the independent variables including intercept have a significant contribution to the regression model except X4: "Sources of Finance for Health Care Services during the Last Hospitalization by Income (%) 96" whose p-value is 0.092442 which is greater than 0.05 level of significance. While for all other independent variables and intercepts in the model, the p-value corresponding to them is less than 0.05 level of significance.

Interpretation

Hence, when constructing a multiple linear regression model with X1, X3, X4, X5, X10, and X13 independent variables, we conclude that the model is significant with respect to the intercept and the following independent variables

X1: "Death Rate (per thousand population) 60 + population 2"
X3: "Mean income from non-agricultural business or self-employed activities by Source (in INR)"
X5: "Agricultural and allied activities (%) 21"
X10: "Sought out-patient care in the past 12 months (%)"
X13: "60–74 Years of Household population (%)".

For this regression model, the Multiple R-squared value is 0.8645 which implies that 86.51% of the variation in the dependent variable is explained by the linear regression model and the Adjusted R-squared of the model is 0.8374.

The Adjusted R-squared of this model obtained using the stepwise procedure after all the assumptions of linearity were satisfied, is slightly greater than the Adjusted R-squared value of the previous model obtained using the Forward Method in which the assumptions of linearity were not satisfied. Also, this new model includes only 6 independent variables while the previous model includes 7 independent variables.

Therefore, this model obtained using the stepwise procedure after all the assumptions of linearity were satisfied is better than the model obtained using the Forward Method in which the assumptions of linearity were not satisfied.

5 Comparison of Models

The new model after all the assumptions of linearity are satisfied has the least number of independent variables which is six while the original full model has the highest number of independent variables which is fourteen and the model using the Forward Selection method before the satisfaction of the assumptions of linearity has seven independent variables. The new model after all the assumptions of linearity are satisfied has the highest number of significant independent variables which is five while the original full model has the least number of significant independent variables which is two and the model using the Forward Selection method before the satisfaction of the assumptions of linearity has four significant independent variables.

The Adjusted R-squared of the new model obtained using the stepwise procedure after all the assumptions of linearity were satisfied, is slightly greater than the Adjusted R-squared value of the previous model obtained using the Forward Method in which the assumptions of linearity were not satisfied and the also from the Adjusted R-squared of the full model.

Therefore, the new model obtained using the stepwise procedure after all the assumptions of linearity were satisfied is better than the model obtained using Forward Method in which the assumptions of linearity were not satisfied since this regression model explains more proportion of the dependent variable with lesser number of independent variables as compared to other two models.

6 Conclusion

In conclusion, the regression model developed in the research paper helps in mortality prediction in all age groups. By incorporating independent variables related to health, income, and demographics, the regression model provides valuable insights to the government authorities in policy-making as to how certain factors, including socioeconomic factors such as income sources, affect the mortality rate in all age groups. This will further help assess the effectiveness of the various policies and initiatives taken by the Government aimed at improving public health.

The research paper does not study and includes the potential confounding variables that may affect the regression models constructed in the paper. The paper also does not include limitations of the LASI dataset. Future studies can explore the relationships between other independent variables and dependent variables related to the welfare of the elderly demographic in India. Additionally, future studies can investigate potential confounding variables that may impact the relationships between the independent and dependent variables. Finally, future studies can compare LASI data with data from other longitudinal aging and retirement surveys to gain a more comprehensive understanding of aging and well-being in a global context.

References

1. Government of India (2024) Open government data (OGD) platform India. https://data.gov.in. Accessed 29 Jan 2024
2. Shrestha N (2020) Detecting multicollinearity in regression analysis. Am J Appl Math Stat 8(2):39–42. https://doi.org/10.12691/ajams-8-2-1
3. Yuan Y, Wang M, Zhu Y, Huang X, Xiong X (2020) Urbanization's effects on the urban-rural income gap in China: a meta-regression analysis. Land Use Policy 99:104995
4. Moitra M, Santomauro D, Degenhardt L, Collins PY, Whiteford H, Vos T, Ferrari A (2021) Estimating the risk of suicide associated with mental disorders: a systematic review and meta-regression analysis. J Psychiatr Res 137:242–249
5. Rana GS, Shukla A, Mustafa A, Bramhankar M, Rai B, Pandey M, Mishra NL (2022) Association of multi-morbidity, social participation, functional and mental health with the self-rated health of middle-aged and older adults in India: a study based on LASI wave-1. BMC Geriatr 22:675. https://doi.org/10.1186/s12877-022-03349-0
6. Mustafa A (2022) Examining correlates of life satisfaction among Indian older adults using household fixed-effect approach. Discov Soc Sci Health 2:25. https://doi.org/10.1007/s44155-022-00028-8

Semi Supervised Flood Damage Detection Using Satellite Images

Manish Nadella, Garapati Venkata Krishna Rayalu, Menta Sai Akshay, S. K. Eswar Sudhan, V. V. Sajith Variyar, V. Sowmya, and Ramesh Sivanpillai

Abstract Flood mapping using satellite data poses significant challenges for disaster response teams. In addressing these challenges, our study introduces a deep learning framework designed to map flood extents in turbid water bodies, thereby advancing practical applicability. The proposed model is a semi-supervised architecture based on ResNet-18, fine-tuned for binary classification of satellite images into flooded or non-flooded categories. Acknowledging the low availability of training samples, our approach leverages a semi-supervised methodology, capitalizing on the benefits of unlabeled data.We explore the challenges of detecting images flooded with turbid water on the model. For the initial training phase, the model undergoes rigorous training on distinct datasets comprising Sentinel-2A images with varying dimensions, specifically $128 \times 128 \times 3$ and $256 \times 256 \times 3$ pixels. Subsequently, the pre-trained model undergoes transfer learning on a flood dataset from Louisiana, enhancing its adaptability to diverse flood scenarios and satellite sources.The capacity to automatically map floods solves significant problems for emergency response teams.

Keywords Flood mapping · Satellite imagery · Semi-supervised learning · Transfer learning · Turbid water · ResNet-18 · Disaster response · Binary classification

1 Introduction

In the continuous confrontation against monsoon-induced floods, disaster response teams use satellite photography to provide essential aid to impacted communities [1]. This technical alarm, is critical for designating the enormous expanse of waterlogged

M. Nadella · G. V. K. Rayalu · M. S. Akshay · S. K. Eswar Sudhan · V. V. Sajith Variyar (✉) · V. Sowmya
Amrita School of Artificial Intelligence, Amrita Vishwa Vidyapeetham, Coimbatore, India
e-mail: vv_sajithvariyar@cb.amrita.edu

R. Sivanpillai
Wyoming GIS Center, School of Computing, University of Wyoming, Laramie, WY 82071, USA

A. C. Frery et al. (eds.), *Computational Intelligence and Data Analytics*, Lecture Notes on Data Engineering and Communications Technologies 236,
https://doi.org/10.1007/978-981-96-0451-7_11

regions, serves as the crux in their supply. However, the significant task comes not only in detecting flood-affected locations, but also in democratizing this crucial information for persons without a technology background. The several satellite datasets, each with its own set of complications, necessitate a level of expertise that goes beyond typical disaster response approaches.

As our research explores flood mapping using satellite images [2], we convey the need for comprehensive solutions that bridge the gap between technological complexity and practical application. Dealing with multiple databases, each with its own complexities, requires a skill set not typically associated with emergency response teams [3]. The thorough cleaning and analysis of large datasets demand significant time and effort, exacerbated by varying spatial resolutions [4].In this intricate environment, building a resilient model capable of handling anomalies is crucial. The model's ability to tolerate different characteristics in satellite datasets is essential for maintaining quality and dependability [5] in flood mapping results.

A pivotal aspect of our research involves developing a semi-supervised deep learning architecture [6]. This architecture, initially trained on images from a single satellite, aims to transcend individual satellite specifications during testing, showcasing adaptability to diverse satellite datasets. Additionally, we focus on creating a semi-supervised framework specifically tailored to map flood images featuring turbid water bodies, thereby enhancing the model's practical utility in real-world flood crisis scenarios.

Our research faces the challenge of limited training data, with a significant portion of available online data comprising clear water flood images rather than turbid water conditions [7]. This imbalance affects the model's understanding of turbidity, requiring careful consideration to prevent misinterpretation as barren land or vegetation. Consequently, evaluating the deep learning model's performance under sparse information, particularly concerning turbidity's influence, is crucial. Furthermore, we investigate how turbidity impacts the model's reliability in detecting floods, acknowledging its adverse effects on image classification [8]. Our study aims to assess the model's resilience in analyzing flood photos from diverse satellite sources, leveraging transfer learning to ensure adaptability to varying images captured by different satellites, thereby enhancing its effectiveness in flood mapping applications [9].

Our research follows a methodology involving data pre-processing steps such as splitting, cropping, and augmentation. We use a semi-supervised learning approach with a modified ResNet-18 model for binary flood classification.Moreover, transfer learning is applied to assess the model's performance on the Louisiana flood 2016 dataset.

2 Related Works

In Wieland et al. [6] the authors evaluated CNN architectures like U-Net and DeepLab-V3+ for semantic segmentation of water bodies in high-resolution satellite and aerial data, considering different encoder backbones, near-infrared bands, slope information, data augmentation, and noisy training data.

In Foroughnia et al. [7] the authors combined unsupervised and supervised classification methods with SAR and multi-spectral data, introducing a novel unsupervised approach (CThS) that handles water patch heterogeneity.

In Edpuganti et al. [10] the authors examined the effect of data quality on DeepLabV3+ training, showing improved accuracy with higher-quality data and stressing the need for analyzing open-source data quality.

In Vandaele et al. [8] the authors explored deep transfer learning for water segmentation and prediction in river camera images, leveraging pre-trained networks and investigating water level estimation.

In Li et al. [9] the authors provided a comprehensive review of water body classification approaches, highlighting the potential of deep learning and proposing open benchmarks.

In Reddy et al. [5] the authors investigated the effect of input image size and batch size on DeepLabV3+ performance for water body segmentation.

While these works address various aspects of water body segmentation and flood mapping, there is still a need to focus on the incorporation of turbid water conditions, the effectiveness of models under turbidity, and the adaptability to diverse satellite sources [4]. Additionally, the importance of data quality, transfer learning, and the integration of different data sources in a semi-supervised deep learning framework for flood mapping with limited training samples and unlabeled data is an aspect that requires further exploration.

3 Dataset

The original dataset has 2,185 images from Sentinel-2A with corresponding masks.

3.1 Augmented Dataset for Semi-supervised Model

The dataset-I comprises of turbid images from Sentinel-2A resized to dimensions of $128 \times 128 \times 3$ pixels. In the training subset of this dataset, there are a total of 1513 training images, while the testing subset consists of 966 images.We labelled the images accordingly as to bring the semi-supervised approach, images labeled as flooded for training are 84, images labeled as non-flood for training are 104 and we had 1325 unlabelled images. Figure 1 illustrates a sample image.

The dataset-II comprises of turbid images from Sentinel-2A resized to larger dimensions of $256 \times 256 \times 3$ pixels. Within the training subset of this dataset, there are 1407 training images, and the testing subset includes 419 images.We labelled the images accordingly as to bring the semi-supervised approach, images labeled as flooded for training are 62, images labeled as non-flood for training are 62 and we had 1283 unlabelled images. Figure 1 illustrates a sample image.

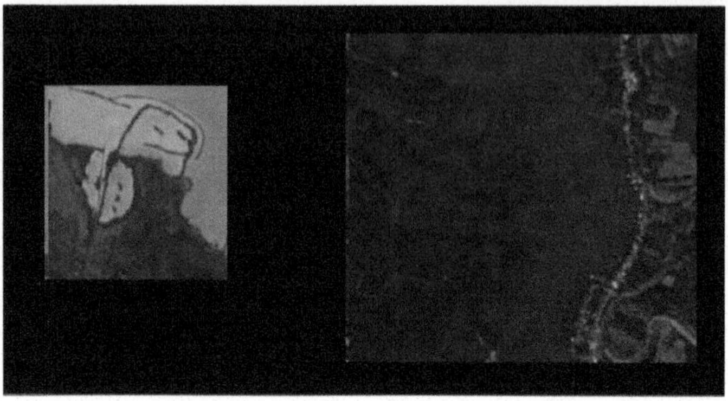

Fig. 1 Sample images from the augmented dataset. (left image: flooded image of size 128 × 128, right image: non-flooded image of size 256 × 256)

3.2 Augmented Dataset for Transfer Learning Model

The dataset-I comprises of satellite images of flood affected regions of louisiana which were taken from kaggle. These images were resized to dimensions of 128 × 128 × 3 pixels. In the training subset of this dataset, there are a total of 870 training images, while the testing subset consists of 342 images.We labelled the images accordingly as to bring the semi-supervised approach, images labeled as flooded for training are 54,images labeled as non-flood for training are 55 and we had 761 unlabelled images.

The dataset-II comprises of images from the same source but where resized to larger dimensions of 256 × 256 × 3 pixels. Within the training subset of this dataset, there are 1314 training images, and the testing subset includes 342 images.We labelled the images accordingly as to bring the semi-supervised approach, images labeled as flooded for training are 64, images labeled as non-flood for training are 66 and we had 1184 unlabelled images.

4 Methodology

4.1 Data Pre-processing

Data Splitting The dataset is partitioned into training and testing sets using a function that creates dedicated folders for each. Random shuffling introduces variability, with approximately 80% of images assigned to the training set and the remaining 20% to the test set.

Data Cropping The script incorporates a function to crop images into sub-images of a specified size from the bottom-right corner, ensuring continuity through overlap. This iterative cropping process generates smaller sub-images from the original dataset for subsequent processing.

Image Size Standardization A function is employed to resize cropped images to a standardized size (256 $\times$ 256 pixels). To guarantee compatibility, another function ensures that sub-image dimensions are multiples of the specified size. Augmented data is created for both training and testing sets and these sets are stored separately. Further data augmentation is achieved by reapplying the cropping function on these sets.

Resulting Structure The final dataset structure encompasses distinct folders for original images, training and testing datasets, and augmented versions of these datasets. This organized dataset is now ready for subsequent stages, such as training and evaluating machine learning models.

4.2 *Training Semi-supervised Model*

Figure 2 serves as an exemplary diagram illustrating the proposed methodology.After the data pre-processing we convert 32-bit images to the standardized 24-bit RGB format that streamlines the data preparation process, ensuring compatibility with subsequent neural network operations.

In the neural network architecture, ResNet-18 serves as the foundational deep learning model. Specifically, ResNet-18 is renowned for its effectiveness in image classification tasks. However for this binary classification task at hand distinguishing between flooded and non-flooded scenarios custom modifications are implemented to tailor the model to the specific requirements of this application.

The primary adaptation involves alterations to the final fully connected layer, the original layer is designed for multi-class classification tasks, typically found in tasks with more than two classes. In the context of binary classification, where there are only two classes (flooded and non-flooded), the structure of the output layer is adjusted to accommodate this binary nature.

Consequently, the original fully connected layer is replaced with a new linear layer configured to have two output nodes. This modification transforms the architecture to produce a binary classification output, where each node corresponds to one of the two classes.

The model is then trained with the two datasets separately with 80% of the data allocation for training and the rest 20% for testing.This model is then used on the validation dataset which generates the predictions and returns a structured JSON file which consists predicted labels for which image from the test dataset.

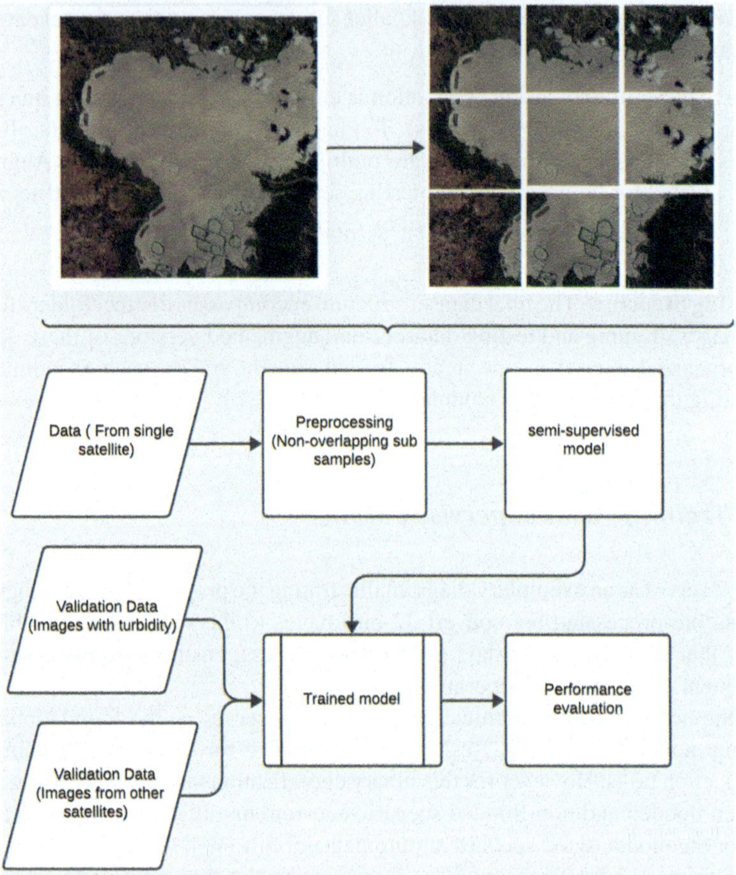

Fig. 2 Block diagram for semi-supervised approach for flood water classification

4.3 *Transfer Learning Approach*

The semi-supervised model which was trained for the purpose of predicting flooded and non-flooded regions from sentinel 2A satellite images is now tested over the Lousiana flood 2016 data after the pre-processing. The weights of model trained on sentinel 2A are trained on small subset of the new dataset and used for testing. This process is carried out for a series of hyperparameters [11].

Table 1 Semi-supervised model trained on images of size 128 × 128

Parameters				Metrics			
Batch size	Learning rate	Epoch	Optimizer	Accuracy (%)	Precision	Recall	F1-score
64	0.01	100	SGD	80.02	0.96	0.80	0.87
16	0.01	100	SGD	76.60	0.91	0.81	0.86
16	0.01	100	ADAM	78.77	0.96	0.79	0.86
128	0.01	100	SGD	72.36	0.93	0.74	0.82
128	**0.01**	**100**	**ADAM**	**81.46**	**0.96**	**0.82**	**0.88**
64	0.001	100	SGD	65.01	0.96	0.62	0.76
64	0.001	100	ADAM	80.74	0.93	0.84	0.88

5 Results and Discussions

5.1 Semi-supervised Learning

Images of size 128 × 128 In Table 1 the results consistently demonstrate the performance of the ADAM optimizer over the traditional SGD (Stochastic Gradient Descent) across multiple evaluation metrics, including accuracy, precision, recall, and F1-score. The F1-score, which provides a balanced assessment of precision and recall, further emphasizes the overall effectiveness of ADAM in achieving a harmonious trade-off between true positives and false positives/negatives. This suggests that the ADAM optimizer may be a favorable choice for optimizing the training of deep learning models, particularly in scenarios where a balance between precision and recall is crucial. The ADAM optimizer, being a first-order gradient-based optimization technique, combines the advantages of adaptive lower-order moment estimates, making it efficient and easy to implement. However, it is important to note that the performance of the ADAM optimizer can be influenced by the specific characteristics of the dataset and the task at hand, and careful hyperparameter tuning may be necessary for optimal results. Figure 3. illustrates the performance of the model on sample images extracted from the test dataset. It is evident from the figure that the model adeptly distinguishes between turbid water and ground, accurately classifying the condition of the images.

Figure 3. showcases the output of sample images classified by the model trained on images sized 128 × 128, utilizing the ADAM optimizer with a batch size of 16 via a semi-supervised approach. In each case, the left side represents a flooded image, while the right side depicts a non-flooded image.

Fig. 3 Sample of images classified by the model trained on images of size 128 × 128 using semi-supervised approach

Images of size 256 × 256 The results in Table 2 clearly show that ADAM outperforms SGD in several key aspects like accuracy, precision, recall, and F1-score. The most impressive part is the significant improvement in F1-score, indicating that ADAM strikes a great balance between getting things right and not missing out on important details. This is especially cool when dealing with bigger images, suggesting that ADAM handles complicated data really well. The F1-score indicates that ADAM effectively balances the trade-off between precision, which avoids false positives, and recall, which prevents false negatives. Essentially, ADAM performs well across different applications, particularly those with extensive and intricate image datasets. Figure 4. demonstrates the performance of the model trained with a batch size of 128 using the ADAM optimizer. Despite the challenging scenario where the water in the flooded image closely resembles the ground, the model successfully classifies the images into the correct classes.

Figure 4. presents the output of sample images classified by the model trained on images sized 256 × 256. This model was trained with the ADAM optimizer and a batch size of 16 using a semi-supervised approach. In each instance, the left side represents a flooded image, while the right side depicts a non-flooded image.

Table 2 Semi-supervised model trained on images of size 256 × 256

Parameters				Metrics			
Batch size	Learning rate	Epoch	Optimiser	Accuracy (%)	Precision	Recall	F1-score
64	0.01	100	SGD	88.30	0.81	0.42	0.55
16	0.01	100	SGD	86.87	0.84	0.30	0.44
16	0.01	100	ADAM	89.97	0.96	0.43	0.60
128	0.01	100	SGD	73.74	0.35	0.64	0.46
128	**0.01**	**100**	**ADAM**	**98.32**	**0.94**	**0.95**	**0.95**
64	0.001	100	SGD	87.35	0.95	0.28	0.44
64	0.001	100	ADAM	98.09	0.95	0.93	0.94

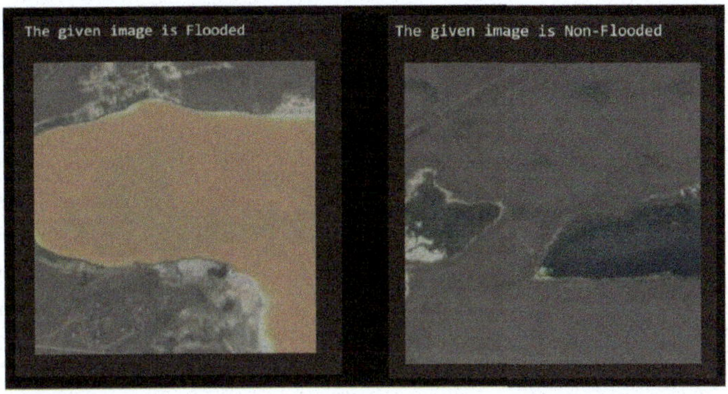

Fig. 4 Sample of images classified by the model trained on images of size 256 × 256, using semi-supervised approach

Table 3 Transfer learning approach–model trained on images of size 128 × 128

Parameters				Metrics			
Batch size	Learning rate	Epoch	Optimiser	Accuracy	Precision	Recall	F1-score
64	0.01	100	SGD	78.75	0.43	0.76	0.83
16	0.01	100	SGD	98.75	0.92	0.98	0.99
16	**0.01**	**100**	**ADAM**	**98.82**	**0.92**	**0.95**	**0.96**
128	0.01	100	SGD	91.25	0.90	0.88	0.93
128	0.01	100	ADAM	98.75	0.99	0.98	0.99
64	0.001	100	SGD	59.37	0.81	0.45	0.62
64	0.001	100	ADAM	97.5	0.97	0.96	0.98

For the semi-supervised approach with the ADAM optimizer and a learning rate of 0.001, applied to images sized 256 × 256, we achieved a true positive score of 79, a true negative score of 708, a false positive score of 25, and a false negative score of 154.

For the semi-supervised approach using the ADAM optimizer with a batch size of 128, applied to images sized 128 × 128, we obtained a true positive score of 3453, a true negative score of 68, a false positive score of 3, and a false negative score of 5.

5.2 *Transfer Learning*

Images of size 128 × 128 In Table 3, while ADAM and SGD showed comparable performance for a batch size of 16, with minimal differences in accuracy, precision, and recall, ADAM exhibited a slightly better F1-score across other configurations,

Table 4 Transfer learning approach–model trained on images of size 256 × 256

Parameters				Metrics			
Batch size	Learning rate	Epoch	Optimiser	Accuracy	Precision	Recall	F1-score
64	0.01	100	SGD	85.67	0.78	0.80	0.87
16	**0.01**	**100**	**SGD**	**89.18**	**0.82**	**0.8**	**0.90**
16	0.01	100	ADAM	81.57	0.73	0.79	0.85
128	0.01	100	SGD	79.82	0.81	0.61	0.76
128	0.01	100	ADAM	67.54	0.70	0.64	0.67
64	0.001	100	SGD	56.72	0.71	0.28	0.40
64	0.001	100	ADAM	79.82	0.76	0.88	0.82

indicating a better balance between precision and recall. This performance underscores the effectiveness of ADAM in the Transfer Learning approach when applied to images of size 128 × 128. Figure 5 displays the performance of the model trained using a batch size of 16 with the ADAM optimizer, employing the transfer learning approach.

Figure 5. illustrates the output of sample images classified by the model trained on images sized 128 × 128, utilizing the ADAM optimizer with a batch size of 128, and employing the transfer learning approach. In each case, the left side represents a flooded image, while the right side depicts a non-flooded image.

Images of size 256 × 256 In Table 4, the choice between ADAM and SGD depends on the specific priorities. ADAM demonstrated higher precision and recall, while SGD achieved a higher accuracy. Figure 6 illustrates the performance of the model trained using a batch size of 16 with the SGD optimizer.

Figure 6 represents the output of sample images classified by the model trained on images sized 256 × 256. This model was trained with the ADAM optimizer and a batch size of 128 using the transfer learning approach. In each instance, the left side depicts a flooded image, while the right side portrays a non-flooded image.

For the transfer learning approach with a batch size of 128, using images sized 128 × 128, we achieved a true positive score of 40, a true negative score of 118, a false positive score of 0, and a false negative score of 2.

For the transfer learning approach with a batch size of 16, using images sized 256 × 256, we achieved a true positive score of 100, a true negative score of 179, a false positive score of 63, and a false negative score of 0.

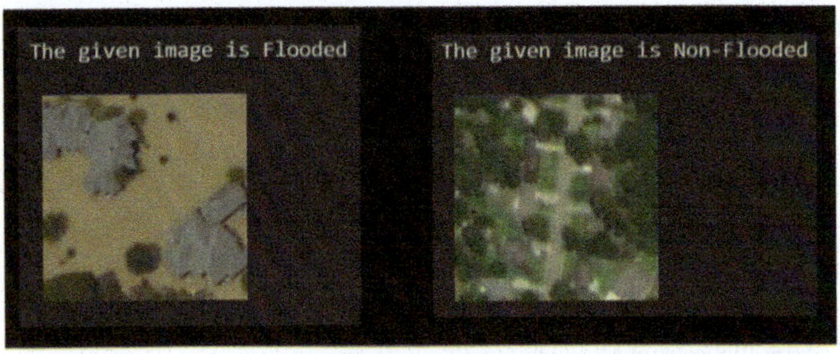

Fig. 5 Sample of images classified by the model trained on images of size 128 × 128, using transfer learning approach

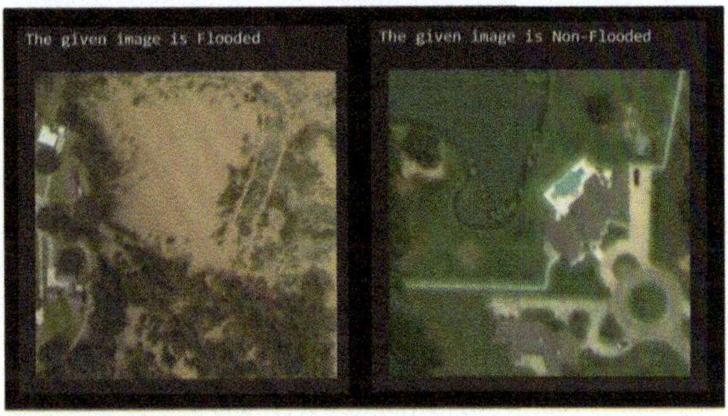

Fig. 6 Sample of images classified by the model trained on images of size 256 × 256, using transfer learning approach

6 Conclusion and Future Works

Our study proposes a robust deep learning framework for mapping flood extent from satellite imagery, addressing the challenges of turbid water bodies. The semi-supervised technique demonstrates adaptability to different flood events and data sources, with the customized ResNet-18 architecture and transfer learning approach performing well in terms of accuracy and F1 score. While highlighting the critical role of artificial intelligence in unlocking the life-saving potential of Earth observation data for disaster management, our study emphasizes the need for more diverse datasets, loss function customization, data augmentation, integration of terrain and hydrological data, higher resolution scaling, near real-time applicability testing, model interpretability research, and user-friendly interfaces to facilitate real-world adoption and aid response teams during disasters.

References

1. Rebecca A, Hamilton L, Södergård B, Liverani M (2021) The role of emergency medical teams in disaster response: a summary of the literature. Nat Hazards 110:1417–142
2. Jain P, Schoen-Phelan B, Ross R (2020) Automatic flood detection in sentinei-2 images using deep convolutional neural networks, pp 617–623
3. Tanim AH, McRae CB, Tavakol-Davani H, Goharian E (2022) Flood detection in urban areas using satellite imagery and machine learning. Water 14(7)
4. Bhadra T, Chouhan A, Chutia D, Bhowmick A, Raju PLN (2020) Flood detection using multispectral images and sar data, pp 294–303
5. Reddy OC, Kumar ID, Sathvika P, Sajith Variyar VV, Sowmya V, Sivanpillai R (2023) Effect of hyperparameters on deeplabv3+ performance to segment water bodies in rgb images. In: The international archives of the photogrammetry, remote sensing and spatial information sciences, vol XLVIII-M-3-2023, pp 203–209
6. Wieland M, Martinis S, Kiefl R, Gstaiger V (2023) Semantic segmentation of water bodies in very high-resolution satellite and aerial images. Remote Sens Environ 287:113452
7. Fatemeh Foroughnia SM, Menenti AM, Lindenbergh RC (2022) Evaluation of sar and optical data for flood delineation using supervised and unsupervised classification. Remote Sens 14(15)
8. Vandaele R, Dance S, Ojha V (2021) Automated water segmentation and river level detection on camera images using transfer learning, pp 232–245. (03, 2021)
9. Li Y, Dang B, Zhang Y, Zhenhong D (2022) Water body classification from high-resolution optical remote sensing imagery: achievements and perspectives. ISPRS J Photogramm Remote Sens 187:306–327
10. Edpuganti A, Akshaya P, Gouthami J, Sajith Variyar VV, Sowmya V, Sivanpillai R (2023) Effect of data quality on water body segmentation with deeplabv3+ algorithm. In: The international archives of the photogrammetry, remote sensing and spatial information sciences, vol XLVIII-M-3-2023, pp 81–85
11. He K, Zhang X, Ren S, Sun J (2016) Deep residual learning for image recognition. 110(1111):770–778

Technological Alternative to Support the Diagnosis of Sarcopenia Through the Application of Machine Learning Techniques

Cristian Castillo-Olea, Clemente Zuñiga, and Guadalupe Flores

Abstract An engineering application project is being presented, in which various machine learning techniques have been applied. These techniques involve the use of classification algorithms and data analysis techniques. Methods were designed using the Python programming language and available libraries for data analysis and machine learning. The purpose of this technological solution is to offer a first-level support alternative in the Mexican healthcare system for the diagnosis of sarcopenia in older adults. Sarcopenia currently has known diagnoses by physicians, but some of these methods are costly, and not all hospitals have access to them. However, there is a possibility of finding certain secondary variables that are related to the disease and can be classified in a software tool that helps primary care physicians recognize risk characteristics in patients in a timely manner. This would ensure that patients who may have sarcopenia receive appropriate attention and medication. The data used in this project comes from the General Hospital of Tijuana, where we have anonymized data from 100 individuals.

Keywords Sarcopenia · Machine learning · Data analysis

1 Introduction

The information that exists on the diagnoses of sarcopenia and the different techniques for the diagnosis of this disease are used by specialists in the subject, however, currently in hospitals in Mexico and around the world doctors attending to patients

C. Castillo-Olea (✉)
Autonomus University of Baja California, Tijuana, Baja California, Mexico
e-mail: castillo.cristian@uabc.edu.mx

C. Castillo-Olea · G. Flores
Universidad La Salle, Mexico City, Mexico

C. Zuñiga
Hospital General de Tijuanma, Tijuana, Baja California, Mexico

A. C. Frery et al. (eds.), *Computational Intelligence and Data Analytics*, Lecture Notes on Data Engineering and Communications Technologies 236,
https://doi.org/10.1007/978-981-96-0451-7_12

who may have symptoms or conditions that may indicate a risk of suffering from the disease sarcopenia, but these symptoms and conditions may go unnoticed by first level doctors because some hospitals do not have the techniques, methods or appropriate equipment for timely diagnosis. The diagnosis of a disease can help to take preventive measures and provide appropriate treatment to the patients, this would result in savings in health care expenses for both patients and hospitals, in addition to the monetary benefits obtained by a timely diagnosis using machine learning also benefits the daily lives of people thanks to early diagnosis. Sarcopenia is a condition characterized by the loss of muscle mass, strength, and function that occurs as a natural part of the aging process [1–3].

It can significantly impact an individual's quality of life, resulting in reduced mobility, an increased risk of falls, and overall weakness. In Mexico, the diagnosis of sarcopenia, like in many countries worldwide, is typically conducted by healthcare professionals trained in dealing with age-related conditions, such as geriatricians or general practitioners. Sarcopenia is usually diagnosed based on a combination of clinical assessment, physical examination, and certain diagnostic criteria. Typically, the diagnosis of sarcopenia involves evaluating an individual's muscle mass, muscle strength, and physical performance. Common diagnostic tools used in Mexico and other countries include Dual-Energy X-ray Absorptiometry (DXA) scans to measure muscle mass, handgrip strength tests, and physical performance assessments, such as the Short Physical Performance Battery (SPPB) or the Timed Up and Go test (TUG) [4, 5]. It's worth mentioning that the diagnosis of sarcopenia may vary somewhat between healthcare providers and institutions in Mexico. Guidelines from international organizations, such as the European Working Group on Sarcopenia in Older People (EWGSOP) or the Asian Working Group for Sarcopenia (AWGS) [6], can provide a framework to standardize the diagnosis and management of sarcopenia. However, it's important to note that each healthcare professional or institution may have their approach to diagnosing and managing sarcopenia. There are different techniques used in machine learning such as supervised, unsupervised, reinforcement, evolutionary, and deep learning [7–10].

2 Data Collection

Data collected during the evaluation included age, dryness, living situation, caregiver, medical history, and admission diagnosis. Information was also collected from instruments related to geriatric assessment, which included, Katz scale, Lawton Brody scale, minimal examination of mental health status, geriatric depression scale, delirium assessment through the CAM confusion assessment method, nutritional status was assessed through a minimal examination of complete nutritional status, A score of less than 17 was considered as malnutrition status, the algorithm for the assessment of sarcopenia was also used, comorbidity was calculated using the Charlson index [11], and quickSOFA was used to assess disease severity [12].

Information on 70 variables was collected from 100 patients. This information was collected in 2022 and is non-personally identifiable information.

3 Methods

All implementations were done with Python 3.8.12 using Pandas version 1.3.3 for data analysis and manipulation, Keras version 2.4.3 and eli5 version 0.11.0 for neural networks as well as Sklearn version 1.0.1 [13] for classification algorithms and evaluation.

For the selection of the data, two machine learning techniques were used: decision tree and cross-validation. In addition to the section of the datasets, which includes those variables or conditions with greater relation to the target, the decision tree was used together with the Rank tool, which evaluates and assigns a score to the variables taking into account the correlation between classes. A study was conducted on patients over 60 years of age who were admitted to the acute geriatric care unit of the Hospital General de Tijuana from November 1, 2021, to May 2, 2022. To carry out the study, a statistical sample was calculated using Freeman's formula [$n = 10 * (k + 1)$] [14], in which a 20% increase was added to compensate for possible losses, which in the end resulted in a total of 100 patients. For this dataset, inclusion parameters were used, the parameters that were applied for this inclusion were such as patients admitted to the geriatrics service, older than 60 years, with cognitive ability to follow simple instructions, with complete clinical history at the time of admission, who had signed the informed consent and who had been hospitalized for more than 72 h. Exclusion criteria were also applied to patients discharged before 72 h, patients with delirium in the first 24 h of hospitalization, patients with respiratory distress, patients with pacemakers, patients with terminal illness, and patients who did not agree to participate in the study. The data set included exclusion criteria, which were applied to all patients who were absent during follow-up, or who decided to leave the study, or from whom the necessary data could not be obtained. All patients underwent an initial assessment that included a bioelectrical impedance (BIA) [15, 16] measurement within the first 24 h of admission. At that time, they were asked if they wished to participate in the study and, if so, signed an informed consent form. Patients who met a criterion for sarcopenia on admission were also excluded, and the analysis included patients who did not meet this diagnosis on admission. Each patient was followed up daily and reassessed for sarcopenia at 72 h and then at 7 days. None of the study patients remained in the hospital for more than 14 days. One month after discharge, an appointment was scheduled and a complete evaluation was performed. If patients were unable to keep this appointment, a telephone consultation was made to verify survival and, in the case of death, the date of death was recorded. Having each of the datasets, a trial run of the classifiers used was carried out: k-Nearest Neighbors (kNN), Tree Decision (TD), Support Vector Machine (SVM), Random Forest (RF), Multi-layer Perceptron Neural Network (MLPNN), Logistic Regression (LR), and AdaBoost (AB) [11, 17–20].

Table 1 Metrics and expressions for analysis

Metric	Expression
Area under curve (AUC)	–
Classification accuracy (CA)	$\text{Accuracy} = \frac{\text{True Positive} + \text{True Negative}}{\text{True Positive} + \text{False Positive} + \text{False Negative} + \text{True Negative}}$
F1	$F1 = \frac{2 * \text{Precision} * \text{Recall}}{\text{Precision} + \text{Recall}}$
Precision	$\text{Precision} = \frac{\text{True Positive}}{\text{True Positive} + \text{False Positive}}$
Recall	$\text{Recall} = \frac{\text{True Positive}}{\text{True Positive} + \text{False Negative}}$

Evaluation criteria

For its evaluation, it was decided to take into account five parameters that are shown in Table 1 with its algebraic description.

4 Results

The dependent variable *seguimiento1m* was used as an indicator of the level of sarcopenia one month after admission to the hospital or if the person died. The variables identified are listed in Table 2.

The best classifiers were Gradient Boosting with an accuracy of 77%, Random Forest with an accuracy of 86%, and XGBoost with an accuracy of 83% (Table 3).

The results for the dependent variable "Follow-up1m", which represents the evaluation of the patients after one month, its statistical analysis showed that 68 patients left the hospital alive, 4 of them died before the month of discharge representing 5.9, 29.4% were alive but functionally deteriorated and 44 or 64.7% were still alive and without any functional deterioration. In the validation of the machine learning algorithms, it was observed that the Random Forest classifier model had the best accuracy result with 86.6% and the 10 variables that had the greatest weight in the result of the dependent variable, as shown in Table 4.

The algorithm showed the independent variable HxHTA as the most relevant when running the RandomForest model, followed by the independent variables SizeIng and Gender. This means that these three characteristics are the most important for predicting the independent variable Tracking1m according to the results of the Random Forest machine learning model.

Table 2 Variables

Variable	Etiqueta
Genero	Sexo
Edad	Edad en años
Camina	Auxiiar de la marcha
Cuidador	Cuidador
KatzP	Katz premorbido
Dependbasica	Katz menor de 5
LawtonP	Lawton premorbido
VelmarchaP	Velocidad marcha premorbido
Depre	GDS > 6
MNAP	MNA Premorbido
Malnutrición	Mna < 17
HxDM	Historia de DM
HxHTA	Historia de HTA
HxICC	Historia de ICC
HxEVC	Historia de EVC
HxIRC	Historia de IRC
HxEPOC	Historia de EPOC
HxParkinson	Historia de Parkinson
HxDepresion	Historia de Depresion
Charlston	Indice de Charlston
qSOFA	qSOFA
DxIngreso1	Dx Ingreso CIE 10
PesoIng	Peso al ingreso
TallaIng	Talla en m
IMCIng	IMC al Ingreso
AngFase	Angulo de fase
MMAIng	Masa muscular apendicular Ingreso
IMMAIng	Indice masa muscular apendicular Ingreso
BFPIng	Baja fuerza de prension Ingreso
PrenIng	Fuerza prension Ingreso
BajaMMIng	Baja Masa Muscular Ingreso
SarcopeniaIng	Sarcopenia al ingreso
Peso3d	Peso al 3er día
IMC3d	IMC al 3er dia
AngFase3d	Angulo de fase al tercer dia
MMA3d	Masa muscular apendicular 3er dia
IMMA3d	Indice de masa muscular apendicular 3er dia

(continued)

Table 2 (continued)

Variable	Etiqueta
BFP3d	Baja Fuerza de presnsion 3er dia
Pren3d	Fuerza de prension 3er dia
BajaMM3d	Baja Masa Muscular 3er dia
Sarcopenia3d	Sarcopenia al 3er dia
Peso7d	Peso al 7o dia
IMC7d	IMC al 7o dia
AngFase7d	Angulo de Fase 7 dia
MMA7d	Masa Muscular Apendicular 7 dia
IMMA7d	Indice de Masa Muscular Apendicular 7 dia
BFP7d	Baja Fuerza de Prension 7 día
Pren7d	Fuerza Prensión 7 dia
BajaMM7d	Baja Masa Muscular 7 dia
Sarcopenia7d	Sarcopenia al 7 dia
Seguimiento1m	Evolucion 3 meses

Table 3 Results of machine learning models on dependent variable dataset "seguimiento1m"

Algoritmo	Presicion
XGBoost()	0.833333333
Ada Boost()	0.6
Naive Bayes Classifier()	0.4
K-Nearest Neighbors Classfier()	0.733333333
Support Vector Machine()	0.4
Decision Tree Classifier()	0.766666667
Logistic Regression()	0.433333333
RandomForestClassifier()	0.866666667
Gradient Boost()	0.771184996

Table 4 Highest performing variables in random forest somewhat-rhythm with dependent variable Se guimiento1m

Variable	Resultado
Genero	0.051651505
Origen	0.048159908
Residencia	0.041670366
KatzP	0.057511256
Dependbasica	0.043227782
HxParkinson	0.053026191
Charlston	0.040858505
qSOFA	0.041572703
HxHTA	0.087898107
HxICC	0.058288591

5 Conclusion

The data analyzed in this paper confirms that machine learning is a powerful tool to diagnose the level of sarcopenia when patients are admitted to the hospital, based on their most common symptoms, as well as the respective comorbidities analyzed for each patient. In this study, we were able to identify variables and measurements at one month of hospitalization. Our future work will focus on combining the different features for better classification performance and applying deep learning techniques for classification. According to the information analyzed the model with the best accuracy is Random Forest with 86%.

References

1. Mostajo NIC (2023) Gama Glutamil transferasa sérica como marcador biológico de sarcopenia en ancianos con fractura. Revista Medica 28(1):13–19. https://doi.org/10.58296/rm.v28i1.53
2. Valisoltani N et al (2024) Association of phase angle with sarcopenia and muscle function in patients with COPD: a case-control study. BMC Pulm Med 24(1). https://doi.org/10.1186/s12890-023-02814-9
3. Wang Q, Wei T, Zhang Z, Xu C, Qiu Y, Xiao W (2024) Assessing causality between inflammatory bowel diseases with frailty index and sarcopenia: a bidirectional Mendelian randomization study. Eur J Med Res 29(1). https://doi.org/10.1186/s40001-023-01614-5
4. Alemán-Mateo H, Valenzuela RER (2014) Skeletal muscle mass indices in healthy young Mexican adults aged 20–40 years: implications for diagnoses of sarcopenia in the elderly population. Sci World J 2014:1–5. https://doi.org/10.1155/2014/672158
5. Alemán-Mateo H, López-Teros MT, Valenzuela RER, Ramírez-Torres M, Urquídez-Romero R (2020) Sarcopenia: influence of regional skeletal muscle cutoff points and fat-free mass in older Mexican people—a pilot study. Curr Gerontol Geriatr Res 2020:1–9. https://doi.org/10.1155/2020/8037503
6. Almeida N et al (2023) Energy and nutrient intake by people with and without sarcopenia diagnosed by the European Working Group on Sarcopenia in older people: a systematic review and meta-analysis. Nutr Rev. https://doi.org/10.1093/nutrit/nuad154
7. Carvalho DV, Pereira EM, Cardoso JS (2019) Machine learning interpretability: a survey on methods and metrics. Electronics 8(8):832. https://doi.org/10.3390/electronics8080832
8. Panch T, Szolovits P, Atun R (2018) Artificial intelligence, machine learning and health systems. J Glob Health 8(2). https://doi.org/10.7189/jogh.08.020303
9. Abasi AK, Khader AT, Al-Betar MA, Naim S, Alyasseri ZAA, Makhadmeh SN (2020) A novel hybrid multi-verse optimizer with K-means for text document clustering. Neural Comput Appl 32:17703–17729
10. Kim M-K, Kim D, Kang H, Park SW, Kim SH, Yoo J (2024) A machine learning model for prediction of sarcopenia in patients with Parkinson's disease. PLoS ONE 19(1):e0296282. https://doi.org/10.1371/journal.pone.0296282
11. Yotruangsri T, Phinyo P, Buawangpong N, Nantsupawat N, Angkurawaranon C, Pinyopornpanish K (2024) Survival analysis of older adults with dementia: predicting factors after unplanned hospitalization in Maharaj Nakorn Chiang Mai Hospital. BMC Geriatr 24(1). https://doi.org/10.1186/s12877-023-04558-x
12. Remelli F et al (2021) Predictive value of geriatric-quickSOFA in hospitalized older people with sepsis. BMC Geriatr 21(1). https://doi.org/10.1186/s12877-021-02182-1

13. Mac Carthy T et al (2024) Automatic Urticaria Activity Score (AUAS): deep learning-based automatic hive counting for urticaria severity assessment. JID Innov 4(1):100218. https://doi.org/10.1016/j.xjidi.2023.100218
14. Muller M, De Vries E, Tomson A, McLachlan C (2023) An introduction to gender affirming healthcare: what the family physician needs to know. S Afr Fam Practi 65(1). https://doi.org/10.4102/safp.v65i1.5770
15. Sasaki R et al (2024) Advancing liver disease sarcopenia prediction: a chest-radiograph-based model for older adults. Res Sq. https://doi.org/10.21203/rs.3.rs-3782222/v1
16. Xu X, Qian Y, Jin K, Chen J, Fu J, Chen C, Zhu Z (2024) The impact of helicobacter pylori infection on low skeletal muscle mass risk in Chinese women over 40: a cross-sectional analysis. Front Cell Infect Microbiol. 13. https://doi.org/10.3389/fcimb.2023.1289909
17. Okada M, Nagata K, Watanabe N, Tada M (2024) Computational learning analytics to estimate location-based self-regulation process of real-world experiences. IEEE Trans Learn Technol 17:445–461. https://doi.org/10.1109/tlt.2023.3262598
18. Jiménez-García J et al (2024) An explainable deep-learning architecture for pediatric sleep apnea identification from overnight airflow and oximetry signals. Biomed Signal Process Control 87:105490. https://doi.org/10.1016/j.bspc.2023.105490
19. Lim YD, Zhao P, Guidoni L, Likforman J-P, Tan CS (2024) Predictive modelling of optical beams from grating structure using deep neural network. J Light Technol 42(2):696–703. https://doi.org/10.1109/jlt.2023.3319692
20. Intelligenza Artificiale per la cura del diabete mellito: opportunità e prospettive future. Il Diabete 35(4) (2024). https://doi.org/10.30682/ildia2304g

Deep Learning—Applications

Automated Crime Detection in CCTV Footage Using Machine Learning

John Adithya, Payam Mortezaei, Khaled Aljawawde, Mohamed Mohamed, Rita Zgheib, and Tamer Mohamed

Abstract Prevention and early detection are more effective than addressing issues after severe symptoms have manifested. Security and safety are essential pillars that provide stability and growth to a community. The installation of CCTV cameras in public places serves to enhance these pillars by offering continuous surveillance and monitoring. This study proposes a sophisticated system for the early detection of crimes using real-time analysis of CCTV footage through advanced machine learning algorithms. The system utilizes an Inflated 3D ConvNet (I3D) model from Google DeepMind, specifically trained to detect kinetic motion, to analyze footage in real time. To train the model, a substantial dataset comprising 35,000 classified clips of violent and non-violent behavior is employed. By leveraging advanced techniques such as 3D Convolutional Neural Networks (3DCNN) and Improved Temporal Segment Networks (ITSM), the system can accurately detect instances of violence in the footage. Once detected, the system promptly alerts the user, providing real-time notifications of potential threats, thereby significantly enhancing public safety and security.

Keywords Machine learning · 3DCNN · ITSM · Violence detection · Crime detection

1 Introduction

The increase in crime in urban areas poses a significant challenge to law enforcement agencies worldwide. Urban areas in the MENA region tend to have higher crime rates due to various factors, including higher population density and socioeconomic disparities. The COVID-19 pandemic also influenced these trends, with some urban

J. Adithya · P. Mortezaei · K. Aljawawde · M. Mohamed · R. Zgheib (✉) · T. Mohamed
Computer Science Department, Canadian University Dubai, Dubai, United Arab Emirates
e-mail: rita.zgheib@cud.ac.ae

T. Mohamed
e-mail: tamer.mohamed@cud.ac.ae

A. C. Frery et al. (eds.), *Computational Intelligence and Data Analytics*, Lecture Notes on Data Engineering and Communications Technologies 236,
https://doi.org/10.1007/978-981-96-0451-7_13

areas experiencing reductions in certain types of crime, such as noise and air pollution, due to lockdowns and reduced economic activity [1]. Traditional surveillance methods have proven ineffective in detecting and responding to criminal activities in real-time. This research project aims to explore the implementation of machine learning techniques in conjunction with closed-circuit television (CCTV) systems for the purpose of real-time crime detection.

This study delves into integrating machine learning algorithms with CCTV technology to create a program capable of detecting criminal behaviors, such as theft, vandalism, and suspicious activities, as they occur. By doing so, law enforcement agencies can respond more rapidly, potentially preventing crimes and significantly enhancing the safety of our communities.

As mentioned, early detection of criminal activities is significant. In many CCTV camera installations, security personnel must actively monitor the cameras, which can result in human error. Also, many small installations do not have regular security to check their cameras; therefore, the footage is only used to be viewed after the criminal incident. A real-time crime detection system can come in handy in many scenarios and provide another layer of security to notify the authorities earlier and enhance security.

The researchers in [2] conducted a systematic review to highlight the areas that required further research in crime detection using machine learning algorithms. One of the primary areas of concern is the development of interpretable models that can elucidate their decision-making processes, a necessity given the severe consequences of erroneous predictions. Additionally, there is a pressing need for acquiring accurate and high-quality data, which is pivotal for effectively implementing these technologies [3]. Another significant gap is the need for comprehensive studies on the real-world effectiveness of these technologies, highlighting the need for more robust evaluations in practical scenarios and the scalability challenges in larger systems. This emphasis on transparency and accountability is crucial to our proposed research.

This research will embark on the same route to utilize machine learning algorithms to further define the route by identifying criminal activity vectors and solutions to detect violent altercations on video, which can then be used in CCTV cameras. The aim is to use a machine learning algorithm, annotate data, define features, and train a model to detect the crime vectors defined for it. The objective is to develop the system to process the data in real-time, but initially, the goal is to detect the crime on the video footage it has given. The system can further expand with the cloud and be implemented in that environment to utilize its resources more efficiently.

Prevention and early detection are believed to be far more effective than curing the problem when it has shown severe symptoms. Within society, security and safety are essential pillars that provide stability and growth to the community. The installation of CCTV cameras in many public places serves this purpose. Early detection of crimes on CCTV cameras can further improve safety and security. We incorporate real-time detection and machine learning algorithms to create a system that can detect threat factors in real-time. The system will analyze the footage of the CCTV camera in real-time on an I3D model from Google DeepMind trained on detecting kinetic motion. Furthermore, 35,000 classified clips of violence and non-violence are used for training. The system can detect violence in the footage using 3DCNN and ITSM and display it to the user.

2 Literature Review

According to [2], the current state of knowledge in using machine learning and deep learning for crime prediction is marked by significant advancements and promising developments. The integration of advanced deep learning models with technologies like drones and aerial devices has opened new avenues for monitoring and responding to criminal activities. These algorithms are adept at analyzing a variety of data sources, including surveillance camera footage, social media, crime reports, and police records, which allows for a more comprehensive understanding of criminal activities.

The researchers in [4], proposes a violence detection system through video analysis, exploring the capabilities of machine learning and hybrid models in evaluating videos. This evaluation identifies violent activities, enabling the system to issue alerts with a certain degree of accuracy and precision. Researchers [5] emphasize that the Convolutional Neural Network (CNN) model enhances neural network performance, particularly in recognizing violent patterns in videos. He notes that retraining these networks yields improved performance, achievable even with limited GPU resources.

Researchers in [6] discuss the application of the CNN model in developing an automated violence detection system. They highlight how combining CNN with an efficient training algorithm can reduce computational complexity. This aspect is particularly beneficial for CCTV systems, which are generally not designed to handle complex algorithms.

In paper [7], the authors address the efficiency and accuracy of violence detection in CCTV cameras. They advocate for the use of Long-Short-Term Memory (LSTM) networks in feature extraction and classification over time. The authors suggest that machine learning-based methods for violence detection could integrate various sensor technologies and network implementations, potentially enhancing the response time of human operators in large-scale systems.

Authors of [8] developed a real-time violence detection system utilizing both CNN and LSTM. They acknowledge the advantages of video detection in today's world, given the abundance of devices capable of recording and transmitting videos in camera systems. A detection system's effectiveness, speed, precision, and accuracy across different video formats and sources is pivotal in determining its utility in violence detection.

In [9], a system leveraging Cloud Computing and Machine Learning is presented for enhancing criminal detection in CCTV footage. This system is divided into three modules targeting specific criminal activities, including known criminal detection via HAAR Cascade, live crime detection, and vehicle theft, to improve suspect identification and reduce false alarms. The study explores various algorithms like HAAR,

Eigen Faces, and CNNs, highlighting the challenge of balancing speed and accuracy in face recognition. The findings suggest that the proposed system could significantly benefit crime detection and prevention efforts, recommending Microsoft Azure for face recognition for its accuracy and speed. The authors proposed further validation and exploration into its utility for locating missing persons as future work.

Authors in [10], conducted a comprehensive review of violence detection methods, analyzing 90 research papers published between 2015 and 2021. They used CNN model for detecting violent activities, including preprocessing steps, model enhancement, and violence detection methods. And they proposed a system showed improved performance compared to existing algorithms like Support Vector Machine and Random Forest.

Another proposed approach in [11], employs the Grassmann algorithm for face detection, aiming to improve reliability in outlier detection and face recognition, as indicated by experimental findings. The methodology of the research is motion detection and face detection using the Grassmann algorithm using pre-trained models for background segmentation, classification, and multi-object tracking. The suggested Grassmann algorithm exhibits superior performance with a lower erroneous denial rate in face detection systems.

In [12], researchers present a comprehensive review of machine learning applications in object detection and classification with specific applications to digital forensics. The paper analyzes a wide range of publications between 2007 and 2019, highlighting the convolutional neural network as a state-of-the-art algorithm for machine learning-based digital forensics. This study is instrumental in understanding the frameworks, applications, and performance of machine learning in enhancing the capabilities of digital forensics in video surveillance systems, thereby aiding in crime detection and analysis.

The paper [13], discusses the increasing need for intelligent systems that can automatically detect crimes from CCTV feeds in real-time. The authors focused on object recognition and the identification of priority frames for crime detection. A variety of techniques and algorithms, notably the YOLO (You Only Look Once) algorithm, have been explored for their ability to detect crimes. The authors highlight the challenges faced in the current research landscape, including the difficulty of real-time crime detection and the limitations of existing datasets. To overcome these challenges, the paper suggests focusing research efforts on developing more sophisticated algorithms that can handle the complex nature of real-time video data, as well as on creating more comprehensive and diverse datasets that better represent the variety of crimes that occur in different settings.

The study [14] proposes a novel deep learning-based surveillance system designed to detect the presence of handheld firearms and bladed weapons in CCTV footage, with the aim of alerting authorities to potential threats before an incident occurs. The authors compare different deep learning-based object detection techniques, including YOLO (You Only Look Once), SSD (Single Shot Multibox Detector), and Faster R-CNN (Region-Based Convolutional Neural Networks), to identify which method offers the best balance of precision and speed for real-time application. The research concludes that YOLOv5, in particular, provides an optimal combination of mean

Table 1 Previous implementations

Technologies used	Results and findings	References
OpenCV, YOLO, CNN	Improved suspect identification, reduced false alarms	[15]
3DCNN	Effective detection of violent crimes	[8]
2D and 3D CNNs	2D CNN achieved 91.2% classification accuracy	[16]
CNN, LSTM	Real-time violence detection in crowd-based footage	[17]
Hybrid 2D and 3D CNN	High accuracy in training (99.93%) and validation (84.5%)	[18]

average precision (mAP) and inference speed, making it the most suitable choice for implementing their proposed surveillance system. The system's ability to detect weapons in real-time and alert authorities preemptively represents a significant step forward in the use of AI for enhancing public safety and crime prevention.

The Table 1 below presents a curated overview of recent advancements in the application of machine learning and 3D Convolutional Neural Networks (3DCNN) for crime detection in CCTV video footage.

3 System Design

3.1 Model Features and Capabilities

The proposed system is designed to detect violent altercations from video footage and alert the administrator promptly. Figure 1 presents a use case diagram for a system that leverages AI and CCTV data for real-time crime detection. The primary actors interacting with the system include the client, who accesses analytics and views live CCTV feeds, and the police officer, who receives generated reports.

The main functionalities of the system encompass configuring it, viewing live CCTV feeds, searching historical data, detecting and classifying crimes, generating reports, viewing crime reports, and sending alerts to the police. The system design focuses on the system's ability to detect violent interactions and automatically alert the police. Additional features like report generation and archiving are extended functionalities that enhance the system's overall utility and effectiveness. By integrating these features, the system aims to improve public safety and security through timely and accurate detection and reporting of violent incidents, facilitating prompt law enforcement response.

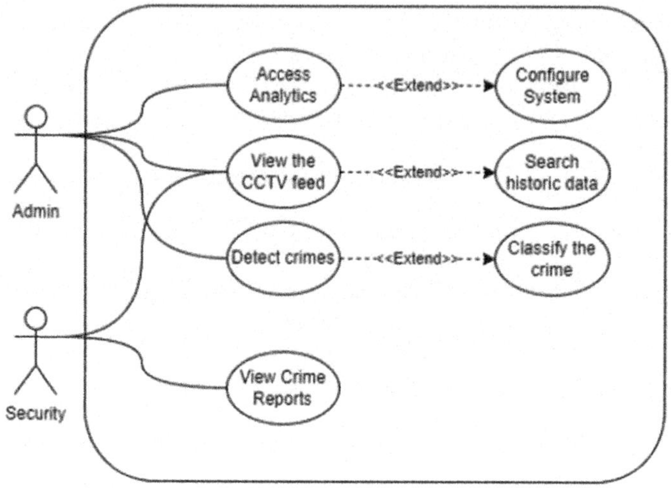

Fig. 1 Real-Time crime detection using AI and CCTV data

The system utilizes a sophisticated technology stack designed for robust performance in real-time crime detection. The core component is a Python-based model leveraging the I3D model from Google DeepMind. The backend interface is also developed in Python, ensuring seamless integration and processing capabilities. The frontend is built using Angular and Angular Material UI, providing an intuitive and responsive user interface.

The implementation incorporates several key libraries, including Keras, TensorFlow, Numpy, OpenCV, and i3d_inception (Google DeepMind), which collectively enhance the model's functionality and performance. Table 2 provides a comprehensive overview of the system's technological components. The model employs various features for detecting violent incidents, such as motion patterns, spatial configurations, temporal patterns (rapid changes over time), color intensity, facial expressions, and environmental context. These features enable the system to accurately identify and classify violent interactions, ensuring timely alerts and responses. By integrating these advanced technologies and methodologies, the system offers a powerful tool for enhancing public safety through real-time crime detection and reporting.

3.2 Methodological Framework

In Fig. 2, the flowchart depicts a process that starts with system initialization, which includes loading models and configuring settings. Following initialization, the system captures and processes CCTV data. It then assesses whether a crime is being detected. If a crime is undetected, the process goes back to capturing and processing new CCTV data. The data is sent to a crime detection algorithm for analysis if a crime is detected.

Table 2 Technology stack

Component	Description	Benefits
Python [19]	A high-level programming language known for its readability and versatility in data processing and machine learning	Ease of use, extensive libraries, and large community support make Python ideal for rapid development in ML and data processing
I3D Model (Inflated 3D ConvNet) [20]	A deep learning model for video action recognition that captures spatial and temporal features	Pre-trained on a large dataset, I3D is effective in understanding motion patterns, crucial for video analysis tasks
Backend (Python)	The server-side application that handles business logic, database interactions, and performance	Python's support for asynchronous tasks and real-time data processing makes it suitable for backend integration with ML models
Frontend (Angular & Angular Material UI) [21]	Client-side framework and UI components for building responsive and interactive web interfaces.	Angular provides a structured framework for dynamic web apps, and Material UI offers a suite of well-designed components
Keras [22]	A high-level neural networks API, capable of running on top of TensorFlow, for fast experimentation	Keras simplifies neural network development and is designed for easy and fast prototyping
TensorFlow [23]	An end-to-end open-source platform for machine learning with a comprehensive, flexible ecosystem of tools	TensorFlow provides a scalable and flexible framework for training and deploying ML models
NumPy [24]	A library for scientific computing with Python that provides support for large, multi-dimensional arrays and matrices	NumPy is foundational for numerical computations in Python, providing efficient array operations
OpenCV [25]	A library focused on real-time computer vision and image processing, widely used in academia and industry	OpenCV is instrumental in video data preprocessing and computer vision tasks required before ML processing
i3d_inception (Inception-v1) [26]	An adaptation of the Inception architecture to the I3D framework for effective feature extraction from video data	Leverages temporal and spatial feature extraction capabilities of the Inception model for analyzing video sequences

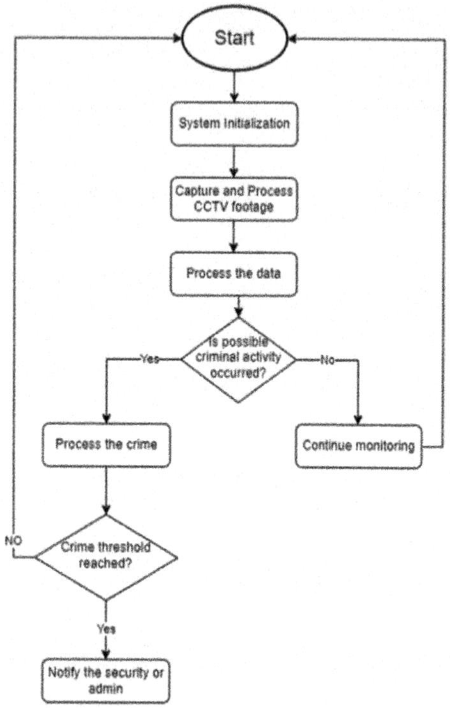

Fig. 2 Real-Time crime detection using AI and CCTV data

After this, there's a decision point to determine if the detected activity has reached a certain crime threshold. If the threshold is not reached, the data is sent to the cloud for further processing, presumably to refine the detection or for storage. If the crime threshold is reached, the data is reported to the police office, and the system's workflow ends.

Preprocessing is critical for improving the quality of data, which includes handling missing values, smoothing noisy data, removing outliers, and resolving inconsistencies. High-quality data is vital for accurate and reliable analytical results [27, 28]. In the context of computer vision research, we present a Python script leveraging the OpenCV library for comprehensive data preprocessing. It includes functions for video frame looping to extend video sequences, center cropping of images to maintain critical information while resizing, resizing images while preserving aspect ratios, and pixel value normalization to ensure consistent and stable model training. These preprocessing steps play a pivotal role in enhancing the accuracy of the results.

The system initializes by loading the required machine-learning models and performing the basic configuration. The input used is a continuous stream of CCTV footage. These frames undergo preprocessing by handling missing values, smoothing noisy data, removing outliers, resolving inconsistencies, center cropping images to maintain critical information, resizing images while preserving aspect ratios and normalizing pixel values.

Each frame from the input clip is individually pre-processed and added to a list. Once all frames are processed, the list is converted into a multi-dimensional array suitable for the model to process. The model analyzes various features such as motion patterns, spatial configurations, object placement, temporal patterns, color intensity, facial expressions, and environmental context.

The model then predicts outcomes for the processed clip, with the batch size set to the number of frames in the clip. Each frame is predicted to depict either violent activity or a normal scenario. The predictions are extracted from the result, which is in a batch. Finally, the function returns the first set of predictions from this batch, corresponding to the video clip's classification results.

Multiple predictions are combined to decide on the most likely category or class for a given data point. It starts by averaging the predictions to get a single array of probabilities. Then, it picks the class with the highest average probability. If the probability is too low, it either sticks with this class or switches to another one, depending on which probability is above the threshold. In the end, the function gives us the chosen class and the accuracy of this choice. Hence, the final decision is made with a certain level of confidence.

If the system is able to assess a level above the threshold and classify it as violence, it shows up on the interface to signal that violence has been detected. This process is continuous as long as a video stream is being pushed into the system as input to be processed.

4 Experimentation and Results

To evaluate the proposed real-time crime detection system, a series of tests were carried out utilizing a comprehensive dataset comprising of 35,000 classified clips exhibiting both violent and non-violent scenarios.

4.1 Implementation

The solution uses Python for various tasks, including machine learning model implementation, data processing, feature identification, and video categorization. The model performs violence detection from Kaggle datasets by analyzing motion patterns, spatial configurations, object placement, temporal patterns, color intensity, facial expressions, and environmental context.

A Python script runs an Angular-based UI on localhost, allowing users to examine specific videos for violence detection. The interface displays findings for violence in frames, with customizable tolerance settings. The model's properties include motion patterns, spatial configurations, item placement, and temporal patterns. The implementation uses several Python libraries, including:

Table 3 Classification results

Metric	Value (%)
Precision	89.58
Recall	84.71
Accuracy	91.6

1. Keras: a high-level neural networks API, written in Python and capable of running on top of TensorFlow.
2. TensorFlow: open-source machine learning framework developed by the Google Brain team.
3. Numpy: a library for the Python programming language, adding support for large, multi-dimensional arrays and matrices, along with a large collection of high-level mathematical functions to operate on these arrays.
4. OpenCV: (Open Source Computer Vision Library) is an open-source computer vision and machine learning software library.
5. I3D_Inception: is not a standard library but a widely adopted 3D video classification network. It uses 3D convolution to learn spatiotemporal information directly from videos.

4.2 Results

The trained model exhibited promising results in the detection of violent incidents in CCTV footage. The subsequent table provides a summary of the evaluation metrics:

The precision of 89.58% indicates that when the model predicts an incident as violent, it is correct approximately 90% of the time. This high precision is crucial for minimizing false positives, ensuring that non-violent incidents are not mistakenly flagged as violent, which could lead to unnecessary interventions or alerts. The recall rate of 84.71% suggests that the model successfully identifies about 85% of all actual violent incidents. While this is a strong result, it also highlights that there is room for improvement in detecting every violent event, as approximately 15% of true incidents might go undetected. The overall accuracy of 91.6% reflects the model's ability to correctly classify both violent and non-violent incidents, underscoring the model's robustness and reliability in varied surveillance scenarios (Table 3).

4.3 Future Developments

As the current system continues to evolve, future developments will concentrate on analyzing live video streams in real-time. This enhancement aims to address various applications, including security and surveillance systems, where the prompt identification of violent incidents is crucial. Furthermore, future rounds will detect, label, and analyse the observed violence. The approach aims to discriminate between physical altercations, acts of vandalism, and other forms of criminal activities, hence providing insights for more effective intervention tactics. This classification capacity improves decision-making for security systems and agencies.

The model's expanded functionalities could be used in content moderation for online platforms. Social media and video-sharing platforms which can aim to create safer online environments by automatically recognizing and flagging violent content in real time.

As technology evolves, ethical considerations and appropriate deployment take precedence. Achieving a balance between public safety and private rights is critical to ensuring the beneficial and responsible use of machine learning for violence identification in CCTV cameras.

5 Conclusion

In conclusion, this study not only addresses the limitations of traditional surveillance systems but also presents a practical and effective solution for real-time criminal detection. The findings pave the way for further advancements in the integration of machine learning and security technology, leading to safer and more secure communities. Looking ahead, the primary focus for enhancing this study lies in the real-time analysis of live video streams, offering significant potential to advance the system's capabilities. This development aligns with the growing demand for early detection of violent situations, thereby contributing to improved security measures across various domains.

References

1. Abouzid M, El-Sherif DM, Al Naggar Y, Alshehri MM, Alothman S, El-Seedi HR, ... Amer T (2022) Investigating the current environmental situation in the Middle East and North Africa (MENA) region during the third wave of COVID-19 pandemic: urban versus rural context. BMC Public Health 22(1):177
2. Mandalapu V, Elluri L, Vyas P, Roy N (2023) Crime prediction using machine learning and deep learning: a systematic review and future directions. IEEE Access 11:60153–60170. https://doi.org/10.1109/ACCESS.2023.3286344
3. Vijeikis R, Raudonis V, Dervinis G (2022) Efficient violence detection in surveillance. Sensors 22(6):2216. https://doi.org/10.3390/s22062216
4. Patel M (2021) Real-Time violence detection using CNN-LSTM. arXiv:2107.07578
5. Jena A, Mohanty SN, SKB S (2020) Detection of violence in videos using hybrid machine learning techniques. Int J Adv Sci Technol 29(3):5386–5392. http://sersc.org/journals/index.php/IJAST/article/view/6040
6. Kisan S, Sahu B, Jena A, Mohanty SN (2019) Detection of violence in videos using hybrid machine learning techniques. Int J Adv Sci Technol 29(03):5386–5392

7. Vijeikis R, Raudonis V, Dervinis G (2022) Efficient violence detection in surveillance. Sensors 22(6):2216. https://doi.org/10.3390/s22062216
8. Sai ANHD, Prasad KS (2023) Machine learning software for the detection of violence from CCTV live footage. J Image Process Artif Intell 9(3):12–18. https://doi.org/10.46610/joipai.2023.v09i03.002
9. Yadav A, TripathiAditya AT, Poojary T, Jeswani J (2020) Criminals as well as crime detection using machine learning and OpenCV. Int Res J Mod Eng Technol Sci 03(04):Art no 2582–5208
10. SD J et al (2022) Real-time based violence detection from CCTV camera using machine learning method. IEEE. https://doi.org/10.1109/i4tech55392.2022.9952805
11. Pandiaraja P, Saarumathi R, Parashakthi M, Logapriya R (2023) An analysis of abnormal event detection and person identification from surveillance cameras using motion vectors with deep learning. IEEE. https://doi.org/10.1109/icears56392.2023.10085466
12. Oladipo F, Ogbuju E, Alayesanmi FS, Musa AE (2020) The state of the art in machine learning-based digital forensics. Social science research network, Jan 2020. https://doi.org/10.2139/ssrn.3668687.
13. Katkar NY, Garg VK (2022) Detection and tracking the criminal activity using network of CCTV cameras. In: 2022 3rd international conference on smart electronics and communication (ICOSEC), Oct 2022. https://doi.org/10.1109/icosec54921.2022.9952104
14. Boukabous M, Azizi M (2023) Image and video-based crime prediction using object detection and deep learning. Bull Electr Eng Inf 12(3):1630–1638. https://doi.org/10.11591/eei.v12i3.5157
15. Reddy NVR, Priya P, Aryan S, Ajay P (2023) Crime detection system with machine learning using open CV, YOLO, and CNN. In: 2023 14th international conference on computing communication and networking technologies (ICCCNT), Delhi, India, pp 1–7. https://doi.org/10.1109/ICCCNT56998.2023.10306367
16. Jan A, Khan GM (2022) Deep vigilante: a deep learning network for real-world crime detection. Res J Appl Sci Eng Technol 14(1):1949–1961
17. J, SD et al (2022) Real-time based violence detection from CCTV camera using machine learning method. In: 2022 international conference on industry 4.0 technology (I4Tech), Pune, India, pp 1–6. https://doi.org/10.1109/I4Tech55392.2022.9952805
18. Jayasimhan A, P P (2022) A hybrid model using 2D and 3D convolutional neural networks for violence detection in a video dataset. In: 2022 3rd international conference on communication, computing and industry 4.0 (C2I4), Bangalore, India, pp 1–5. https://doi.org/10.1109/C2I456876.2022.10051324
19. Python Software Foundation (nd) Python. https://www.python.org/
20. Carreira J, Zisserman A (2017) Quo Vadis, action recognition? A new model and the kinetics dataset. In: 2017 IEEE conference on computer vision and pattern recognition (CVPR), Honolulu, HI, USA. IEEE, pp 4724–4733. https://doi.org/10.1109/CVPR.2017.502
21. Angular (nd) Angular. https://angular.io/
22. Keras (nd) Keras Documentation. https://keras.io/
23. TensorFlow (nd) TensorFlow Documentation. https://www.tensorflow.org/
24. NumPy (nd) NumPy Documentation. https://numpy.org/
25. OpenCV (nd) OpenCV Documentation. https://opencv.org/
26. Google DeepMind (nd) Kinetics I3D. GitHub. https://github.com/google-deepmind/kinetics-i3d
27. Nahar J, Promi ZT, Ferdous J, Ishrak F, Khurshid R (2022) Anomalous behavior detection using Spatio temporal Feature and 3D CNN model for Surveillance (Doctoral dissertation, Brac University)
28. Li J, Chen H (2022) Research on abnormal data analysis and data processing method based on network traffic. In: IEEE Xplore, 27 May 2022. https://doi.org/10.1109/iccbe56101.2022.9888172
29. SD J et al (2022) Real-time based violence detection from CCTV camera using machine learning method. In: 2022 international conference on industry 4.0 technology (I4Tech), 23 Sep 2022. https://doi.org/10.1109/i4tech55392.2022.9952805

Automated Detection and Severity Classification of Coffee Leaf Diseases Using Modified ResNet 50 with GradCam Technique

Ch Anuradha, Ganesh Kumar Vathumilli, Dhara Sesha Sai, and Uriti Lalithya Pavan

Abstract Automated systems for detecting plant diseases play a pivotal role in precision agriculture, crucial for maintaining crop health and maximizing yields. This study specifically targets the early identification and severity classification of coffee leaf diseases, focusing on the Arabica variety. Leveraging the superior image classification capabilities of ResNet-50, the methodology introduces a system designed to differentiate between healthy leaves and those experiencing biotic stress such as rust, miner Cercospora, and Phoma. In cases of disease presence, the system further categorizes the severity level. The process commences with the collection and preprocessing of a diverse dataset containing coffee leaf images depicting various health states. Employing data augmentation and visualization techniques, such as the Grad-Cam method, enhances the model's robustness and accurately highlights regions within coffee leaf images for precise classification decisions. A modified ResNet-50 model is utilized for classifying leaf health status and identifying biotic stress, followed by severity grading employing the Grad-Cam technique.

Keywords Grad-Cam · ResNet-50 · Rust · Miner · Phoma · Cercospora · Biotic stress · Severity estimation

1 Introduction

The Western Ghats' hilly regions are where Arabica coffee is most commonly grown in India they include states like Karnataka, Kerala, Tamil Nadu, and portions of Andhra Pradesh. Arabica coffee is a favorite among coffee lovers due to its moderate flavor and scent. But the Arabica type is vulnerable to a range of biotic stress illnesses, similar to those that affect Brazilian coffee crops [1]. Leaf miner, rust, Phoma, and cercospora leaf spot are diseases that can induce defoliation and reduce photosynthesis, which can hurt the ultimate coffee product's yield and quality [2]. Modern

Ch Anuradha · G. K. Vathumilli (✉) · D. Sesha Sai · U. Lalithya Pavan
Department of CSE, Velagapudi Ramakrishna Siddhartha Engineering College, Vijayawada, Andhra Pradesh, India
e-mail: ganeshkumarvattumilli2003@gmail.com

A. C. Frery et al. (eds.), *Computational Intelligence and Data Analytics*, Lecture Notes on Data Engineering and Communications Technologies 236,
https://doi.org/10.1007/978-981-96-0451-7_14

technologies like artificial intelligence and computer-aided diagnosis systems, which have been adopted in other areas, may also help Indian coffee growers by making it easier to accurately identify and grade the severity of diseases that affect Arabica coffee plants. This will allow for prompt interventions to maximize crop health and yield.

Government organizations and coffee growers have been working hard in recent years to prevent coffee illnesses by using manual detection techniques. But despite these initiatives, misdiagnosis is still common, leading to incorrect treatment and consequent environmental damage [3]. Moreover, to maintain the long-term sustainability and financial viability of coffee growing, it is necessary to improve the crop output of coffee leaves in addition to managing diseases [4]. It is critical to identify coffee leaf diseases early in order to minimize yield losses and maximize crop health. Farmers can reduce economic losses and prevent widespread infection by detecting diseases early on and taking appropriate action, such as using crop varieties resistant to the disease or carefully dousing crops with fungicides.

This study suggests using cutting-edge technology, notably ResNet-50, a powerful convolutional neural network, to address the problems associated with coffee leaf illnesses. ResNet-50, a deep learning model that is widely used for image classification, will be used to create a strong system that can identify and classify common coffee leaf diseases Automated Detection and Severity Classification of Coffee Leaf Diseases Using Modified ResNet 50 with GradCam Technique including rust, miner Cercospora, and Phoma in order to determine their severity. With the use of ResNet-50's abilities, the system seeks to precisely detect these illnesses in the early stages so that prompt action may be taken to lessen their effects on crop health and yield. Moreover, the research integrates the GradCAM visualization method to approximate the severity of these illnesses, improving the accuracy of diagnostic evaluations. The goal of the project is to give coffee growers a complete toolkit for proactive disease management by combining customized ResNet-50 with Grad Cam.

2 Related Works

This section provides a thorough summary of several literature survey studies that are cited in our work. As one of the most consumed drinks in the world, coffee has a significant influence on global agriculture, particularly when it comes to plant diseases that affect coffee plants [5]. Timely detection and management of these illnesses are vital in order to avert the spread of infections and guarantee the caliber and quantity of coffee crops. In order to overcome these issues, it is critical to employ cutting-edge technology for precise and effective illness diagnosis, such as deep learning models like DenseNet, ResNet, and VGG16. The DenseNet model has remarkable F1 scores and MCC metrics, together with its great efficiency.

Using cutting-edge methods such as Grad-CAM, Grad-CAM++, and Score-CAM, the project aims to visualize coffee illnesses in order to improve classification accuracy and pinpoint critical areas for disease detection in coffee plants. The project intends to shed light on the variables impacting classification accuracy and the significance of visualizing the classification process by putting these visualization techniques into practice [6]. The guided strategy achieved a classification accuracy of 94% compared to 74%, greatly outperforming the naïve approach using masked pictures without background noise and tiny patches of images with sick areas.

In the field of agriculture, biotic stressors on crops are quite important, especially in Brazil where coffee production is a major economic driver [7]. These pressures, which have an impact on both productivity and quality, include leaf miner, rust, and leaf spot. They represent serious challenges to coffee farms. A unique strategy that makes use of deep learning techniques has been created to effectively identify and measure these biotic stressors on coffee leaves in order to overcome these difficulties. This cutting-edge technology provides a more dependable and effective solution than conventional image processing methods, which represents a significant breakthrough. This approach is important because it gives experts and farmers a strong tool for managing and identifying plant diseases early on.

Cutting-edge technology like artificial intelligence (AI), machine learning (ML), and deep learning (DL) must be used in the agriculture industry, especially when diagnosing coffee plant diseases [8]. These technologies provide a viable way to address the difficulties in precisely recognizing and distinguishing between different types of leaf diseases in coffee plants. The integration of digital and physical systems has been made possible by the Fourth Industrial Revolution (FIR), opening the door for the use of DL, ML, and AI in agriculture. Farmers may increase their output and efficiency using these technologies, resulting in better results with less work. The advancement of automated methods for identifying plant diseases not only improves precision but also tackles the drawbacks of manual inspection, such as mistakes.

The agricultural industry places a great deal of attention on the creation of a model for biotic stress categorization and severity assessment on coffee leaves using semantic segmentation [9]. The model offers farmers and professionals a useful and efficient tool to recognize and measure biotic stressors impacting coffee crops, with classification accuracy rates for biotic stresses above 97%. For semantic segmentation, the model integrates the UNet and PSPNet architectures, enabling accurate symptom recognition and quantification on coffee leaves. By allowing users to precisely detect problems affecting coffee plants and obtain feedback on the severity of their symptoms, the development of an Android app improves accessibility even further. This invention raises productivity and quality in coffee production while also enhancing crop management techniques.

Diverse techniques have been suggested to identify and quantify the extent of damage to plant leaves, such as the use of artificial neural networks, fuzzy c-means, k-means, and evolutionary algorithms. Preprocessing of the images, feature extraction, and training for damaged area categorization are all steps in the process [10]. To achieve precise detection, leaves must be segmented using several color spaces, such as YUV and YCgCr. Neural networks are trained using supervised learning algorithms such as Extreme Learning Machines and Backpropagation. For creating a mobile solution, Mean Stack, MongoDB, Express, AngularJS, and Node.Js are featured. Experiments, analysis of the data, and plans for the work are also included in this article. The development of automated techniques for identifying and evaluating plant leaf damage is the significance of this research.

The objective of the research is to improve the classification accuracy of Arabica coffee leaf disease by applying sophisticated neural network architectures, including InceptionResNetV2, DensNet169, MobileNetV2, and ResNet50. Optimizing illness detection is the goal of this research, which involves examining a dataset of 5000 photographs divided into five categories. The efficacy of the InceptionResNetV2 and DensNet169 models in identifying leaf diseases is demonstrated by their excellent accuracy. The fact that MobileNetV2 was chosen for web application development because of its small size emphasizes how the study findings are put to use. Farmers rely on the system's precise identification and treatment recommendations for Arabica coffee leaf diseases to minimize crop loss and increase yield [11]. The importance of this research is in its ability to use cutting-edge technology to completely transform disease control in agriculture.

3 Proposed Architecture

An outline of the dataset, proposed system architecture, and methodology algorithms, for the designed system is presented here.

3.1 Dataset Preparation and Preprocessing

The dataset, known as the Arabica dataset, initially comprised 58,000 images categorized into five classes: rust, healthy, Cercospora, miner, and phoma. To mitigate overfitting issues, we pruned the dataset to 30,000 images. Each image is sized 128 × 128 pixels. However, it's worth noting that the images contain noise. The first step in preparing the Arabica dataset for coffee leaf disease identification was to resize each image to 128 by 128 pixels, which was a standard dimension. Consistency within the dataset was guaranteed by this uniformity, which was essential for further research. The photos were resized and then turned into grayscale, which reduced them to a single channel while keeping all the information needed for analysis. This translation preserved important information for illness diagnosis while assisting in reducing computing complexity.

Histogram equalization was used after the image was converted to grayscale to improve the contrast and general quality. This method maximized the dynamic range and efficiently stretched out the histogram by redistributing intensity values around the image. Histogram equalization enhanced the visual clarity and distinguishability of disease-relevant characteristics, hence facilitating correct identification. This painstaking process was essential to improving the interpretability of the images and getting them ready for additional processing [12].

Furthermore, Gaussian blur was used to reduce noise and improve image clarity even further. Noise reduction without sacrificing important characteristics was accomplished by applying Gaussian blur within a predetermined range and computing threshold values separately for each image. By maintaining essential information for illness diagnosis, the preprocessing strategies improved image quality while maintaining a cautious balance.

3.1.1 Proposed Architecture

The suggested system of the adapted ResNet-50 model for coffee leaf disease categorization and severity estimate is shown in Fig. 1. The model's base is created by loading a ResNet-50 model that has already been trained. ResNet-50 is well known for being an efficient model for picture classification tasks because it has been trained on the large ImageNet dataset. The foundation of the architecture is this pre-trained ResNet-50, which contains a plethora of information about feature extraction from different visual patterns. The pre-trained layers are then frozen to avoid having their weights changed while the layer is being trained. The ability of the model to effectively generalize to the particulars of the coffee leaf disease classification problem is ensured by the preservation of previously learned features. Custom layers will be added to the architecture to enhance the pre-trained ResNet-50 model's capabilities. These extra layers are designed especially to address the complexities involved in classifying coffee leaf diseases. Convolutional layers give the model the ability to extract hierarchical features from the input images, enabling it to identify diseases by capturing both high-level abstractions and low-level details.

In addition, downsampling the feature maps with max-pooling layers reduces computational complexity and highlights the most important characteristics. The pre-trained ResNet-50 backbone combined with the custom layers creates an extensive architecture that can accurately identify coffee leaf illnesses. Grad cam is applied to the output to overcome the pre-trained model outputs' inaccuracy, which clarifies

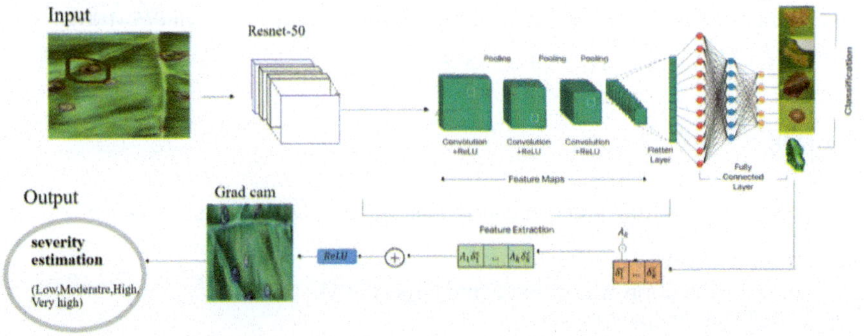

Fig. 1 Proposed architecture for classification and severity estimation

the complexities of the custom layers. A max-pooling layer is placed after each of the two convolutional layers. The first convolutional layer uses 256 filters, each with a 3 × 3 kernel size, and the Rectified Linear Unit (ReLU) activation function. This layer is essential for identifying complex characteristics and patterns in the input photos. The feature maps are then downsampled using a max-pooling layer, which essentially reduces their spatial dimensions while preserving the most important data for later processing phases. The procedure is then repeated with an additional 128-filter convolutional layer, which improves the model's ability to extract and represent features.

3.1.2 Severity Estimation of Coffee Leaves

The comprehensive methodology for visualizing and quantifying the severity of coffee leaf diseases using a customized ResNet-50 and Grad-CAM technique. Initially, the last convolutional layer of the customized ResNet model is extracted to serve as the foundation for generating the Grad-CAM visualization. Subsequently, a custom function,getgradcam(), is defined to compute the Grad-CAM heatmap, leveraging the activations of the last convolutional layer and the predicted class probabilities obtained from the model.

Within the getgradcam() function, the input image undergoes preprocessing before being fed through the customized ResNet model to obtain relevant activations and predicted class probabilities. TensorFlow's gradient tape is then utilized to compute the gradients necessary for calculating neuron importance weights(0.7 and 0.3). Following this, the Grad-CAM heatmap is generated, post-processed, and superimposed onto the original image using the jet colormap, providing a visual representation of regions crucial for disease classification.

This involves converting the superimposed image to the YCbCr color space, extracting relevant components, and applying Otsu's thresholding method to segment the image. By calculating the area occupied by the detected regions and comparing it to the total image area, the severity of the detected disease is quantified as a percentage. These percentages are used for categorizing stressed images into four levels low (0–25), moderate (25–50), high (50–75), and very high (75–100).

4 Algorithm

The process of building and evaluating a model involves three main tasks:

1. Image Preprocessing Algorithm.
2. Disease Detection Algorithm using Grad-CAM.
3. Model Evaluation Algorithm.

Algorithm 1 Image Preprocessing Algorithm

Step 1: Gather a dataset of images containing examples of healthy, rust, miner Cercospora, and Phoma, ensuring each image is labeled or categorized correctly.
Step 2: For each image in the dataset:
  Check if the image size is not 128 × 128 pixels.
  If not, resize the image to 128 × 128 while maintaining the aspect ratio:
    Compute the aspect ratio of the original image:
    $\text{aspect\_ratio} = \frac{\text{original\_width}}{\text{original\_height}}$ ▷ (1)
    If the original width is greater than the original height:
      $\text{height} = \frac{128 \times \text{original\_height}}{\text{original\_width}}$ ▷ (2)
    If the original height is greater than the original width:
      $\text{width} = \frac{128 \times \text{original\_width}}{\text{original\_height}}$ ▷ (3)
Step 3: Convert each RGB image into a grayscale image to reduce complexity.
Step 4: Apply histogram equalization to enhance the contrast of the grayscale images:
  Transformation function: $T(v) = \frac{c(v) - c_{\min}}{MN - c_{\min}} \times (L - 1)$ ▷ (4)
Step 5: Apply the transformation function to each pixel value in the grayscale image.
Step 6: Apply Gaussian filtering to remove noise and smooth the edges of the grayscale images:
  Gaussian function: $G(x, y) = \frac{1}{2\pi\sigma^2} \times e^{-\frac{x^2+y^2}{2\sigma^2}}$ ▷ (5)
Step 7: Save the preprocessed images with their respective labels/categories.
Step 8: Repeat steps 2 to 6 for all images in the dataset.

Algorithm 2 Disease Detection Algorithm using Grad-CAM

Step 1: Load pre-trained ResNet-50 model with specified input shape.
Step 2: Define custom layers for further feature extraction and classification. Combine the pre-trained model with the custom layers to create the final model..
Step 3: Compile the model with a specified optimizer, loss function, and metrics.
  Loss function: $L(y, y_1) = -[y \cdot \log(y_1) + (1 - y) \cdot \log(1 - y_1)]$ ▷ (6)
  Update rule: $\theta(t + 1) = \theta(t) - \text{rate} \cdot \frac{m(t)}{\sqrt{v(t)} + \text{epsilon}}$ ▷ (7)
Step 4: Get the last convolutional layer from the model that extracts the outputs of the last convolutional layer and model predictions.
Step 5: Compute gradients of the predicted class concerning the last convolutional layer output.
  Algorithm:
  Compute gradients: $weights = \text{tf.reduce\_sum}(pooled\_grads \cdot \text{last\_conv\_output}, \text{axis} = -1)$ ▷ (8)
Step 6: Convert Grad-CAM to a heatmap and superimpose it on the original image.
Step 7: Convert the superimposed image to YCbCr color space and extract the blue-difference chroma component from the YCbCr image.
Step 8: Apply Otsu's thresholding to the Cb channel to segment regions of interest.
Step 9: Calculate the area and severity of the detected regions in the segmented image.
  Area: $area = \text{cv2.countNonZero}(thresholded\_image)$ ▷ (9)
  Severity: $Severity = \left(\frac{area}{\text{total\_pixels}}\right) \times 100$ ▷ (10)

Algorithm 3 Model Evaluation Algorithm

1: **Step 1:** Use the preprocessed images from the testing set to evaluate the trained model performance.

2: **Step 2:** To evaluate model performance, compute evaluation metrics like accuracy, precision, recall, and F1 Score.

3: Accuracy: $\text{Accuracy} = \frac{\text{number of correctly predicted samples}}{\text{total number of samples}}$ ▷ (11)

4: Precision: $\text{Precision} = \frac{\text{True Positives}}{\text{True Positives+False Positives}}$ ▷ (12)

5 Results and Outputs

This part presents and discusses the comparison of different approaches, strategies, and results at every stage of our investigation. Upon evaluation of the test dataset, the model achieved an impressive accuracy of 95.4%, showcasing its robust performance in accurately classifying images shown in Fig. 2.

In order to track the model's performance on untested data, a section of the dataset is usually set aside as a validation set throughout the training phase. The model's performance on the validation set is assessed after every training session, and the validation accuracy is calculated. By using this data, overfitting can be avoided and hyperparameters can be adjusted. As demonstrated in Fig. 3, your example resulted

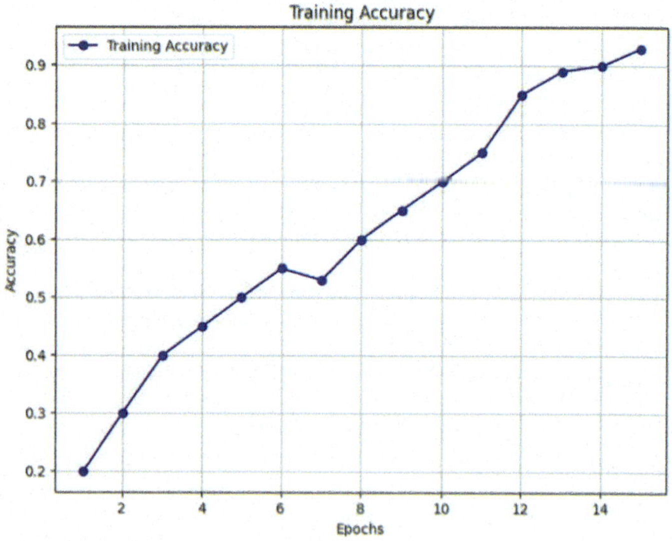

Fig. 2 Training accuracy of customized ResNet-50

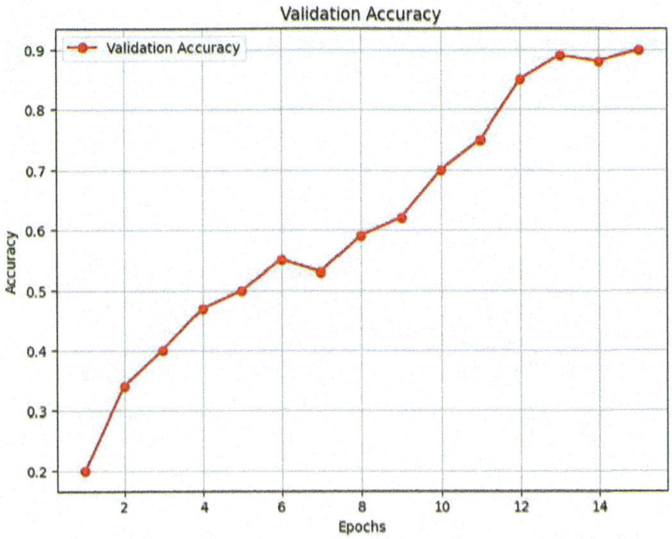

Fig. 3 Training accuracy of customized ResNet-50

in a validation accuracy of 93.3%, demonstrating the model's strong generalization to new data and its ability to accurately categorize and estimate the severity of coffee leaf illnesses.Upon closer examination, the model's predictive skills for distinct classes exhibit both strengths and weaknesses, which facilitates the identification of areas that require further development in subsequent versions of the model.

In addition to rust, Cercospora, and healthy cases, the accuracy heat map also shows the performance of a classification model across 15 epochs for the classes of miner disease and phoma. Interestingly, the model is more accurate at identifying Cercospora, rust, and healthy cases. Because they resemble rust sickness, the phoma and miner groups have somewhat poorer accuracy. Rust and phoma have traits in common, which causes these classes to be misclassified and have lower accuracy. Despite these difficulties, Fig. 4 provides insightful information about the model's advantages and disadvantages, helping to direct future adjustments to increase classification accuracy for all classes.

Our model's propensity to incorrectly identify cases as belonging to the phoma and miner classes is a significant flaw. The underlying similarities between rust, miner, and phoma diseases are the cause of these incorrect classifications, which cause uncertainty during the classification process. Reduced accuracy is the result of the model's incapacity to distinguish between these groups, particularly for Phoma and Miner. To improve the model's overall performance and reliability, this challenge must be addressed. Grad-CAM is a sophisticated approach that is used to identify regions shown in Fig. 5 accurately. Otsu thresholding is also used to segment images efficiently, identifying healthy and unhealthy areas based on differences in intensity. Moreover, the incorporation of YCbCr color space analysis improves the differen-

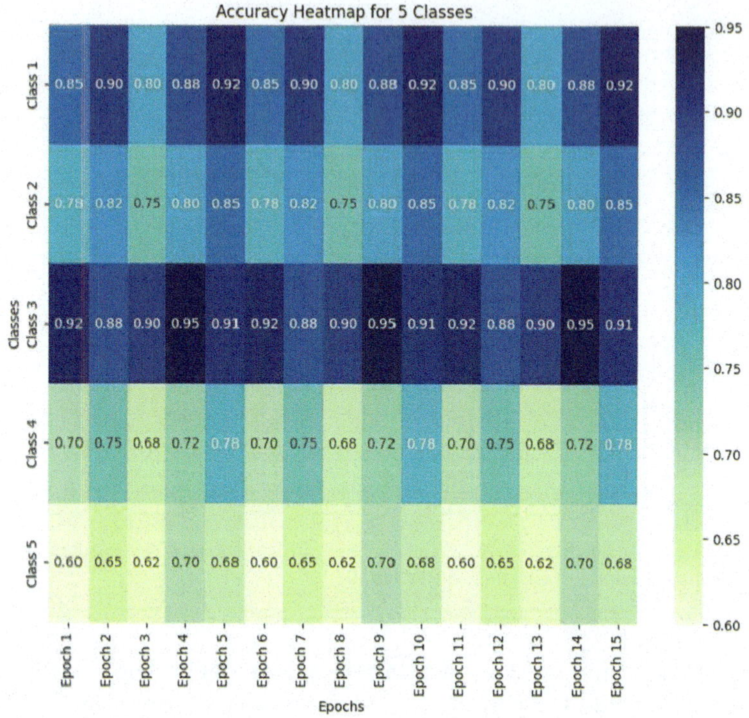

Fig. 4 Accuracy heatmap for five classes

tiation between various color channels, making it easier to identify and measure disease-affected regions precisely. I produced a CSV file with severity estimates after testing one thousand photos.

To evaluate accuracy, these predictions were then contrasted with ground truth labels. Notably, the comparison showed a 92% accuracy rate in determining the severity of the condition. As demonstrated by a 92% accuracy rate when compared to ground truth labels, the severity estimation system achieves a high level of accuracy by integrating various techniques. In addition to enhancing disease identification, this all-encompassing strategy allows for a more sophisticated comprehension of disease severity, which facilitates more informed decision-making regarding disease management tactics.

Table 1 represents the comparison of our proposed models with the existing models from which our proposed system Customized ResNet-50 model is considered as the best model for Severity Estimation.

Fig. 5 Visualizing disease severity with Grad-CAM

Table 1 Performance comparison with existing models

Model name	Accuracy (%)	References
DNN	77	[6]
ResNet-50	86.51	[7]
MobileNet-v2	92.80	[5]
Customized ResNet-50	92.5	–

6 Conclusion and Future Work

With a focus on the Arabica variety specifically, our work aims to address the pressing demand for automated systems in precision agriculture, notably in the early identification and severity classification of coffee leaf diseases. Our methodology, which makes use of ResNet-50's robust image classification capabilities, presents a comprehensive system that can differentiate between healthy leaves and those that are affected by biotic stresses including rust, miner Cercospora, and Phoma. Additionally, our system efficiently classifies the disease severity level when it occurs, giving farmers important information they need to take prompt corrective action. Our model shows robustness and accuracy in identifying regions within coffee leaf images for precise classification decisions by carefully gathering and preprocessing a diverse dataset with images of coffee leaves representing different health states, along with sophisticated techniques like data augmentation and Grad-Cam visualization. With the addition of the Grad-Cam technique for severity grading and a modified ResNet-50 model for leaf health status classification and biotic stress identification, our system provides a comprehensive solution for early disease diagnosis and control in coffee crops.

By improving the design and training methodologies of our model, we hope to address the misclassification of Phoma and mining illnesses in subsequent work. To help the model distinguish these diseases from other leaf conditions, one method would be to gather more data that is particularly related to these illnesses. Furthermore, to improve the discriminatory strength of the model, we intend to investigate more sophisticated deep learning approaches like ensemble methods and transfer learning.

References

1. Edgardo JGM et al (2022) An app to assist farmers in the identification of diseases and pests of coffee leaves using deep learning. Inf Process Agric 9.1:38–47
2. Latif MA et al (2023) Enhanced classification of coffee leaf biotic stress by synergizing feature concatenation and dimensionality reduction. IEEE Access
3. Martins B, Deroco G, de Lourdes Bueno Trindade Galo M, Vieira BS (2017) Detecting and mapping root-knot nematode infection in coffee usingremote sensing measurements. IEEE J Sel Top Appl Earth Observ Remote Sens 10.12:5395–5403
4. Kondo T et al (2023) Prediction of short-term mortality of cardiac care unit patients using image-transformed ECG waveforms. IEEE J Transl Eng Health Med 11:191–198
5. Hitimana E et al (2023) An intelligent system-based coffee plant leaf disease recognition using deep learning techniques on rwandan arabica dataset. Technologies 11.5:116
6. Yebasse M et al (2021) Coffee disease visualization and classification. Plants 10.6:1257
7. Esgario, JGM, Krohling RA, Ventura JA (2020) Deep learning for classification and severity estimation of coffee leaf biotic stress. Comput Electron Agric 169:105162
8. Manso GL et al (2019) A smartphone application to detection and classification of coffee leaf miner and coffee leaf rust. arXiv:1904.00742
9. Aufar Y, Abdillah MH, Romadoni J (2023) Web-based cnn application for arabica coffee leaf disease prediction in smart agriculture. Jurnal Resti (Rekayasa Sistem Dan Teknologi Informasi) 7.1:71–79
10. Nawaz M et al (2024) CoffeeNet: a deep learning approach for coffee plant leaves diseases recognition. Exp Syst Appl 237:121481
11. Martins GD, de Lourdes Bueno Trindade Galo M, Vieira BS (2017) Detecting and mapping root-knot nematode infection in coffee crop using remote sensing measurements. IEEE J Sel Top Appl Earth Observ Remote Sens 10.12:5395–5403
12. Mahapatra D, Ge Z, Reyes M (2022) Self-supervised generalized zero-shot learning for medical image classification using novel interpretable saliency maps. IEEE Trans Med Imaging 41(9):2443–2456

Comparative Analysis of DL Models for Early Detection of COVID-19 Using Cough Audio Data

J. Sreedutt Ram, P. Sidharth, Sukrith Sunil, S. S. Poorna, and K. Anuraj

Abstract Early detection of the SARS-CoV-2 virus in human mucus is now an urgent need to settle down the COVID-19 outbreaks. These global pandemic outbreaks are hard to stop without massive testing. One promising solution is non-intrusive collection of data in the form of cough sounds and using various speech processing techniques for classifying diseased samples from healthy ones. In this paper we present a non-invasive, quick and efficient pre-screening methodology using cough sounds for the detection of COVID-19, employing cepstral domain analysis viz. Mel-Frequency Cepstral Coefficients (MFCC), Gammatone Frequency Cepstral Coefficients (GFCC) and Human Factor Cepstral Coefficients (HFCC) using architectures such as 1-D Convolutional Neural Network (CNN), Recurrent Neural Network (RNN), Long Short-Term Memory (LSTM) and Bidirectional Long Short-Term Memory (BiLSTM). Subsequently, we analyse the impact of feature fusion technique for the classification of cough audio data as COVID-19 affected or not affected categories and we have found that the model with feature selection using Spearman Rank Correlation on combined cepstral features gave highest accuracy 72% with LSTM.

Keywords COVID-19 · MFCC · GFCC · HFCC · CNN · RNN · LSTM · BiLSTM

1 Introduction

The current COVID-19 situation has given rise to an urgent demand for early identification of the presence of the SARS-CoV-2 virus. One promising method is to gather data in a non-invasive manner using cough sounds for COVID-19 detection. Recently, most of the research works related to COVID-19 detection have focused

J. S. Ram · P. Sidharth · S. Sunil · S. S. Poorna (✉) · K. Anuraj
Department of Electronics and Communication Engineering, Amrita Vishwa Vidyapeetham, Amritapuri, India
e-mail: poornass@am.amrita.edu

A. C. Frery et al. (eds.), *Computational Intelligence and Data Analytics*, Lecture Notes on Data Engineering and Communications Technologies 236,
https://doi.org/10.1007/978-981-96-0451-7_15

on the non-invasive mode of disease detection. In this work, we present a detailed analysis comparing various cepstrum-based approaches for COVID-19 detection using cough audio datasets. We delve into the use of advanced deep learning models like CNN, RNN, LSTM and BiLSTM, focusing on features like MFCC, GFCC and HFCC. Our research entails the use of publicly available datasets through crowd-sourcing to create a reliable benchmark dataset for accurately diagnosing COVID-19. We put forth a COVID-19 detection framework that relies on Deep Learning (DL) and is centred around audio, specifically cough sounds. The results are validated through extensive analysis.

The proposed model will offer a non-invasive and efficient pre-screening method for COVID-19 detection by utilizing cough sounds by using cepstral features—MFCC, GFCC and HFCC—with deep learning architectures like 1-D CNN, RNN, LSTM and BiLSTM. This approach highlights the strength of cepstral features and the capacity of LSTM to handle long sequential data.

2 Literature Survey

In recent years, several studies have proposed cough sounds as one of the detection methodologies for methodologies for COVID-19. However, as cough sound is considered one of the major things in most respiratory diseases, it still poses some challenges to differentiate the cough sounds of COVID-19 patients from other diseases. In [1], the authors use a CNN to recognize COVID-19 from cough sounds. Support vector machines (SVMs) and random forests are employed in [2] to categorize COVID-19 using three types of audio features derived from cough sound: used three types of audio features: MFCC, GFCC and HFCC. They tested this method using four cough datasets viz. NoCoCoD, Cambridge, Virufy and Coswara. They also utilized recurring neural network (RNN) to classify time–frequency spectrograms of cough sounds. In [3], authors used K-nearest neighbours (KNN) and decision tree models to detect COVID-19 utilizing statistical features viz. mean, standard deviation and the energy of the coughs. They also made use of MFCC as a technique to extract features from cough sounds. Hence audio features like MFCC, GFCC and HFCC can play a crucial role in detecting COVID-19.

In [4], the authors introduced an innovative approach using a multi-channelled CNN to autonomously diagnose COVID-19 based on respiratory sounds. An innovative approach to respiratory sound analysis is addressed in [5]. Authors classified MFCC derived from cough sounds using CNN for COVID-19 detection. The paper also discussed the dataset collection and labelling process, underscoring the importance of data handling and pre-processing in developing effective models for COVID-19 diagnosis through respiratory sounds. Cost-effective and non-contact methods for COVID-19 screening through the analysis of respiratory sounds, marking a significant stride in the ongoing battle against the pandemic. Pahar et al. studied various datasets for identifying COVID-19 positive coughs using Residual-based Neural Network Architecture (Resnet50) and yielded an area under the ROC curve (AUC) of 0.98 [6]. The study revealed that COVID-19 positive coughs showed distinct characteristics, and the length were 15–20% shorter than non-COVID coughs.

MFCCs highlight inherent resilience against noise. This distinctive quality enables them to filter out extraneous information while preserving the perceptually crucial features of speech, positioning them as indispensable tools for various applications. MFCCs appeared as a cornerstone in speech and audio analysis, showing their prowess across diverse tasks like emotion classification [7], speaker recognition [8] and even cough detection.

MFCCs are an impressive tool for researchers and developers working across various domains of speech and audio analysis. MFCCs have proved their ability to distinguish coughs from background noise and speech with the help of spectral and temporal patterns of cough sounds [9]. Unlike normal methods, GFCC helps in finding the small changes in whispers of spectral information hidden within [10]. This biomimetic ability is utilized in diverse applications, ranging from pinpointing COVID-19 in noisy coughs to amplifying children's speech recognition [11].

GFCC's similarities to the human auditory system help to represent sound's spectral quality and reduction in unwanted noise. GFCC's capability in noise reduction helps in diagnosing Parkinson's disease, exceeding traditional methods [12]. MFCCs exhibit limitations in filter bandwidth causing issues in noisy settings. Human-factor cepstral coefficient's (HFCC) superior noise robustness enables them to overcome these limitations of MFCCs by decoupling filter bandwidth from filter spacing [13].

Usage of wider filters in HFCC results in a remarkable 7 dB improvement in signal-to-noise ratio compared to MFCCs [14]. This enhanced robustness directly helps in superior bird species recognition accuracy in real-world environments [15]. This study was conducted in a noisy urban environment, revealed higher recognition accuracy using HFCC compared to MFCC. The adaptability and computational efficiency of HFCCs, comparable to MFCCs make them a compelling choice for practical applications. Their superior noise robustness proved efficacy in real-world settings, and potential for optimization paved the way for significant advancements in multiple fields.

In this work, we proposed a non-invasive, contactless DL-based COVID-19 detection methodology using cough audio datasets. Here DL based models viz. CNN, LSTM, RNN and BiLSTM are the algorithms adopted to classify cepstral features viz. MFCC, GFCC and HFCC derived from cough audio datasets. Further, we analysed the effect of feature fusion in cough sound classification. The rest of the paper follows the organization as given below.

Section 2 gives the description of methodology adopted in this paper. Experimental setup followed is elaborated in the next section. A detailed discussion of obtained results is given in Sect. 4, followed by conclusion.

3 Methodology

Figure 1 shows the block level representation of COVID-19 detection system. The proposed work aims at classifying cough audio samples to potentially show the presence or absence of COVID-19. The cough audio data is pre-processed subjected to extraction of features in the cepstral domain. These features include MFCCs, HFCCs and GFCCs, which can bring out the discriminative nature in the cough audio by modelling the human perception system using various filtering techniques in the spectral domain. These cepstral features are further concatenated. Feature selection strategies viz. Recursive Feature Elimination (RFE) and Spearman Rank Correlation method and were independently applied to the concatenated features. In order to carry out comparative analysis and study the effectiveness of these features individually and as combinations, various Deep Learning techniques viz. CNN, RNN, LSTM and BiLSTM are employed to classify COVID-19 "positive" or "negative" categories.

3.1 *Dataset*

Our analysis is carried out using the Covid-19 Cough Sounds Kaggle dataset [16] created from the samples collected from COVID-19 Sound Resource Allocation (COSWARA) [17] and Virufy [18] datasets. Both these datasets have cough and respiratory voice samples, with crowdsourced and clinically validated data, well known for AI-based COVID-19 detection. To mitigate the class imbalance problem

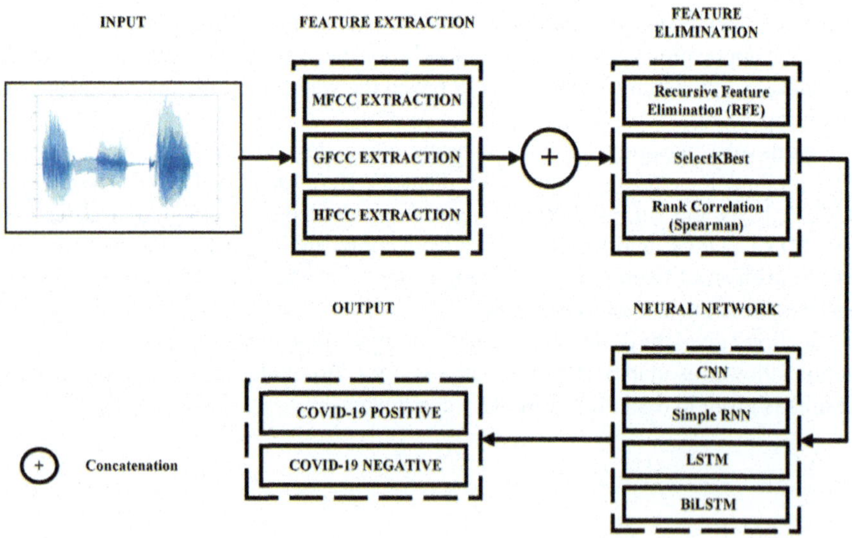

Fig. 1 Block diagram

Table 1 Statistics of dataset for COVID-19 cough audio

Label	Samples	Total audio duration (min)	Average per sample (min)
COVID-19 positive	150	11.228	0.074
COVID-19 negative	150	16.407	0.109

with the dataset, 150 cough audio samples of people who tested positive, and 150 samples of people negative for COVID-19 are used in our analysis. The statistics of cough Audio samples used in our analysis is summarized in Table 1.

3.2 Feature Extraction

Cepstrum-based features from audio have demonstrated their effectiveness in various applications involving speaker identification, speech emotion identification, lung disease detection, etc. [19–21]. These cepstrum-based features include MFCCs, HFCCs and GFCCs, which can bring out the discriminative nature in the audio by modelling the human perception system using various filtering techniques in the spectral domain. We have used these 3 features for COVID-19 detection from cough audio data in our work viz. MFCCs, GFCCs and HFCCs.

MFCC are representations of energy distribution in the power spectrum of speech signal. They are obtained by breaking the audio signal into short overlapping frames. Then, Fast Fourier Transform (FFT) is applied to each frame to obtain the spectral variations. To replicate the response of human auditory system, we convert the spectral scale to Mel scale as given in Eq. (1), where f_{linear} is the linear frequency scale and f_{mel} is the Mel scale frequency. This is accomplished by using a series of triangular filters that overlap one another. The individual triangular filter in Mel filter bank MFCC_i is given in Eq. (2), where $f(i)$ is the frequency of the ith filter bank. The power spectrum of the signal is passed through these filters, and subsequently, Discrete Cosine Transform (DCT) is applied to the obtained log-filter bank energies, resulting in a set of MFCC coefficients. These coefficients contain the spectral characteristics of the audio signal in a compact manner, making them useful tool for audio-based classification tasks.

$$f_{\text{mel}} = 2595 \log_{10}\left(1 + \frac{f_{\text{linear}}}{700}\right) \tag{1}$$

$$\text{MFCC}_i(k) = \begin{cases} 0, & \text{if } k < f(i-1) \\ \dfrac{k - f(i-1)}{f(i) - f(i-1)}, & \text{if } f(i-1) \le k \le f(i) \\ \dfrac{f(i+1) - k}{f(i+1) - f(i)}, & \text{if } f(i) \le k \le f(i+1) \\ 0, & \text{if } k > f(i-1) \end{cases} \tag{2}$$

The computation of HFCC [22] follows the same way as MFCC, but with a difference that though the filter spacing follows a Mel scale, the bandwidth of the filter banks is spaced according to the human auditory system, following an Equivalent Rectangular Bandwidth (ERB) as given in Eq. (3).

$$\mathrm{ERB} = 6.23f_c^2 + 93.39f_c + 28.52 \tag{3}$$

where f_c is the centre frequency, HFCC is capable of capturing audio-based features more effectively with noise reduction, than normal MFCC. This is achieved by wider filter banks, enhancing noise removal by adjusting filter spacing, which further improves the signal-to-noise ratio.

GFCC [23] is similar to MFCC but are found using the filter bank design known as the gammatone filter bank. This filter bank is created by analysis of human auditory system, and is based on a set of overlapping bandpass filters, obtained as the product of gamma distribution and single tone sinusoid as given in Eq. (4).

$$h(t) = h_0 t^{n-1} e^{-2\pi Bt} \cos(2\pi f_c t + \emptyset) \tag{4}$$

where h_0 is the amplitude factor, n is the order of the filter, B is the duration of impulse response and Ø represents the phase. The audio signal is passed through the gammatone filter bank, and the outputs of each filter are passed through a non-linear compression stage to imitate the response of cochlea in the ear, followed by DCT to obtain the GFCCs. These coefficients capture the spectral characteristics of the audio signal, with a focus on capturing subtle spectral details effectively. These features are analysed individually as well as in combination for COVID-19 detection.

3.3 Feature Selection

Feature selection is done to find and retain the most appropriate features from the dataset while it discards the irrelevant and redundant ones. By selecting the most informative features, we can improve model performance, reduce overfitting and enhance computational efficiency. In this work, to identify the features based on their relevance, Recursive Feature Elimination (RFE) and Spearman's Rank Correlation methods were employed.

RFE-based feature selection uses recursion to eliminate the least relevant features, utilizing a ranking metric. The initially uses all features for training and ranks the features as per their relevance, and based on the rank, the least relevant features are removed. This is done until the required number of features are obtained on there is no improvement to the model. In Spearman's rank correlation, the relation between two variables is evaluated based on correlation measure. The correlation coefficient takes the values between −1 and +1, which gives negative and positive correlations respectively. A zero value for rank correlation coefficients specifies that there is no correlation between the features.

3.4 *Classification Techniques*

In order to analyse the effectiveness of individual features, feature combinations and the effect of feature selection, four distinct deep learning architectures are known to handle sequential data are employed. This includes 1-D-CNN [24–27], RNN, LSTM and BiLSTM. All these models employed an 80:20 ratio for the train test split. The analysis carried out in this work used an accuracy measure of COVID-19 prediction as the performance metric.

4 Experiment Setup

Our dataset contains audio files with a sampling rate of 22.05 kHz with the largest track duration of 29.95 s and a minimum track duration of 0.25 s The dataset held files of format MPEG Audio Layer 3 (.mp3) which was converted to Waveform Audio File (.wav) format for maintaining better audio quality using Python package pyhub. In the feature extraction phase, diverse techniques are employed to extract informative features from audio data, leveraging prominent libraries such as Librosa in Python. A Hamming window function is applied to each speech samples with a duration of 30 ms frame and 15 ms overlap. FFT is utilized to compute the magnitude spectrum. 21-Mel frequency filter banks are employed, followed by logarithmic transformation and Discrete Cosine Transform to obtain a set of 21 MFC Coefficients. GFCC features are computed using SPafe library and same number of coefficients were employed for our analysis. For obtaining HFCC coefficients the parameters h_0 was chosen as 1, $\emptyset$ as 0 and n as 4.

The Deep Learning architectures were independently trained and evaluated using four different audio feature types: MFCC, GFCC, HFCC and concatenated feature which have undergone feature fusion, with 80:20 train-test split. We perform feature selection operations on the concatenated dataset, Recursive Feature Elimination (RFE) using Random Forest Classifier and Spearman's rank correlation, which is done by the scikit-learn library. These techniques are applied to each feature type to pick 21 optimal feature coefficients.

5 Result and Discussion

The experimental outcomes present a trend in the accuracy of evaluated neural network architectures across various audio feature types, highlighting the substantial impact of feature selection methods. The BiLSTM architecture consistently had the highest accuracy along various scenarios, underscoring its versatility across diverse feature sets. HFCC features proved to be informative, consistently delivering better accuracy along all architectures. The implementation of feature selection techniques,

Table 2 Accuracies (%) of different models and feature selection methods are compared

Model	CNN	Simple RNN	LSTM	BiLSTM	Feature selection optimization used
MFCC	61	62	50	63	No
GFCC	55	55	53	58	No
HFCC	62	62	55	67	No
MFCC + GFCC + HFCC	65	67	58	62	Recursive feature elimination (RFE)
MFCC + GFCC + HFCC	62	64	72	52	Rank correlation (Spearman)

such as Recursive Feature Elimination (RFE) and Rank Correlation, helped in accuracy enhancement. LSTM achieved accuracies of 50, 53 and 55% for MFCC, GFCC and HFCC, respectively, and accuracies of 58 and 72% for concatenated features using RFE and Rank Correlation respectively. CNN was able to attain accuracies of 61, 55 and 62% for MFCC, GFCC and HFCC and accuracies of 65 and 62% for concatenated features using RFE and Rank Correlation, respectively. Simple RNN achieved accuracies of 62, 55 and 62% for MFCC, GFCC and HFCC, and accuracies of 67 and 60% for concatenated features using RFE and Rank Correlation, respectively. BiLSTM demonstrated accuracies of 63, 58 and 67% for individual feature types. These findings help us to understand how to approach neural network architecture selection and feature optimization for achieving optimal audio classification outcomes. The findings are listed in Table 2.

6 Conclusion and Future Work

The experiments show that HFCC features consistently performed, exhibiting slight variation inaccuracies across all the deep learning architectures analysed. Hence HFCC is considered robust and effective for COVID-19 detection, for the dataset considered. Further feature selection using Spearman Rank Correlation on combined cepstral features, gave the highest accuracy with LSTM. The accuracy of 72%, highlights the effectiveness of the feature selection technique as well as the learning capacity of LSTM architecture to handle long sequential concatenated data. Our future work aims to analyse the effectiveness of COVID-19 classification using these features, employing attention mechanisms on the aforementioned DL models.

The effectiveness of a model is dependent on the dataset used, and the lack of diversity in cough sound samples limited our generalizability. The computational complexity of deep learning architectures like LSTM and BiLSTM encounters challenges when deployed in resource-constrained environments. There is also a risk of overfitting because of the model's complexity and the size of the dataset. Also, the model's real-world, real-time performance has not been tested.

References

1. Rao S, Narayanaswamy V, Esposito M, Thiagarajan JJ, Spanias A (2021) COVID-19 detection using cough sound analysis and deep learning algorithms. Intell Decis Technol 15(4):655–665
2. Chowdhury NK, Kabir MA, Rahman MM, Islam SMS (2022) Machine learning for detecting COVID-19 from cough sounds: an ensemble-based MCDM method. Comput Biol Med 145:105405
3. Kapoor T, Pandhi T, Gupta B (2022) Cough audio analysis for COVID-19 diagnosis. SN Comput Sci 4(2):125
4. Lella KK, Pja A (2022) Automatic diagnosis of COVID-19 disease using deep convolutional neural network with multi-feature channel from respiratory sound data: cough, voice, and breath. Alex Eng J 61(2):1319–1334
5. Bansal V, Pahwa G, Kannan N (2020) Cough classification for COVID-19 based on audio mfcc features using convolutional neural networks. In: 2020 IEEE international conference on computing, power and communication technologies (GUCON). IEEE, pp 604–608
6. Pahar M, Klopper M, Warren R, Niesler T (2021) COVID-19 cough classification using machine learning and global smartphone recordings. Comput Biol Med 135:104572
7. Vimal B, Surya M, Sridhar VS, Ashok A (2021) MFCC based audio classification using machine learning. In: 2021 12th international conference on computing communication and networking technologies (ICCCNT). IEEE, pp 1–4
8. Tiwari V (2010) MFCC and its applications in speaker recognition. Int J Emerg Technol 1(1):19–22
9. Mashika M, van der Haar D (2023) Mel frequency cepstral coefficients and support vector machines for cough detection. In: International conference on human–computer interaction. Springer Nature Switzerland, Cham, pp 250–259
10. Swain BK, Khan MZ, Chowdhary CL, Alsaeedi A (2023) SRC: superior robustness of COVID-19 detection from noisy cough data using GFCC. Comput Syst Sci Eng 46(2)
11. Ankita S, Shahnawazuddin S (2023) Studying the effect of frame-level concatenation of GFCC and TS-MFCC features on zero-shot children's ASR. In: International conference on speech and computer. Springer Nature Switzerland, Cham, pp 140–150
12. Boualoulou N, Drissi TB, Nsiri B (2023) Comparison of feature extraction methods between MFCC, BFCC, and GFCC with SVM classifier for Parkinson's disease diagnosis. In: International conference on IoT based control networks and intelligent systems. Springer Nature Singapore, Singapore, pp 231–247
13. Skowronski MD, Harris JG (2002) Human factor cepstral coefficients. J Acoust Soc Am 112(Suppl 5):2279–2279
14. Skowronski MD, Harris JG (2004) Exploiting independent filter bandwidth of human factor cepstral coefficients in automatic speech recognition. J Acoust Soc Am 116(3):1774–1780
15. Wielgat R, Zieliński TP, Potempa T, Lisowska-Lis A, Król D (2007) HFCC based recognition of bird species. In: Signal processing algorithms, architectures, arrangements, and applications SPA 2007. IEEE, pp 129–134
16. https://www.kaggle.com/datasets/pranaynandan63/covid-19-cough-sounds
17. Bhattacharya D, Sharma NK, Dutta D, Chetupalli SR, Mote P, Ganapathy S, Chandrakiran C et al (2023) Coswara: a respiratory sounds and symptoms dataset for remote screening of SARS-CoV-2 infection. Sci Data 10(1):397
18. Chaudhari G, Jiang X, Fakhry A, Han A, Xiao J, Shen S, Khanzada A (2020) Virufy: global applicability of crowdsourced and clinical datasets for AI detection of COVID-19 from cough. arXiv preprint arXiv:2011.13320
19. Poorna SS, Jayachandran Nair G (2019) Multistage classification scheme to enhance speech emotion recognition. Int J Speech Technol 22:327–340
20. Yerramreddy DR, Marasani J, Gowtham PSV, Yashwanth S, Poorna SS, Anuraj K (2023) Speaker identification using MFCC feature extraction: a comparative study using GMM, CNN, RNN, KNN and random forest classifier. In: 2023 Second international conference on trends in electrical, electronics, and computer engineering (TEECCON). IEEE, pp 287–292

21. Alghamdi NS, Zakariah M, Karamti H (2024) A deep CNN-based acoustic model for the identification of lung diseases utilizing extracted MFCC features from respiratory sounds. Multimed Tools Appl 1–33
22. Benba A, Jilbab A, Hammouch A (2017) Using human factor cepstral coefficient on multiple types of voice recordings for detecting patients with Parkinson's disease. IRBM 38(6):346–351
23. Adiga A, Magimai M, Seelamantula CS (2013) Gammatone wavelet cepstral coefficients for robust speech recognition. In: 2013 IEEE international conference of IEEE region 10 (TENCON 2013). IEEE, pp 1–4
24. Renjith S, Manazhy R (2023) A comparative analysis of ISLRS using CNN and VIT. In: International conference on iot based control networks and intelligent systems. Springer Nature Singapore, Singapore, pp 1–9
25. Raja AK, Sugandhi C, Nymish G, Havish NS, Rashmi M (2023) Convolutional neural network based virtual mouse. In: 2023 4th international conference for emerging technology (INCET). IEEE, pp 1–7
26. Megalingam RK, Aryan K, Gopika A, Jogesh G, Kunnambath AR, Kota AH (2024) Deep learning approach to identify pests in coconut trees. In: 2024 3rd international conference for innovation in technology (INOCON). IEEE, pp 1–6
27. Charan MGKS, Poorna SS, Anuraj K, Praneeth CS, Sai Sumanth PG, Gupta CVSP, Srikar K (2022) Sign language recognition using CNN and CGAN. In: Inventive systems and control: proceedings of ICISC 2022. Springer Nature Singapore, Singapore, pp 489–502

Deepfake Detection Using Deep Learning and Evolutionary Optimization

Ashina Gaur, Bhawna Jain, Aarti Kushwaha, Anushka Dahiya, and Simran Saigal

Abstract The advancement of technology and Artificial Intelligence (AI) has posed challenges to online security. While AI has numerous benefits, it comes with its fair share of shortcomings. Deepfakes refer to manipulated media like video, image, and audio. The existence of the Deepfake technology is a notable example of how AI has the potential to pose risks to society. Deep Learning (DL) models have proven effective in differentiating genuine images from synthesized ones. When there is a limited amount of training data, transfer learning is employed wherein a pre-trained DL model is used. Though DL models are predominantly used, there is a need to eliminate features that are redundant from the feature set. To achieve that, in this work, the functionality and workings of the Particle Swarm Optimization (PSO) algorithm for feature selection have been studied. First, features have been extracted using a pre-trained DL model. Thereafter, optimal features are derived from the extracted features by implementing the PSO algorithm. This feature subset is then used to perform classification. The model has been evaluated on the OpenForensics public dataset while using, on average, 44.18% of the feature set. Classification was performed using an artificial neural network (ANN). The selected features using PSO drastically improved the modelâŁ™s performance from an accuracy of 63.4% (without PSO) to 79.1%.

Keywords Evolutionary algorithm · Particle swarm optimization · Deepfake detection

A. Gaur (✉) · B. Jain · A. Kushwaha · A. Dahiya · S. Saigal
Department of Computer Science and Engineering, Indira Gandhi Delhi Technical University for Women, Delhi, India
e-mail: ashinagaurigdtuw@gmail.com

A. C. Frery et al. (eds.), *Computational Intelligence and Data Analytics*, Lecture Notes on Data Engineering and Communications Technologies 236,
https://doi.org/10.1007/978-981-96-0451-7_16

1 Introduction

Deepfakes are generated by manipulating media through the use of artificial intelligence, resulting in videos, images, or audio that appear highly realistic but are entirely fabricated. DeepFake technology gained prominence around the year 2017, but in the year 2019, DeepFakes hit the mainstream when realistic-looking synthesized videos began to surface on the internet, raising concerns about their misuse [1]. The increased accessibility of this technology raises the potential for its misuse. The detection of deepfakes is essential to ensuring the privacy of users online by identifying and preventing the spread of misleading content.

A pre-trained model is a model that has been trained to solve a comparable problem using an extensive benchmark dataset. Given the massive resources needed to train the deep learning models, the popular concept of transfer learning is used in the field of DL [2]. Various DL models structured on the Convolutional Neural Network (CNN) architecture, trained on a huge dataset are used for feature extraction using transfer learning, like VGG16 (Visual Geometry Group 16) [2], ResNet (Residual Network) [3], Inception (GoogLeNet) [4], and Xception [5]. The deeper layers capture high-level features and complex patterns. Features can be extracted by directly fetching the output from one of the layers.

Classifiers experience the "curse of dimensionality" when datasets have a large number of features [6]. Algorithms become more computationally complex as dimensions increase, requiring more resource allocation. The risk of overfitting is magnified, with models prone to capturing noise rather than the underlying patterns in the data [7].

Feature selection stands out as a key strategy for mitigating the challenges posed by the curse of dimensionality, as mentioned above. Feature selection here involves the identification and retention of the most consistent and relevant variables while discarding irrelevant or redundant ones.

The aim is, therefore, to discover a set of features that considerably improve a model's predictive performance. Feature selection methods achieve this by using evaluation criteria.

The task of selecting the optimal feature subset from a dataset is recognized as a computationally challenging problem. This is attributed to the non-deterministic polynomial-time (NP) nature of the problem, signifying that the time required to solve it grows exponentially with the size of the input.

For a dataset having 'n' features, the total count of possible subsets (including the empty set as well as the full set) is given by the power set, comprising 2^n elements.

$$\text{Number of subsets} = 2^n \tag{1}$$

Since it's not feasible to check every possible combination of features, we rely on heuristic approaches to handle this problem more efficiently. Heuristics are like greedy algorithms or metaheuristic optimization techniques [8] that might not always give the absolute best solution but offer practical and quicker ways to solve problems [9].

Particle Swarm Optimization (PSO) algorithm [10] is an evolutionary optimization method inspired by nature. The fundamental idea it is based on is to mimic the collective movement of particles in a search space to discover the best solutions. This work aims to classify actual and synthesized images using the PSO feature selection approach. The suggested method begins by extracting the image features by applying the pre-trained CNN, ResNet50 [10] on the dataset to extract the image features. The PSO evolutionary algorithm is then used to obtain a suitable set of features from the extracted feature set that delivers the best results. Finally, the obtained feature set is used to categorize the images as real or fake by use of classifiers. This article analyzes experiments with a publicly available dataset, named OpenForensics. An ANN and SVM are used as classifiers. The objective function utilizes a Random Forest (RF) ML model to evaluate the fitness of the subset of features used. The results are evaluated based on the evaluation parameters: precision, accuracy, recall, and F1 score.

The following is the arrangement of the paper's sections: Sect. 2 provides an overview of the related work; Sect. 3 contains preliminary information; Sect. 4 elaborates on the feature selection methodology in fake face image classification; Sect. 5 explores the findings after the experimental analysis; and Sect. 6 concludes the paper.

2 Related Work

In this section, a summary has been provided of some relevant research on the application of evolutionary optimization and deep learning to detect deepfakes.

Saraswathi et al. [11] have proposed a method of detection of fake videos using CNN wherein the ResNet model was used to extract facial features, which were then passed through a Long Short-Term Memory (LSTM) classifier with sequential input. The final prediction was made based on CNN architecture and its layers with an overall accuracy of 90.37%. This methodology offered a deep learning framework to identify deepfakes. The models were trained using a database of videos that were taken out of a mixed dataset.

Yadav et al. [12] put forward an approach of Deepfake detection that used LSTM as the deep learning neural network and Inception-ResNet-V2 Network utilizing CNN. The working sequence included pre-processing the dataset, and extraction of features using ResNet architecture [13] followed by evaluation and prediction of the model. This approach leads to an accuracy of 98.37% on training data and 87.06% on validation data.

To detect deepfake images through feature ensemble learning [14], Zhang et al. proposed a heterogeneous methodology. They extracted gray gradients, texture features and spectral features from genuine and deepfake images, then integrated them with a feature vector smoothing process, and finally adopted a back-propagation neural network to train a deepfake detector with feature vectors. Experiments showed that this approach using proGAN, SAGAN, DCGAN, BEGAN, and EBGAN showed better detection results compared to several other deepfake detectors with the feature vector. Experimental results showed that this approach using the models proGAN, SAGAN, DCGAN, BEGAN, and EBGAN achieved better detection accuracy compared with several state-of-the-art deepfake detectors.

Patel et al., worked on improving deep-CNN (D-CNN) algorithm for the detection of deepfakes to increase accuracy [15]. The images were transformed and passed through the D-CNN model. Adam Optimizer as well as binary cross-entropy were utilized to enhance the results of the D-CNN. The given model yielded an accuracy of 98.33% in AttGAN, 99.33% in GDWCT, 99.17% in StarGAN, and 94.67% in StyleGAN2.

Mohiuddin et al. [16] made a hierarchical selection plan (HFS) for deepfake video detection. They extracted both hand-crafted and DL features from the inputs. Subsequently, HFS was employed to select the most effective features. Each stage of this process incorporated a hybrid feature selection approach that combines Gray Wolf Optimization, a population-based metaheuristic model, and the Vortex Search algorithm, a singular solution-based metaheuristic model. The model's performance was assessed across the publicly accessible datasets: FaceForensics++, Deepfake Detection Challenge (DFDC), and Celeb-DF (V2) yielding accuracy scores of 99.35%, 99.16%, and 85.67%, respectively.

Das and Sebastian [17] performed a comparative analysis of non-sequential CNN-based Deepfake Video/images detection strategy along with feature selection analysis. This model collected features utilizing Principal Component Analysis (PCA) for the selection of features and Support Vector Machine (SVM) classification in conjunction with three convolutional models: VGGNet, Efficient16, and ResNet. Accuracy using this model came out to be 96.5%.

Inspiration has been taken from these approaches and methodologies used by researchers in developing systems for detecting deepfake videos and images. This study integrates PSO to enhance the effectiveness of feature extraction on images for deepfake detection. This is part of our effort to increase the efficiency and reliability of deepfake detection systems.

3 Preliminaries

Particle Swarm Optimization (PSO), proposed by Eberhant and Kennedy [18], is a renowned optimization algorithm that is built on the idea of observing the accumulative behavior of decentralized systems. These systems include a group of individuals(agents) collectively aiming to achieve a goal. The agents not only interact with their immediate local environment but also engage in interactions that extend beyond their immediate surroundings. PSO is a heuristic algorithm therefore it does not guarantee finding the global optimum but it is effective in finding good solutions by quickly exploring solution spaces for a wide range of optimization problems.

There are two major equations in the PSO algorithm, velocity and the particle equation. Each particle computes its velocity using the global best and the personal best at a particular position in any iteration.

The velocity equation in PSO is given by:

$$v_{ij}(t+1) = w \cdot v_{ij}(t) + c_1 \cdot r_1 \cdot (p_{best}(t) - x_{ij}(t)) + c_2 \cdot r_2 \cdot (g_{best}(t) - x_{ij}(t)) \tag{2}$$

where $v_{ij}(t+1)$ represents the velocity of the particle i in iteration j at time $t+1$, w is the inertia weight that balances exploration and exploitation, c_1 and c_2 are the acceleration coefficients, r_1 and r_2 are two random numbers within the range [0,1], $x_{ij}(t)$ is the current position of particle i in iteration j at time t.

The particle equation in PSO is given as:

$$x_{ij}(t+1) = x_{ij}(t) + v_{ij}(t+1) \tag{3}$$

where $x_{ij}(t+1)$ is the latest position of particle i in iteration j at time $t+1$, and $v_{ij}(t+1)$ is the velocity of particle i in iteration j at time $t+1$.

Algorithm 1 Particle Swarm Optimization

Initialize:
Initialize the particle positions x_i and velocities v_i for $i = 1, 2, \ldots,$ num_particles
Initialize the individual best positions $p_i = x_i$ and corresponding fitness values $f_{\text{personal\_best},i} = f(x_i)$
Initialize global best position g_{best} and corresponding fitness value $f_{\text{global\_best}}$
for each iteration until convergence **do**
    for each particle $i = 1$ to num_particles **do**
        Update the velocity: $v_i \leftarrow \omega v_i + c_1 r_1 (p_i - x_i) + c_2 r_2 (g_{\text{best}} - x_i)$
        Update the position: $x_i \leftarrow x_i + v_i$
        if $f(x_i) < f_{\text{personal\_best},i}$ **then**
            Update the personal best value: $p_i \leftarrow x_i$, $f_{\text{personal\_best},i} \leftarrow f(x_i)$
        end if
        if $f(x_i) < f_{\text{global\_best}}$ **then**
            Update the global best value: $g_{\text{best}} \leftarrow x_i$, $f_{\text{global\_best}} \leftarrow f(x_i)$
        end if
    end for
end for

Fig. 1 Sample images corresponding to individual classes taken from the OpenForensics dataset

4 Methodology

This section details the proposed methodology.

4.1 Considered Dataset

A publicly accessible deepfake dataset called OpenForensics is taken into consideration for experimental analysis. The labeled images in the dataset are divided into two categories: real and fake. Figure 1 illustrates a few examples of the two classes taken from the same dataset. A total of 12602 images from the dataset were used for this experiment, with 6301 of those images being real and the rest being fake. To further divide the dataset into training and testing datasets, a test size of 0.2 was used (Fig. 2).

4.2 Stages of Implementation

4.2.1 Feature Extraction Using ResNet50

In this paper, ResNet50 [19] has been used for feature extraction. ResNet-50 is a deep CNN architecture that belongs to the ResNet(Residual Network) family. It was proposed by Kaiming et al. [19]. ResNet-50 is particularly known for its depth and the use of residual blocks, which address the vanishing gradient problem and are

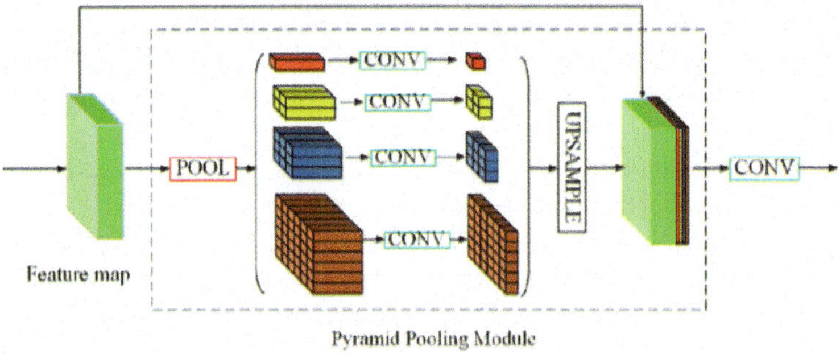

Fig. 2 Diagram illustrating the architecture of the ResNet-50 CNN [18]

used for training very deep neural networks. Often trained on large datasets like ImageNet, it can then be used as a feature extractor in transfer learning [20].

The shape of the features (n_samples, n_features) is the total number of features extracted from each image by the ResNet50 model, where n_samples indicates the number of images in the dataset, and the number of features extracted is denoted by n_features from each image by the ResNet50 model. In this case, it's (n_samples, 2048) because pooling='avg' generates a feature vector of length 2048.

4.2.2 Optimal Feature Subset Selection Using PSO

To reduce the large feature space, PSO [21] evolutionary technique has been used to select the optimal subset of features by evaluating the fitness until termination is met.

Number of particles, num_particles (refer to Algorithm 1) has been initialized to 25 to achieve an optimal outcome. The RF Classifier is utilized for fitness evaluation [22].

Each particle represents a binary array of size (2048), 1 indicating the presence of the *ith* feature in the subset and 0 indicating the absence of the *ith* feature in the subset particle $i = [1_0, 0_1, 1_2, \ldots, 0_{2047}]$ (*Sample Particle*). For each particle, fitness is computed, and subsequently, the velocity and position are updated accordingly. This iterative process persists until the solution converges. For Termination, either iteration exceeds the threshold value or the g_best_fitness value remains the same for >20 consecutive iterations.

4.2.3 Classification

After identifying the best features through PSO, classification has been done to distinguish between genuine and fake images using the following techniques, thereafter, their results have been compared.

1. *Support Vector Machines (SVM)*
   SVM determines which hyperplane, in the context of binary classification, best divides the two classes. To obtain higher-dimensional space from input features, SVM employs a kernel trick that facilitates the process of identifying a separating hyperplane for data classification. The data points are categorized using a non-linear decision boundary and the Sigmoid kernel.
2. *Artificial Neural Network (ANN)*
   ANN [23] is a model that is inspired by the organization and working of the human brain. The two activation functions used are Rectified Linear Unit (ReLU) and Sigmoid, which the selected ANN uses within its internal layers for binary classification.

5 Results

The experiment was simulated on Google Collaboratory using a system with an Intel Core i7-7600U CPU and 16GB of Random Access Memory (RAM).

Figure 4 depicts the performance of the proposed method based on accuracy, precision, F1-score, and recall on the testing dataset, contrasting it with the performance observed without feature selection using PSO. For the testing dataset, the accuracy obtained with the ANN classifier was 79.1%. As for the training dataset, the accuracy and precision obtained were 97.28% and 97.38% respectively.

The performance achieved with the SVM classifier was not as favorable as that observed with ANN. The accuracy in the testing dataset stood at 70.1% when SVM was used as the classifier therefore the results obtained by using ANN have been emphasized. Figure 3 shows the accuracies of the particles, which is nothing but a subset of features or the selected features, over four consecutive generations. The maximum of these accuracies per generation and the previous generations is the global best value, which determines the final selected features. There is improvement in accuracy as the PSO algorithm progresses, thus proving to be an efficient method for feature selection (Table 1).

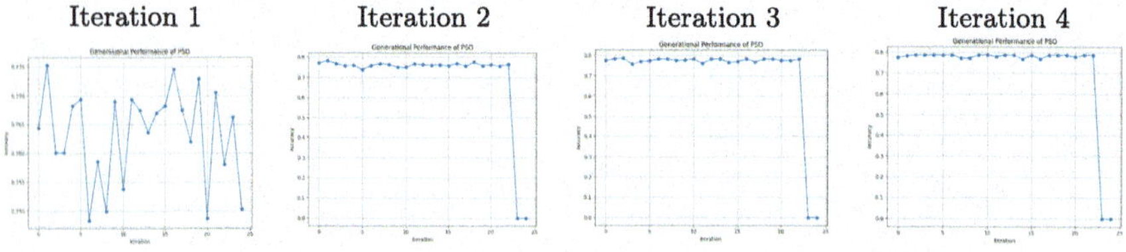

Fig. 3 Accuracy of particles over four consecutive generations from left to right

Table 1 Comparison of selected features for over 10 runs

Parameter	Without PSO	With PSO
Minimum number of features	2048	481
Maximum number of features	2048	1047
Average number of features	2048	904

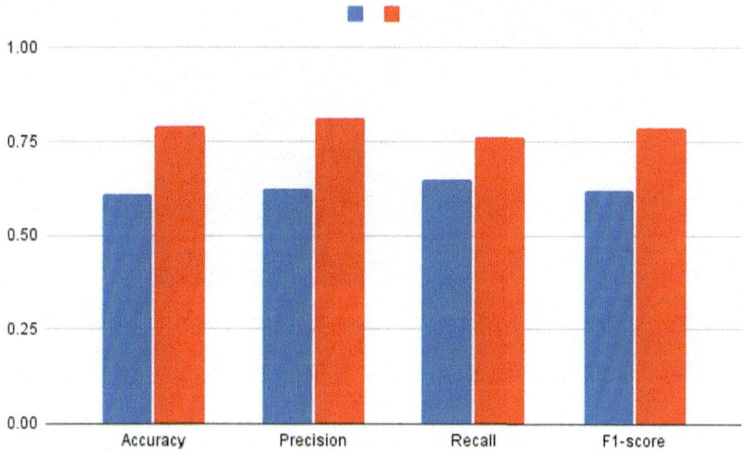

Fig. 4 Improved performance in evaluation metrics after using PSO

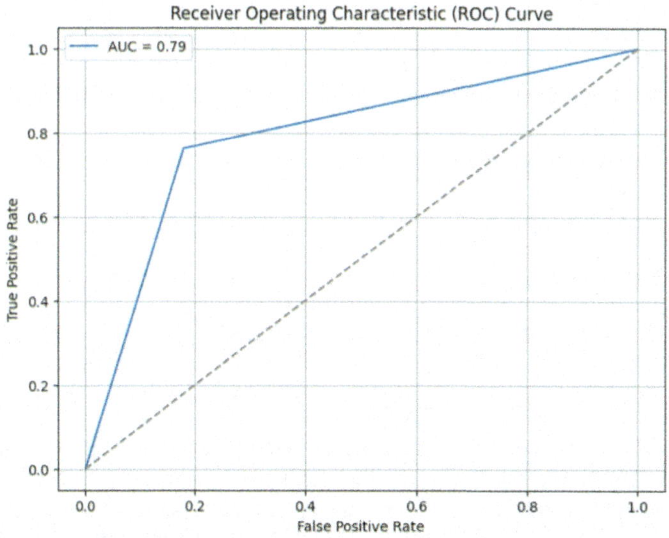

Fig. 5 ROC curve after evaluation using ANN

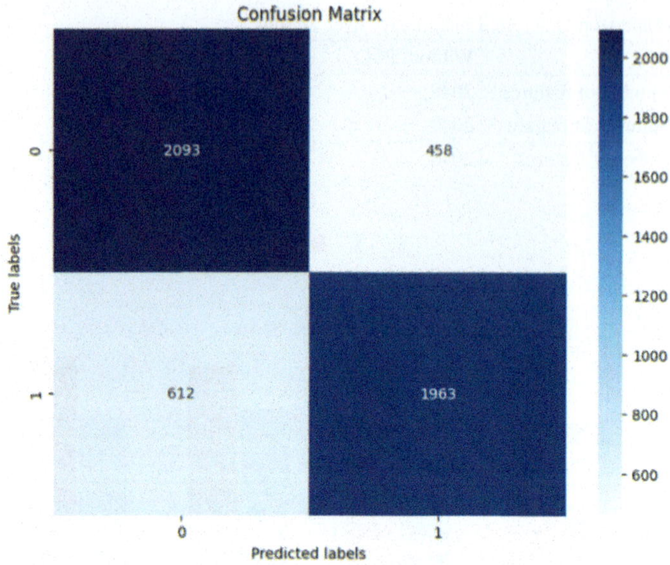

Fig. 6 Confusion matrix after evaluation using ANN

In the worst case, there is a feature percentage reduction of 48.87% and in the average case, the reduction is ≈55.85%. The selected features improved the model's performance from an accuracy of 63.4% (*Without PSO*) to 79.1% with a percentage increase of 24.76.

Further, the performance of the considered method is depicted using the receiver operating characteristic curve (ROC) in Fig. 5, with an AUC of 0.79. This demonstrates a moderate ability to classify deepfakes.

In this experiment, 2093 real images and 1963 fake images were correctly identified by the ANN model. These values represent the true positive and true negative predictions, respectively as illustrated in Fig. 6, using the CM evaluation metric for binary classification.

6 Conclusion

This research paper utilizes particle swarm optimization (PSO), an evolutionary algorithm (EA) feature selection method, for classifying Real-Fake Images. The suggested approach first extracts an image's characteristics via the deep learning model ResNet. The features extracted are optimized using the PSO algorithm to find a subset of features having the approximately best accuracy. Each particle updates its position and velocity at each iteration, assesses the fitness function, and carries out other operations that are typically linear in both the number of particles (N) and the dimensionality (D), PSO's time complexity is estimated to be $O(N \cdot D \cdot T)$ where N: number of particles in the swarm, D: dimensionality of the problem (number of parameters to be optimized), T: maximum number of iterations. In this paper, since RF Classifier is used in the objective function having time complexity $O(B \cdot N \cdot d \cdot F)$, where, B: number of trees in the Forest, d: depth of each tree, F: number of features. If each particle involves evaluating the fitness function using a RF, and you have N particles, the overall time complexity for one iteration would be approximately $O(N \cdot B \cdot N \cdot d \cdot F)$. Ultimately, the ANN classifier uses the elicited feature subset to classify. The experimental evaluations are performed on a public deepfake image dataset, namely OpenForensics which is built on top of Google Open Images. The evaluation parameters considered are classification accuracy, precision, F1 score, recall, AUC - ROC, and confusion matrix. Future work includes exploring and testing with other PSO variants based on parameter tuning techniques such as PSO-Time Varying Acceleration Coefficient (PSO-TVAC), and PSO-Time Varying Inertia Weight (PSO-TVIW) among others. Additional evolutionary algorithms for feature selection can also be used. Different classification algorithms can be used for better accuracy along with ensembling. The effectiveness of the proposed method could also be evaluated on different classifiers, along with its application on diverse image and video datasets.

References

1. Heo YJ, Choi YJ, Lee YW, Kim BG (2021) Deepfake detection scheme based on vision transformer and distillation
2. Marcelino P (2018) Transfer learning from pre-trained models. Towards Sci 10:23
3. Mei Y, Jin H, Yu B, Wu E, Yang K (2021) Visual geometry group-UNet: deep learning ultrasonic image reconstruction for curved parts. J Acoust Soc Am 149(5):2997–3009
4. Li B, Lima D (2021) Facial expression recognition via ResNet-50. Int J Cogn Comput Eng 2:57–64
5. Sam SM, Kamardin K, Sjarif NNA, Mohamed N (2019) Offline signature verification using deep learning convolutional neural network (CNN) architectures googlenet inception-v1 and inception-v3. Procedia Comput Sci 161:475–483
6. Sharma S, Kumar S (2022) The xception model: a potential feature extractor in breast cancer histology images classification. ICT Exp 8(1):101–108
7. Taşkın G, Kaya H, Bruzzone L (2017) Feature selection based on high dimensional model representation for hyperspectral images. IEEE Trans Image Process 26(6):2918–2928
8. Zhang Y, Zheng L, Thing VLL (2017) Automated face swapping and its detection. In: 2017 IEEE 2nd international conference on signal and image processing, IEEE ICSIP, no 17412486, pp 15–19
9. Maier HR, Razavi S, Kapelan Z, Matott LS, Kasprzyk J, Tolson BA (2019) Introductory overview: optimization using evolutionary algorithms and other metaheuristics. Environ Model Softw 114:195–213
10. Rajalaxmi R, Sudharsana R, Preethika D (2023) Gothai: deepfake detection using inception-resnet-v2 network. IEEE ICCMC, no 22888222
11. Vijaya R, Mihir S, Srinivas S, Gadkol A (2022) Detection of synthesized videos using CNN. IEEE ICAISS, no 22540465
12. Yadav P, Jaswal I, Maravi J, Choudhary V, Khanna G, DeepFake Detection using Inception-ResNetV2 and LSTM

13. Mohiuddin S, Sheikh KH, Malakar S, Velásquez JD, Sarkar R (2023) A hierarchical feature selection strategy for deepfake video detection. Latin Am Comput Intell 35(6):9363–9380
14. He K, Zhang X, Ren S, Sun J (2016) Deep residual learning for image recognition, pp 770–778
15. Goodfellow IJ, Pouget-Abadie J, Mirza M, Xu B, Warde-Farley D, Ozair S, Courville A, Bengio Y (2014) Generative adversarial networks
16. Caldelli R, Galteri L, Amerini I, Bimbo AD (2021) Optical flow based CNN for detection of unlearnt deepfake manipulations. Pattern Recogn Lett 146:31–37
17. Li J, Zhang H, Wang C, Wu F (2020) Spaceborne SAR data for regional urban mapping using a robust building extractor
18. ul Huda N, Javed A, Maswadi K, Alhazmi A, Ashraf R (2023) Fake-checker: a fusion of texture features and deep learning for deepfakes detection. Springer (2023)
19. Das A, Sebastian L (2023) A comparative analysis and study of a fast parallel CNN based deepfake video detection model with feature selection (FPC-DFM). IEEE ACCTHPA, no 22888067
20. Mishra M, Srivastava M (2014) A view of artificial neural network
21. Das A, Sebastian L (2023) A comparative analysis and study of a fast parallel CNN based deepfake video detection model with feature selection (FPC-DFM). IEEE ACCTHPA, no 22888067
22. Kunhare N, Tiwari R, Dhar J (2020) Particle swarm optimization and feature selection for intrusion detection system. Sādhanā 45:1–14
23. Zhang Y, Zheng L, Thing VLL (2017) Automated face swapping and its detection. In: 2017 IEEE 2nd international conference on signal and image processing, IEEE ICSIP, no 17412486, pp 15–19

Deep Learning Based Approach for the Detection of Diabetic Retinopathy and Glaucoma

M. Venkata Dasu, K. Purushotham, M. Sravani, A. Pavan Kumar, and T. Siva Kumar

Abstract The diagnosis of diabetic retinopathy has historically depended on specialists analyzing digital retinal images, but these knowledgeable individuals are in short supply. Interest in using computer assistance for monitoring has increased because of this challenge. The goal of our project is to present a novel approach to retinal image analysis. We are concentrating on identifying problems with blood vessels and exudates, which are important markers for comprehending retinal diseases such as diabetes in their early stages and Glaucoma. Our goal in researching these alterations in the blood vessels of the eye is to create a computerized system that will aid in the diagnosis of eye disorders. The goal of this project is to provide a more straightforward and effective method for assessing retinal health. One important objective is to use technology to identify problems more quickly and precisely, particularly when there aren't as many expert observers as possible available. The goal of this research is to develop a trustworthy instrument that will facilitate the early identification and evaluation of retinal disorders, ultimately leading to better outcomes for eye health.

Keywords Diabetic retinopathy · Glaucoma · Deep learning · Segmentation · Gray level co-occurrence matrix (GLCM)

1 Introduction

In the medical field, better patient outcomes are achieved through early disease detection and treatment. Insufficient insulin causes blood glucose levels to rise, which is a condition known as diabetes [1, 2]. The term "eye diseases" includes conditions like diabetic retinopathy, glaucoma, and other conglomerations of eye disorders. People are at risk from all types of ocular disorders. Eye disorders that result in blindness and vision loss are most common in people between the ages of 20 and

M. V. Dasu (✉) · K. Purushotham · M. Sravani · A. P. Kumar · T. S. Kumar
Department of Electronics and Communication Engineering, Annamacharya Institute of Technology and Sciences, Rajampet, Andhra Pradesh, India
e-mail: dassmarri@gmail.com

A. C. Frery et al. (eds.), *Computational Intelligence and Data Analytics*, Lecture Notes on Data Engineering and Communications Technologies 236,
https://doi.org/10.1007/978-981-96-0451-7_17

74. Globally, 2.2 billion people are estimated by the WHO to have one of the many eye conditions [3, 4].

In addition, the 2017, IDF-International Diabetes Federation report claimed that 425 million individuals globally suffer from diabetes. This is expected to rise to 692 million by 2045. It is the fourth most common cause of mortality worldwide and has major medical, social, and financial ramifications that have an impact on public health [5, 6]. The number of Americans estimated to have diabetes in 2020 is 30.5 million (10.5% of the population), and 7.3 million individuals across all age categories remain undiagnosed. Diabetic retinopathy is the most prevalent condition caused by damage to the blood vessels in the retina [7]. Even though diabetic retinopathy is primarily asymptomatic, there are minimal to no clinically apparent microvascular changes and neuronal retinal damage in the early stages of the disease. Consequently, diabetics must have regular eye exams since early detection and management of the condition are crucial [8–10]. A brief description of each of the four types of Diabetic eye disease DED that endanger eye vision can be found in the following subsection. The light-sensitive tissue is located at the back of the eye. Detecting light and relaying that information to the brain is the retina's job. To perceive the objects in the environment, the brain decodes those signals [11].

DR is divided into two phases (a) Advanced DR (b) early DR. Proliferative diabetic retinopathy (PDR) is the term used to describe advanced diabetic retinopathy. In this instance, aberrant blood vessels grow in the retina as a result of the damaged blood vessels leaking the vitreous, a translucent jelly-like fluid that fills the centre of the eye. Because the newly developed blood vessels are blocking the fluid's usual flow, pressure might accumulate inside the eyeball. By harming the optic nerve, which transmits visual information from the eye to the brain, this may result in glaucoma [12].

NPDR-Non-proliferative diabetic retinopathy is the word used to depict the condition in which new blood vessels do not form (proliferate) in the initial stages of DR. The retina's internal blood vessel walls deteriorate as a result of NPDR. The surfaces of the smaller veins give rise to smaller bulges called microaneurysms, which frequently leak fluid and blood into the eye. Large retinal vessels also start to enlarge and alter in diameter. As more blood vessels get occluded, NPDR progresses from mild to severe. Depending on their severity, the retina's nerve fibres may begin to bulge. The disorder known as macular edema, in which the macular portion of the retina often expands, requires medical attention. NPDR is divided into three stages: mild, moderate, and severe [13].

The following are the four stages of diabetic retinopathy:

A. No DR (normal): Diabetes has no effect on the eyes at this stage. Thus, the eye is in good condition.
B. Mild DR: It is the DR's initial stage. Swelling in the tiny blood vessel regions that harm the optic nerve is a hallmark of this stage.
C. Moderate DR: It is also referred to as mild non-proliferative retinopathy, and it is the second stage of DR that results in partial blockage of the retina's blood veins.

D. Severe DR Severe non-proliferative retinopathy, the term used to describe this third stage, results in further obstruction of blood vessels. These affect the parts of the retina that aren't getting enough blood flow.

The retina cannot support the growth of new blood vessels, because there is sufficient blood flow, the damaged blood vessels cannot be replaced by new ones.

Proliferative DR (PDR) can eventually lead to blindness because of the development of blood vessels in the retina because of blood vessel damage caused by diabetes, which badly affects the retina in non-proliferative DR (NPDR) [14].

Machine learning technologies have made a wide range of computer-assisted solutions for the automatic assessment and classification of diabetic retinal health possible. The DR identification procedure uses a variety of characteristic extraction techniques to retrieve pertinent data from the input fundus images. The unique extractor, which is made by hand, needs to be able to adapt to changes in the visual characteristics of different lesions and endure variations in the diabetic retinopathy's state [15, 16].

The eye disease known as glaucoma affects the optic nerve, which is responsible for sending signals from the eye to the brain. The main cause of glaucoma is the eye's incapacity to discharge waste fluid. The optic nerve is harmed by intraocular pressure, which is brought on by excess fluid accumulation. Ten to fifteen percent of those with glaucoma may lose one or both eyes and glaucoma accounts for 10.6–13.5% of blindness in Europe [17, 18]. Approximately 50% of glaucoma patients are misdiagnosed because the ailment doesn't manifest symptoms until considerable damage to the retinal nerve fibre layer has already happened [19].

This research paper is organized into 5 sections. Section 2 reviews relevant literature on diabetic retinopathy disease. This section provides a thorough analysis and discussion of the research and literature currently available on the topic. The primary content of this work is presented in Sect. 3, which provides a detailed introduction to the suggested technique. Section 4 covers the simulation results and finally, conclusion and future scope are discussed in Sect. 5.

2 Literature Review

There are numerous methods for detecting diabetic retinopathy, including the following ones.

In 2023, Aurangzeb et al. [20] created a framework for the methodical creation of AI-enabled glaucoma and diabetic retinopathy diagnostic tools. It determines the optimal parameters for the application of intelligent evolution algorithms to improve the contrast of images of the retinal fundus. Additionally, one drawback of this research is that the suggested deep learning models might not be able to reliably identify extremely thin blood vessels in retinal images, a crucial component in the diagnosis of some eye conditions.

In 2023, Podnar et al. [21] A model developed for a diabetic retinopathy screening programme in Slovenia based on the Relative Significance of Glaucoma-Referral Indicators in Retinal Images: A Cross-Sectional Study. In a cross-sectional study, retinal images in the DRSP were used to accurately refer patients with glaucoma. To validate this, however, more research with a larger sample size is required.

In 2022, Sheng et al. [22] examined an overview of artificial intelligence in diabetic retinopathy and other ocular diseases. A diabetic condition that affects the eyes is called diabetic retinopathy. Diabetic retinopathy results from injury to the blood arteries that supply the retina, the light-sensitive tissue located at the back of the eye. The range of ophthalmic images that can be used for AI analysis goes beyond colour fundus photography. Additionally, despite the encouraging advancements in AI technology, there are still certain difficulties in modern clinical practice. These developments will have a significant impact on clinical care in the future.

In 2021, Tamim et al. [23] suggested a method for accurately diagnosing glaucoma and diabetic retinopathy through retinal fundus images. The decision tree (DT) provides the highest accuracy built on hybrid features and genetic algorithms, with 96.67% for DR, 98.42% for glaucoma, and 96.67% for normal. However, if the image simultaneously exhibits pathological signs of glaucoma and DR, the system will be limited.

In 2023, Arias-Serrano et al. [24] presented "Artificial intelligence-based glaucoma and diabetic retinopathy detection using MATLAB-retrained AlexNet convolutional neural network." It is especially crucial to screen regularly for these diseases to detect them early and stop their progression. However, future developments will be necessary for users who are not proficient in programming languages.

In 2022, Mahapatra et al. [25] provided a unique framework for segmenting retinal vessels utilizing adaptive weighted spatial FCM and an optimally optimized Frangi filter. An efficient method for segmenting the retinal vessels, but the suggested research has a significant flaw in that its sensitivity value is low.

In 2019, Ali et al. [26] examined the Using CUDA Deep Neural Network to Identify Diabetic Retinopathy from Retinal Images. DR will be detected earlier and treated more quickly. However, the accuracy is low.

3 Methodology

The proposed model's block diagram is displayed in Fig. 1.

A. **Acquisition of Input Images**

The input images are gathered from Kaggle, constituting a collection of images from a public database. Kaggle, a platform hosting various datasets, provides a valuable resource for researchers, developers, and data enthusiasts to access and utilize diverse image datasets for tasks such as machine learning, computer vision, and deep learning projects.

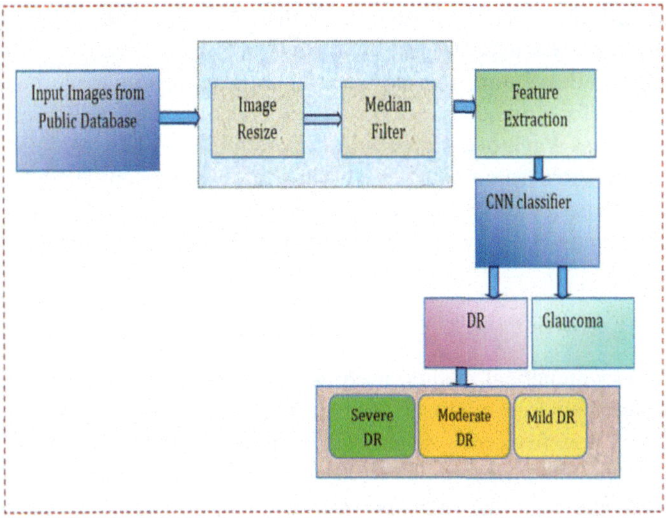

Fig. 1 Proposed block diagram

B. **Pre-processing**

Pre-processing methods like resizing and median filtering are applied to enhance image quality and facilitate more effective analysis.

- Resizing is an essential component of Image Processing. It ensures standardized dimensions for consistent processing.
- Median Filter: Median filtering helps reduce noise and enhance relevant features. The median filter is a type of filter that is used to eliminate noise from signals and pictures. The median filter holds great significance in the field of image processing due to its well-known ability to preserve edges while reducing noise. The first job of the filter is to go through each piece of input data and use the "window" strategy, which is a median function, to filter out every entry. The window is usually slightly more complex across the higher-dimensional signals. The number of windows determines the number of medians, which are divided into even and odd groups.

C. **Otsu Segmentation**

A thresholding method used in image segmentation is called Otsu's method. By maximizing the variance between two classes of pixels in an image, it automatically determines the best threshold and frequently separates the foreground from the background. This aids in binary image segmentation, in which intensity values are used to classify pixels as either background or object [27]. The important steps of Otsu segmentation approach are

- **Histogram Computation**: The distribution of pixel intensities in the image is shown by the histogram.

- **Normalization**: Adjust the histogram so that each bin value corresponds to the likelihood that each intensity level will occur.
- **Class Probabilities and Means**: Determine the means (average intensity) and probabilities (foreground and background) for each class for each threshold value.
- **Between-Class Variance Calculation**: Determine the between-class variance for each threshold value using the probabilities and means determined in the preceding step. The difference in distance between the two classes is measured by the between-class variance. Maximizing this variance is the goal of Otsu's approach.

D. **Feature Extraction (GLCM)**

The GLCM is a numerical method used in computer vision and image processing to analyse the spatial correlations between pixel values in an image. It provides information on how frequently different pixel intensity combinations show up in a specific spatial arrangement. In this work, GLCM is used to extract image features. The special relationship that forms between two pixels in an image that are spaced apart by a predetermined amount is exploited by GLCM. Pixel value pair occurrences in predefined spatial patterns are used to build the GLCM. There are several steps in this computing process. The notation $S(k. l, t, \theta)$ represents the Gray Level Co-Occurrence Matrix (GLCM), where k and l are the pixel values, t is the distance between them, and θ is the orientation. Several parameters can be chosen for analysis depending on your specific needs and the characteristics of the pictures [28].

Several statistical attributes, such as contrast, energy, homogeneity, and correlation, were extracted using this technique. Using the GLCM, a variety of features are retrieved from the images, including contrast, energy, and correlation. Particularly in fields like medical imaging, these properties are helpful for texture characterization, picture processing, and classification.

E. **Classification**

The proposed article aims to achieve accurate classification in the final stage of our work by using Convolutional Neural Network (CNN).

F. **Convolutional Neural Network (CNN)**

One kind of deep neural network that is intended for processing structured grid data, like images, is the convolution neural network (CNN). It is made up of several layers, such as fully connected, pooling, and convolution layers. The architecture of the proposed CNN classifier is shown in Fig. 2.

- **Convolution Layers**: These layers process input data using convolution operations. The network can identify features because convolution filters teach it the spatial hierarchies of patterns.
- **Pooling Layers**: By down sampling, these layers minimize the spatial dimensions of the input. It's common practice to use average or maximum pooling to preserve important data. Layers that are fully connected link each neuron in one layer to

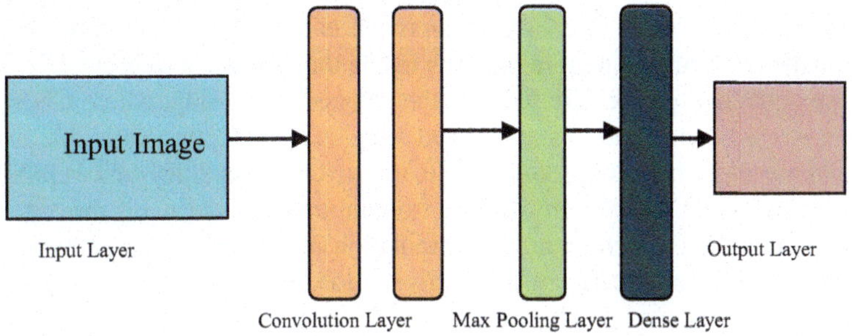

Fig. 2 CNN architecture

every other layer's neuron. They usually occur at the end of the network and record high-level information.

- **Activation Functions**: By introducing non-linearity, non-linear activation functions such as Rectified Linear Unit (RLU) enable the network to learn intricate relationships.
- **Layers of Normalization**: Methods such as Batch Normalization aid in accelerating and stabilizing the training process.

G. **Statistical Parameters**

An effective method for assessing a classifier's effectiveness in binary classification is the confusion matrix. The sorts of errors generated are revealed by the matrix, which neatly separates the actual classes from the classes the classifier predicted. There are typically four entries in the confusion matrix.

- True positive (TP): True positives (TP) are instances that fall inside the positive class and are correctly classified as positive by the classifier.
- True Negative (TN): cases that, despite the classifier's accurate classification of them as negative, belong to the negative class.
- False Positives: cases that belong in the negative class but were inadvertently recognized as positive by the classifier.
- False Negative (FN): Misclassified cases that fall within the positive category.

The statistical parameters such as accuracy, sensitivity, and specificity are used for assessing the performance of the proposed method.

4 Results and Discussions

Input images from public database (Kaggle database) were collected and applied as input to the pre-processing techniques where the image is resized for further processing and the resized image is given as input to the median filter to remove noise in the input image. The GLCM approach is then used for feature extraction, allowing the capturing of texture patterns inside the images.

These collected features are fed into the categorization step, where CNN model is trained to recognize patterns associated with Diabetic Retinopathy disease. This comprehensive methodology attempts to improve the accuracy and reliability of identifying Diabetic Retinopathy disease from input images by using the capabilities of CNN classification technique. The simulation results are obtained for mild and serious DR stages and are discussed in this section.

Mild DR Stage

Figure 3 depicts the input image and its median filtered image. The features are obtained from GLCM and these features are fed to CNN classifier to identify stages of area disease image.

Figure 4 shows a clustered image which simplifies the classification process. Figure 5 illustrates the classification output. The marked region shows the affected part of DR and the area is calculated and given as 30.8553 mm^2. The detected part is extracted and is shown in Fig. 6.

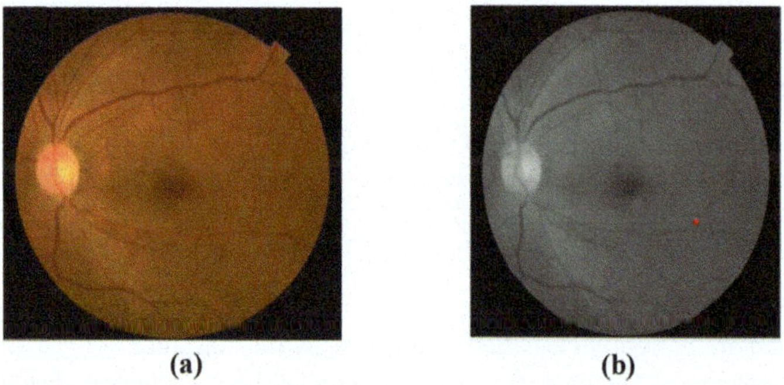

Fig. 3 **a** Input image, **b** median filter image

Fig. 4 Clustered image

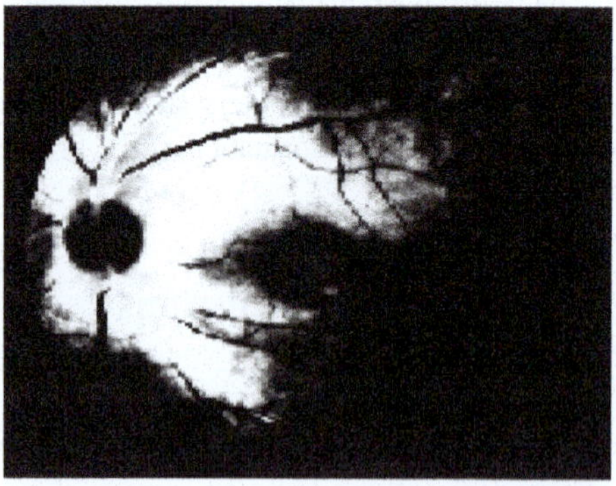

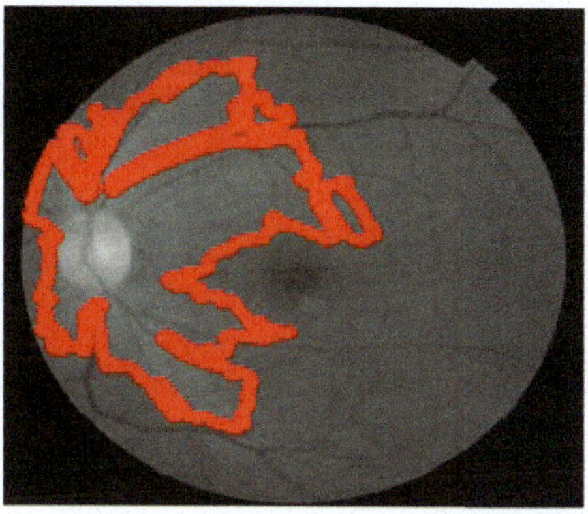

Fig. 5 Localization area: 30.8553 mm^2

Fig. 6 DR detected region in the input image

Severe Stage

Figure 7 portrays the input image alongside its corresponding median filtered image. The median filtering process likely helps to reduce noise and enhance the clarity of the image, making it more suitable for feature extraction.

Features are then extracted from the processed images using Gray Level Co-occurrence Matrix (GLCM) analysis. GLCM is a method commonly used in image processing to capture texture information by analyzing the spatial relationships of pixel intensities. Features extracted from GLCM may include contrast, energy, homogeneity, and correlation among others, which are indicative of different textural patterns present in the image.

These extracted features serve as input to a Convolutional Neural Network (CNN) classifier. CNNs are well-suited for image classification tasks due to their ability to automatically learn hierarchical features directly from the raw pixel data. By training the CNN classifier on the extracted features, the model can effectively identify and classify the stages of the area disease depicted in the input image. This process enables

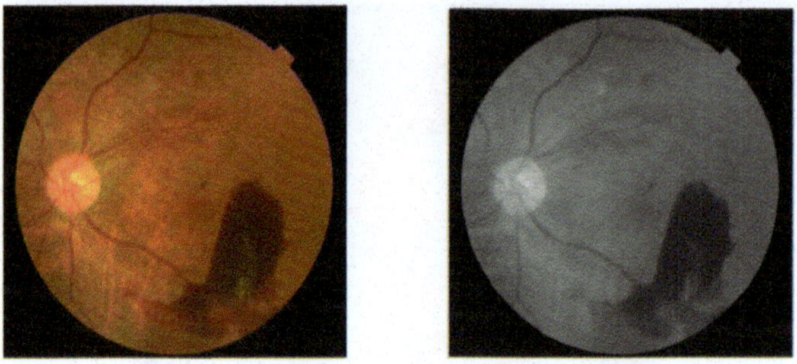

(a) (b)

Fig. 7 **a** Input image, **b** median filter image

automated diagnosis and classification of medical conditions based on image data, aiding healthcare professionals in making accurate and timely assessments.

In the context provided, it seems that Figs. 8 depicts clustered images that aid in simplifying the classification process, likely related to diagnosing a medical condition such as diabetic retinopathy (DR). Figure 9 specifically illustrates the output of the classification process, with a marked region indicating the affected part of DR. The area of this affected region is quantified and provided as 50.2343 mm^2. In mild DR stage, the affected region is 30.8553 mm^2 but in severe stage, the affected area is increased to 50.2343 mm^2. This information is valuable for understanding the extent and severity of the condition, enabling healthcare professionals to make informed decisions regarding treatment and management strategies. The detected part is extracted and is shown in Fig. 10. The graphical plot of obtained parameters is shown in Fig. 11. Table 1 displays the statistical properties of the suggested methodology for mild and severe DR symptoms, including Sensitivity, Specificity, Accuracy, and Peak Signal to Noise Ratio values.

Fig. 8 Clustered image

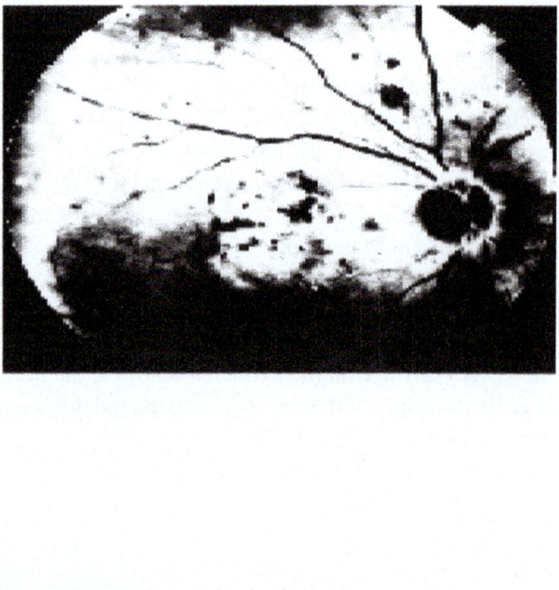

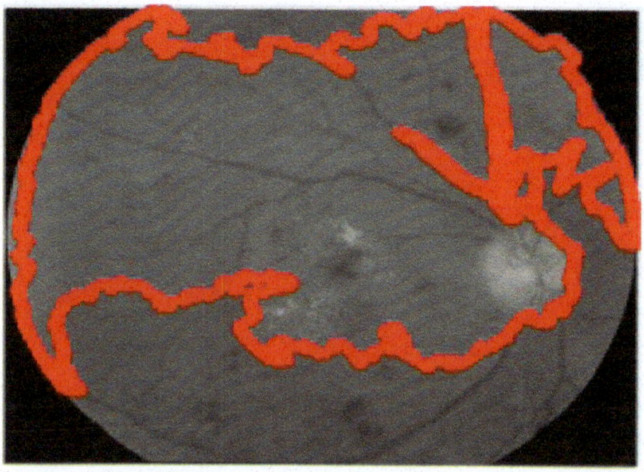

Fig. 9 Localization area: 50.2343 m

Fig. 10 DR detected region in the input image

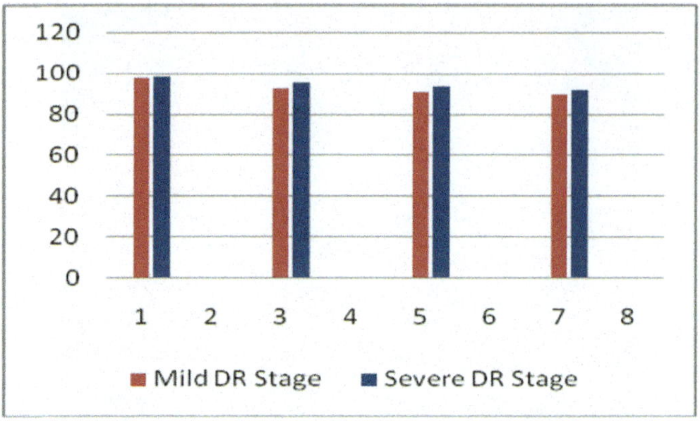

Fig. 11 Statistical parameters in graphical representation

Table 1 Statistical parameters

Sl. No.	Parameters	Mild DR stage (%)	Severe DR stage (%)
1	Accuracy	97.60	99.18
2	Sensitivity	92.43	95.73
3	Specificity	90.72	93.84
4	PSNR	89.82	91.37

5 Conclusion

The proposed method for detecting abnormalities in retinal images leverages input images obtained from a public database, employing pre-processing techniques to enhance their quality. Subsequently, the pre-processed images undergo segmentation using the OTSU method, followed by feature extraction. These features encompass contrasted Gray Level Co-occurrence Matrix (GLCM) features, capturing textural information crucial for abnormality identification.

The extracted features are then inputted into a Convolutional Neural Network (CNN) for the purpose of abnormality identification. Through this process, the method achieves impressive statistical parameters, with a sensitivity of 95.73, specificity of 93.84, and an overall accuracy of 99.18%. These metrics underscore the robust performance of the proposed method in accurately detecting retinal abnormalities. The classification accuracy of the proposed method is better than the existing methods in the literature survey.

Notably, the achieved accuracy of 99.18% surpasses that of existing methods discussed in the literature, indicating the superior efficacy of the proposed approach in retinal abnormality detection. This suggests promising prospects for the application of this method in clinical settings, potentially enhancing the early detection and management of retinal diseases.

References

1. Alyoubi WL, Shalash WM, Abulkhair MF (2020) Diabetic retinopathy detection through deep learning techniques: a review. Inform Med Unlocked 20:100377. ISSN: 2352-9148. https://doi.org/10.1016/j.imu.2020.100377
2. Gómez-Valverde JJ, Antón A, Fatti G (2019) Automatic glaucoma classification using color fundus images based on convolutional neural networks and transfer learning. Biomed Opt Express 10(2):892–913
3. Kamal MM, Shanto MHI, Hossan MM, Hasnat MA, Sultana S, Biswas M (2022) A comprehensive review of the diabetic retinopathy, glaucoma, and strabismus detection techniques based on machine learning and deep learning. Eur J Med Health Sci 4(2):24–40. https://doi.org/10.34104/ejmhs.022.024040
4. Forouhi NG, Wareham NJ (2010) Epidemiology of diabetes. Medicine 38(11):602–606
5. Tabish SA (2007) Is diabetes becoming the biggest epidemic of the twenty-first century? Int J Health Sci 1(2):V–VIII
6. Hacisoftaoglu RE, Karakaya M, Sallam AB (2020) Deep learning frameworks for diabetic retinopathy detection with smartphone-based retinal imaging systems. Pattern Recogn Lett. https://doi.org/10.1016/j.patrec.2020.04.009
7. Yang W, Lu J, Weng J, Jia W, Ji L, Xiao J, Shan Z, Liu J, Tian H, Ji Q (2010) Prevalence of diabetes among men and women in China. N Engl J Med 362(12):1090–1101
8. Hamid S, Safi S, Hafezi-Moghadam A, Hamid A (2018) Early detection of diabetic retinopathy. Surv Ophthalmol 63(5):601–608
9. Tsiknakis N, Theodoropoulos D, Manikis G, Ktistakis E, Boutsora O, Berto A, Scarpa F, Scarpa A, Fotiadis DI, Marias K (2021) Deep learning for diabetic retinopathy detection and classification based on fundus images: a review. Comput Biol Med 135:104599. ISSN: 0010-4825. https://doi.org/10.1016/j.compbiomed.2021.104599
10. Kolb H (2011). Simple anatomy of there Tina by Helga Kolb. https://www.ncbi.nlm.nih.gov/books/NBK11533/. Accessed 31 Jan 2012
11. Early Treatment Diabetic Retinopathy Study Research Group (1991) Grading diabetic retinopathy from stereoscopic color fundus photographs—an extension of the modified Airlie house classification: ETDRS report number 10. Ophthalmology 98(5):786–806
12. Sarki R, Ahmed K, Wang H, Zhang Y (2020) Automatic detection of diabetic eye disease through deep learning using fundus images: a survey. IEEE Access 8:151133–151149. https://doi.org/10.1109/ACCESS.2020.3015258
13. Pathak KC, Shah RB, Tharakan RR, Patel BN, Jariwala DC (2021) Diabetic retinopathy diagnosis and categorization using deep learning—a review. In: 2021 5th international conference on intelligent computing and control systems (ICICCS), Madurai, India, pp 1063–1069. https://doi.org/10.1109/ICICCS51141.2021.9432312
14. Gunasekaran K, Pitchai R, Chaitanya GK, Selvaraj D, Annie Sheryl S, Almoallim HS, Alharbi SA, Raghavan SS, Tesemma BG (2022) A deep learning framework for earlier prediction of diabetic retinopathy from fundus photographs. BioMed Res Int 2022:Article ID 3163496, 15 pages. https://doi.org/10.1155/2022/3163496
15. Xiao D, Bhuiyan A, Frost S, Vignarajan J, Tay-Kearney M-L, Kanagasingam Y (2019) Major automatic diabetic retinopathy screening systems and related core algorithms: a review. Mach Vis Appl 30(3):423–446
16. Quigley HA (1996) Number of people with glaucoma worldwide. Br J Ophthalmol 80:389–393
17. Bourne RR, Jonas JB, Flaxman SR, Keeffe J, Leasher J, Naidoo K, Parodi MB, Pesudovs K, Price H, White RA et al (2014) Prevalence and causes of vision loss in high-income countries and Eastern and Central Europe: 1990–2010. Br J Ophthalmol 98:629–638
18. Cedrone C, Mancino R, Cerulli A, Cesareo M, Nucci C (2008) Epidemiology of primary glaucoma: prevalence, incidence, and blinding effects. Prog Brain Res 173:3–14
19. Aurangzeb K, Alharthi RS, Haider SI, Alhussein M (2023) Systematic development of AI-enabled diagnostic systems for glaucoma and diabetic retinopathy. IEEE Access 11:105069–105081. https://doi.org/10.1109/ACCESS.2023.3317348

20. Podnar B, Albreht T, Cvenkel B (2023) Relative importance of glaucoma-referral indicators in retinal images in a diabetic retinopathy screening programme in Slovenia: a cross-sectional study. Medicina 59:1441. https://doi.org/10.3390/medicina59081441
21. Sheng B, Chen X, Li T, Ma T, Yang Y, Bi L, Zhang X (2022) An overview of artificial intelligence in diabetic retinopathy and other ocular diseases. Front Public Health 10:971943. https://doi.org/10.3389/fpubh.2022.971943
22. Tamim N, Elshrkawey M, Nassar H (2021) Accurate diagnosis of diabetic retinopathy and glaucoma using retinal fundus images based on hybrid features and genetic algorithm. Appl Sci 11:6178. https://doi.org/10.3390/app11136178
23. Arias-Serrano I, Velásquez-López PA, Avila-Briones LN et al (2023) Artificial intelligence-based glaucoma and diabetic retinopathy detection using MATLAB—retrained AlexNet convolutional neural network [version 1; peer review: 2 approved with reservations]. F1000Research 12:14. https://doi.org/10.12688/f1000research.122288.1
24. Mahapatra S, Agrawal S, Mishro PK, Pachori RB (2022) A novel framework for retinal vessel segmentation using optimal improved Frangi filter and adaptive weighted spatial FCM. Comput Biol Med 147:105770. ISSN: 0010-4825. https://doi.org/10.1016/j.compbiomed.2022.105770
25. Ali A, Qadri S, Mashwani WK, Kumam W, Kumam P, Naeem S, Goktas A, Jamal F, Chesneau C, Anam S, Sulaiman M (2020) Machine learning based automated segmentation and hybrid feature analysis for diabetic retinopathy classification using fundus image. Entropy 22(5):567. https://doi.org/10.3390/e22050567
26. Furht B (eds) Encyclopedia of multimedia. Springer, Boston, MA. https://doi.org/10.1007/978-0-387-78414-4_305
27. Otsu N (1979) A threshold selection method from gray-level histograms. IEEE Trans Syst Man Cybern 9(1):62–66
28. Gonzalez RC, Woods RE (2017) Digital image processing. Pearson Education

Deep Learning Based Speckle Noise Removal in Color Images Using Supervised Denoising Convolutional Neural Network

M. Ravichandran, T. Jerry Alexander, P. Ganesan, L. M. I. Leo Joseph, V. Elamaran, S. Jency, and G. Sajiv

Abstract Speckle noise offers a considerable challenge in the turf of digital image processing, exclusively in color images due to its complicated and unexpected nature. Conventional techniques for removing speckle noise frequently fail to properly reduce noise while maintaining image features. Deep learning methods have demonstrated encouraging performance in a number of image restoration challenges in recent years. In this work, we present an innovative method using a supervised denoising convolutional neural network (SDCNN) for speckle noise reduction in color images. The presented SDCNN successfully distinguishes between noise and actual image information by utilizing deep learning to extract complex patterns and characteristics from noisy input imagery. To improve the network's competence, the

M. Ravichandran
Department of CSE, Vel Tech Rangarajan Dr. Sagunthala R&D Institute of Science and Technology, Chennai, India
e-mail: drmravichandran@veltech.edu.in

T. J. Alexander
CSE department, Alpha College of Engineering, Chennai, India
e-mail: csehod@alphagroup.edu

P. Ganesan (✉)
Department of ECE, Vidya Jyothi Institute of Technology, Hyderabad, India
e-mail: gganeshnathan@gmail.com

L. M. I. L. Joseph
Department of ECE, SR University, Warangal, Telangana, India
e-mail: leojoseph.lmi@sru.edu.in

V. Elamaran
Department of ECE, School of EEE, SASTRA Deemed University, Thanjavur, India
e-mail: elamaran@ece.sastra.edu

S. Jency · G. Sajiv
Saveetha School of Engineering, SIMATS, Chennai, India
e-mail: jencys9026.sse@saveetha.com

G. Sajiv
e-mail: sajivg2003.sse@saveetha.com

A. C. Frery et al. (eds.), *Computational Intelligence and Data Analytics*, Lecture Notes on Data Engineering and Communications Technologies 236,
https://doi.org/10.1007/978-981-96-0451-7_18

suggested method develops a multi-scale structure that allows the model to make decisions at various dimensions by capturing contextual information at both the local and global levels. In addition, the proposed method utilizes residual learning to ease the training procedure and mitigate the vanishing gradient issue. The suggested SDCNN offers a reliable and effective method for eliminating speckle noise from color images. This makes it potentially useful in fields like underwater imaging, satellite imaging, medical imaging, and other fields where precise image reconstruction is crucial.

Keywords Speckle noise · SDCNN · Deep learning · PSNR · SSIM · BRISQUE · NIQE · PIQE

1 Introduction

Noise is an unavoidable artifact in digital image processing that degrades image quality and presents serious problems in a number of applications, including photography, remote sensing, medical imaging, and surveillance [1, 2]. Speckle noise is one of the most annoying types of noise; it is especially common in images taken by satellite imaging systems, underwater photography, ultrasound equipment, and synthetic aperture radar (SAR) [3]. Multiplicative noise, or speckle noise, introduces granular patterns that deteriorate image quality and make it more difficult to analyze and comprehend the data [4]. Speckle noise reduction is an important topic for many applications, from surveillance and environmental monitoring to medical diagnostics. Speckle noise mitigation has traditionally been accomplished using a range of filtering methods and model-based approaches. Among these techniques are wavelet-based algorithms, Gaussian filtering, and median filtering, which work on small regions (local neighborhoods) to smooth out noise while trying to conserve image patterns. Although somewhat successful, conventional techniques frequently fail to achieve a compromise between feature retention and noise suppression, resulting in either excessively smoothed pictures or inadequate noise reduction [5].

The discipline of image processing has seen a revolution with the introduction of deep learning, which presents potential answers to the problems associated with noise reduction. Above and beyond the enactment of conventional approaches, CNN in particular has shown impressive skills in learning complicated mappings from noisy images to clean targets. CNN-based methods have demonstrated improved denoising performance by capturing complex picture features and exhibiting more flexibility to noise characteristics through the use of deep learning [6]. In order to remove speckle noise from color images, this research focuses on using deep learning approaches, more especially supervised denoising convolutional neural networks (SDCNN). The objective is to devise a consistent and effectual technique that can precisely eliminate speckle noise while keeping important texture and information in the image. Our methodology seeks to learn directly from data, in contrast to conventional approaches that rely on manually designed features and heuristics, allowing for more flexible and effective noise reduction.

The primary contribution of the proposed method is the implementation of SDCNN architecture made especially for the elimination of speckle noise from color images. This SDCNN learns from a huge dataset of clean and noisy image pairings using a supervised learning framework, which enables it to distinguish complex patterns related to speckle noise from real image information. This supervised approach enables the network to generalize well to unseen data and various noise distributions, leading to improved denoising performance. Additionally, we use state-of-the-art methods like residual learning and multi-scale processing into our SDCNN architecture to improve its denoising performance. We find that our model efficiently suppresses speckle noise at different scales while maintaining image structures and features by taking into account both local and global contextual information. By reducing the vanishing gradient issue and quickening convergence, the residual learning technique makes training easier and boosts the network's overall performance. The proposed method quantifies and evaluates the denoising performance using quality parameters, proving the efficiency and superiority of the proposed work.

The remaining of the article is structured as follows: We examine relevant methods in deep learning based image denoising and speckle noise reduction in Sect. 2. In Sect. 3, the design and technique of our suggested SDCNN are presented. The outcome of the proposed work is debated in Sect. 4. Finally, we conclude with a discussion of our study's results, possible uses, and directions for further research in Sect. 5.

2 Related Work

In image processing, speckle noise reduction is a crucial step, especially for applications like underwater photography, remote sensing, and medical imaging. This topic has been attempted through various strategies all over the years, from more complex model-based solutions to more conventional filtering procedures. This section discusses the available techniques for removing speckle noise and offers an analysis of their advantages, disadvantages, and restrictions. Figure 1 illustrates the various techniques for image denoising.

One of the most popular methods for eliminating speckle noise is median filtering. It works by substituting the adjoining pixel's median value within a predetermined window for each pixel's value. Median filtering reduces sharpness in images by blurring edges and tiny details, despite being easy to use and computationally efficient. In order to reduce noise, Gaussian filtering convolves the image using a Gaussian kernel. Due to its non-Gaussian distribution, it might not be the best option for eliminating speckle noise, even though it works well for Gaussian noise. Particularly in textured areas, Gaussian filtering may result in over-smoothing and a loss of image features. The statistical technique known as Wiener filtering seeks to reduce the MSE

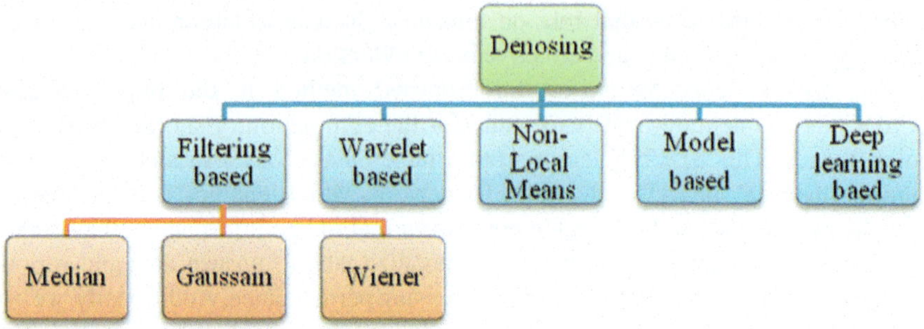

Fig. 1 Techniques for image denoising

between the test and filtered images. It works in the frequency domain and necessitates an understanding of SNR & the noise power spectrum. While Wiener filtering can work well in some situations, non-stationary noise like speckles might cause it to perform badly [7].

Wavelet-based techniques eliminate noise while keeping image features by making use of the multi-resolution capabilities of wavelet transformations. Generally, these techniques break down the image into distinct frequency bands, separately denoise each band, and then put the denoised image back together. Noise reduction often uses wavelet shrinking techniques like soft and hard thresholding [8]. Wavelet-based techniques can provide good results in some situations, but they can also have ringing effects and artifacts, especially when noise levels are high. Non-local means denoising is a patch-based technique that estimates the underlying clean image by making use of similarities between patches in the image. To estimate the noise-free pixel values, it computes weighted averages of comparable patches inside a search window. Speckle noise and other non-Gaussian and Gaussian noise may be effectively eliminated with NLM denoising. It may not be appropriate for real-time applications or massive image databases due to its processing complexity. Model-based techniques present the speckle noise removal issue as an optimization assignment by minimizing a predetermined cost function, with the objective of estimating the clean image. To regularize the optimization process, these algorithms frequently include previous knowledge about noise distribution and picture attributes. Markov random field (MRF) modeling, total variation (TV) regularization, and maximum a posteriori (MAP) estimation are a few examples. Model-based techniques may provide amazing outcomes, particularly when precise noise models are available. However, they might not generalize well to other noise distributions and picture formats, and they could need human parameter adjustment [9, 10].

Recently, deep learning based methods have become effective tools for eliminating speckle noise. In particular, CNNs have revealed amazing abilities to learn intricate mappings from noisy test images to clean destinations. These techniques usually entail training a CNN with a sizable dataset of matched noisy and clean pictures, whereby the network is trained to automatically denoise incoming photos. Convolutional neural networks with supervised denoising (SDCNNs) use supervised learning frameworks; the network is trained on clean, ground truth pictures to direct the denoising process. Compared to conventional techniques, SDCNNs achieve better denoising performance because they are able to retain picture information while capturing complex patterns linked to speckle noise (Table 1).

Table 1 Comparison of various denoising methods

Method	Description	Advantages	Disadvantages
Traditional filtering techniques	Removes noise using simple spatial filters such as median, mean, and Gaussian filters	• Simple and computationally efficient • Easy to implement	• Tends to blur edges and fine details • May result in over-smoothing
Wavelet-based methods	The data disintegrated as multiple bands using wavelet techniques like wavelet shrinking	• Preserves image details and structures • Effective for various noise types	• May suffer from artifacts and ringing effects • Complexity increases with higher noise levels
Non-local means denoising	Uses similarities between image patches (weighted averages) to compute the pristine image	• Removes noise while preserving image structures • Effective for diverse noise types	• Computationally intensive • Limited scalability to large-scale datasets
Model-based approaches	Uses optimization problem based on past information and incorporates MAP, TV, and other features	• Robust against noise variations • Can handle complex noise distributions	• Manual parameter tuning may be required • May require accurate noise models
Deep learning based approaches	Create complex mappings from noisy input pictures to clean targets. Train with a huge dataset and integrate sophisticated designs such as multi-scale processing and residual learning	• Superior real-time denoising performance • Learns intricate patterns associated with noise • Generalizes well to diverse noise distributions	• Requires large annotated datasets for training • Computationally intensive training • May overfit to specific noise distributions

3 Methodology

This section sketches out the recommended methodology for speckle image denoising using SDCNN. The initial stage is creating an extensive dataset with matched clean and noisy color images. The clean images are corrupted with synthetic speckle noise to create noisy images, which mimic real-world situations seen in several imaging modalities. The dataset has been meticulously selected to encompass a range of situations, noise levels, and image attributes, hence guaranteeing the trained model's resilience and capacity to generalize. The architecture of SDCNN is made to efficiently gather and eliminate speckle noise while maintaining important image characteristics. The network consists of several convolutional layers, normalization layers (such as batch normalization), and activation functions (like ReLU). These layers facilitate noise reduction and feature retention by allowing the network to retrieve hierarchical features from the input images. Advanced methods like residual learning and multi-scale processing are integrated into the network architecture to improve denoising performance. The network can gather contextual information at both the local and global levels due to multi-scale processing, and training can be made more reliable and effective via residual learning. The noisy-clean image pairings that are gathered from the dataset are used to train the SDCNN. The network gains the ability to link clean patches to their corresponding noisy input patches during training. The network's parameters are tuned by supervised learning framework in order to lessen the difference among the denoised output and the clean ground truth image. Following training, the trained SDCNN's performance is evaluated by testing and validation data. The denoised images are visually inspected and compared to the noisy input and ground truth clean images to perform qualitative evaluations. In this work, the full reference quality metrics (PSNR, MSE, SSIM, MS-SSIM) and no-reference quality metrics (NIQE, PIQE, and BRISQUE) are computed to assess denoising performance [11] (Fig. 2).

The input layer of SDCNN processes noisy images. The noisy image is commonly represented as a group of pixels, with all pixels denoting a color or intensity value. The SDCNN design includes many convolutional layers to extract features by convolving filters over the input image. Convolutional layers use learnable filters to generate feature maps that capture many characteristics of an image. The number and size of filters may change between the network's convolutional layers. Following each convolutional layer, the feature maps are activated element by element using a rectified linear unit (ReLU) which provides non-linearity into the network by converting negative values to zero, letting it to learn complicated mappings between input and output. Residual learning is a major component of the SDCNN architecture. This is accomplished by utilizing skip connections, which enable the network to directly learn residual functions. This residual link allows the network to learn to anticipate the dissimilarity between the noisy and output image rather than simply predicting the clean result. Batch normalization layers are frequently employed in the SDCNN architecture to increase training stability and convergence. These layers normalize the activations of each layer throughout the mini-batch, eliminating internal covariate

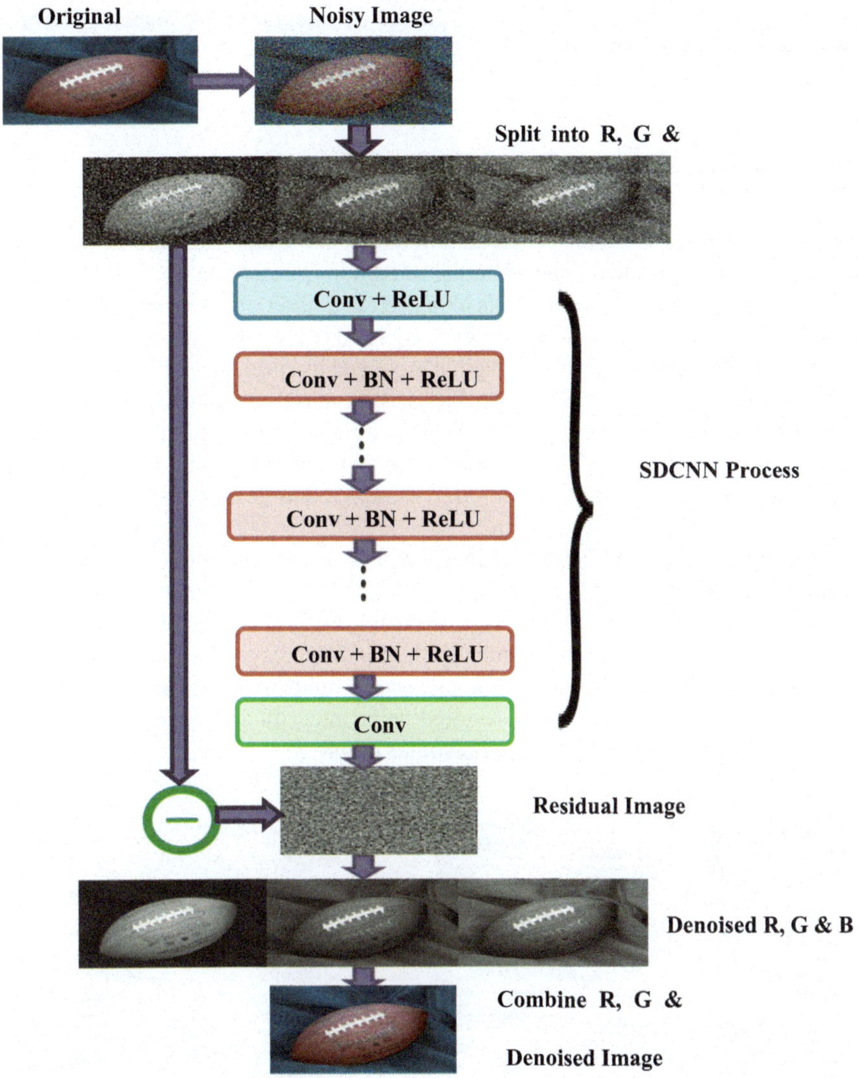

Fig. 2 Proposed method for denoising

shift and speeding up training. In traditional learning, each layer of a neural network straightforwardly absorbs a mapping to the output from the input. As the network becomes deeper, it becomes increasingly difficult to train effectively due to issues like vanishing gradients and degradation in performance. In residual learning, preferred output (mapping) $H(x)$, is the summation of input (x) and residual mapping $F(x)$, i.e., $F(x) = H(x) - x$. The output layer generates a denoised image. It usually comprises of a single convolutional layer followed by linear or no activation function. The output layer creates the final denoised image by integrating the features learnt in previous layers and applying the learned transformations to the input image.

4 Experimental Result and Discussion

This section explains the investigational findings and examines the competence of the suggested technique for speckle noise reduction in color images, which employs SDCNN. We conducted experiments on a diverse dataset of color images corrupted with speckle noise. The end result of the suggested process for denoising is depicted in Fig. 3. Figure 3a represents the original input and its speckle version of variance of 0.1 is illustrated in Fig. 3b. The noisy data is split into three channels as depicted in Fig. 3b through e. The outcome of the process, a clean image after the noise removal, is demonstrated in Fig. 3f. The proposed system applied to the same image with different variances such as 0.2, 0.25, 0.3, and 0.35, and its outcome is portrayed in Fig. 4.

The suggested method's performance was tested using full reference quantitative measures such as the PSNR, SSIM, and no-reference (blind) quality measures such as BRISQUE, NIQE, and PIQE. They measure image quality and resemblance to ground truth images. Visual assessment of the denoised photos demonstrated a significant reduction in speckle noise while preserving image features and structure. The suggested technique successfully restored color integrity and increased perceptual quality, yielding visually appealing outcomes (Table 2).

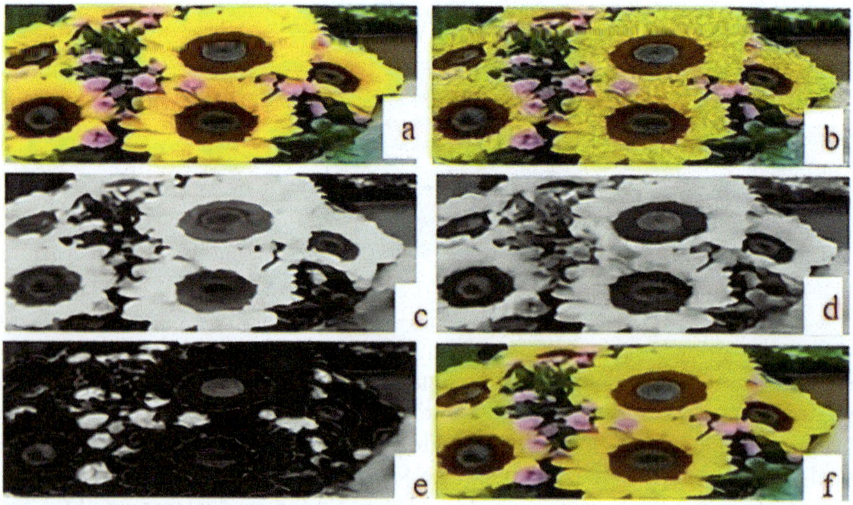

Fig. 3 Outcome of the proposed method for denoising

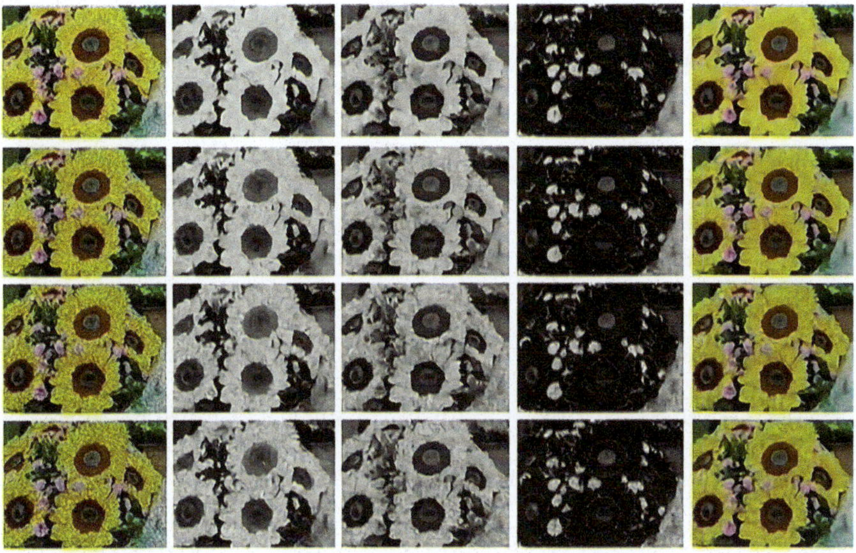

Fig. 4 Outcome for various variances of speckle noise. The process with speckle noise of variances 0.2, 0.25, 0.3, and 0.35 is illustrated in first, second, third, and fourth rows respectively

Blind quality measures evaluate the quality of a picture without access to its source or ground truth [12, 13]. These approaches are especially effective when the unique and perfect image is not existing, such as in real-time image processing applications or when working with images received from unknown sources [14, 15]. A lower score suggests improved perceptual quality. NIQE is a no-reference evaluation method that uses spatial domain statistics to analyze a picture's naturalness. It assesses picture quality by measuring factors such as sharpness, noise, and texture. NIQE has been demonstrated to be strongly correlated with human perceptions of picture quality. BRISQUE analyzes picture quality using natural scene statistics. It uses the distribution of local image gradients and other statistical variables to determine image quality. PIQE is a model of the image formation process and human visual perception that is intended to imitate the perception of image quality by the visual system [16, 17] (Table 3).

We conducted considerable testing and analysis to assess the efficiency of image quality metrics in assessing image quality and offering significant insights into the perceived integrity of denoised images. Furthermore, our findings emphasize the significance of using both types of quality criteria in the evaluation of denoising approaches. While complete reference measures serve as a comparison point for ground truth imagery, no-reference measures provide vital insights into the perceptual quality of denoised images in real-world applications. Based on our experimental findings and analysis, we conclude that combining complete reference and no-reference image measurements provides a comprehensive framework for evaluating the perceptual quality of denoised images (Figs. 5 and 6).

Table 2 Proposed method evaluation using full reference image quality measures

Image quality metrics	Test and noisy	Test and denoised	Test and noisy	Test and denoised	Test and noisy	Test and denoised
	Var = 0.1		**Var = 0.2**		**Var = 0.25**	
MSE	0.0207	0.0059	0.0392	0.0108	0.0571	0.0162
PSNR	16.8336	22.3280	14.0648	19.6523	12.4351	17.9031
SSIM	0.8277	0.9292	0.7338	0.8908	0.6620	0.8554
MSSSIM (R)	0.8419	0.9254	0.7784	0.8758	0.7303	0.8245
MSSSIM (G)	0.8521	0.9291	0.7973	0.8864	0.7534	0.8483
MSSSIM (B)	0.9598	0.9697	0.9359	0.9470	0.9160	0.9287
		Var = 0.3		**Var = 0.35**		
MSE		0.0732	0.0216	0.0845	0.0255	
PSNR		11.3525	16.6478	10.7292	15.9307	
SSIM		0.6044	0.8227	0.5688	0.8023	
MSSSIM (R)		0.6924	0.7825	0.6659	0.7581	
MSSSIM (G)		0.7127	0.8091	0.6895	0.7869	
MSSSIM (B)		0.9016	0.9192	0.8844	0.9066	

Table 3 Proposed method evaluation using no reference image quality measures

Quality metrics	Test image	Noisy image	Denoised image	Noisy image	Denoised image	Noisy image	Denoised image
		Variance = 0.1		**Variance = 0.2**		**Variance = 0.25**	
PIQE	33.0871	58.4100	31.8222	64.7611	31.2943	66.5323	36.7232
NIQE	4.5931	17.3042	4.3665	15.1171	4.6422	19.6203	4.3659
BRISQUE	37.5922	44.2355	34.6348	44.3919	34.0785	44.2296	28.8423
		Variance = 0.3		**Variance = 0.35**			
PIQE	33.0871	68.7496	40.5206	70.6519	41.6159		
NIQE	4.5931	21.1232	4.8105	17.6549	4.9406		
BRISQUE	37.5922	45.7512	28.1170	48.3718	28.2620		

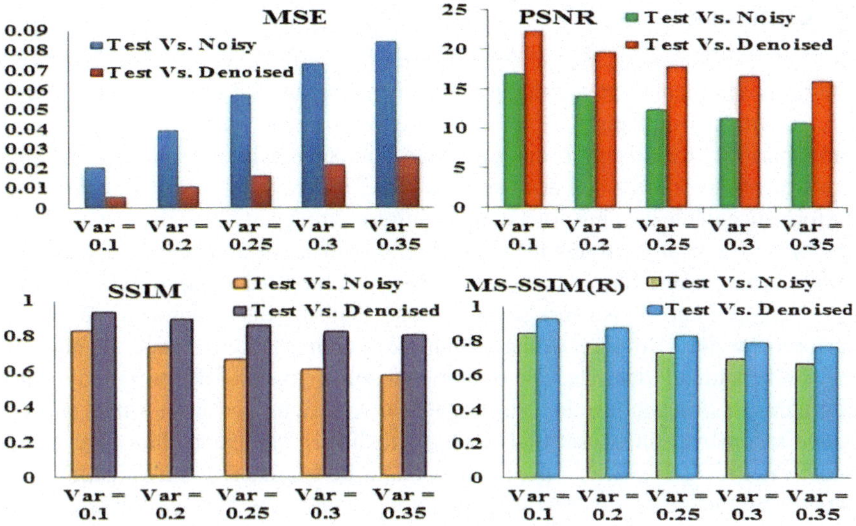

Fig. 5 Assessment based on full reference quality measures

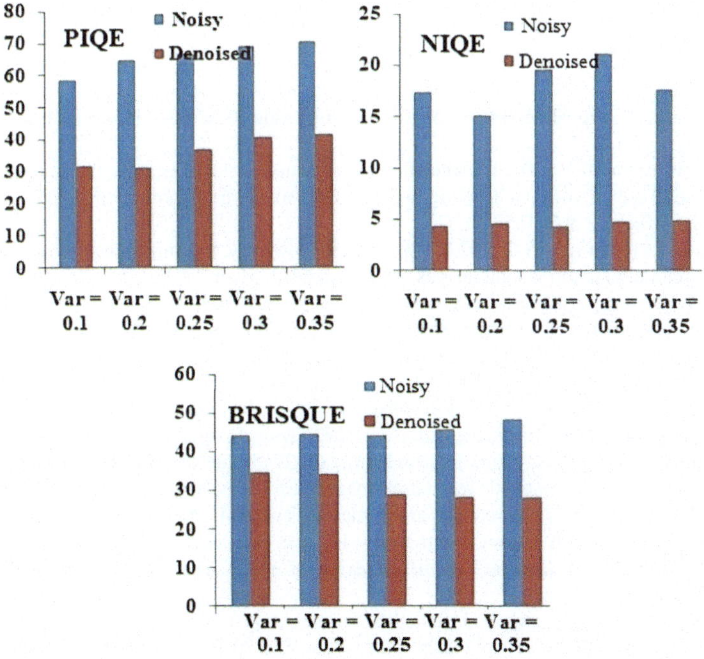

Fig. 6 Assessment based on no-reference quality measures

5 Conclusion

The proposed method portrayed an innovative approach for speckle noise removal in color images utilizing SDCNN. The proposed method has shown the efficacy and superiority over the most current image denoising techniques through thorough experimentation results and analysis. Our SDCNN provides outstanding denoising performance while keeping crucial image information by utilizing the power of deep learning, specifically through the creation of a multi-scale architecture and a supervised learning framework. Residual learning is incorporated to improve training even more and speed up the network's convergence. Our suggested approach consistently performed well for different variances of noise as demonstrated by experimental findings on benchmark datasets. The exceptional performance of the proposed method highlights its potential use in a variety of real-world circumstances where superior image reconstruction is essential, such as medical imaging, satellite images, and underwater imagery. The proposed work might be extended by investigating other network architectures, incorporating attention mechanisms, or modifying the model to fit certain noise distributions.

References

1. Ai J et al (2022) A trilateral filter for video SAR speckle noise reduction. IEEE Geosci Remote Sens Lett 19:1–5
2. Ganesan P, Rajini V (2013) Segmentation and denoising of noisy satellite images based on modified fuzzy C means clustering and discrete wavelet transform for information retrieval. Int J Eng Technol 5(5):3858–3869
3. Lv X-G, Li F, Liu J, Lu S-T (2022) A patch-based low-rank minimization approach for speckle noise reduction in ultrasound images. Adv Appl Math Mech 14(1):155–180
4. Bedi AK, Sunkaria RK (2022) Ultrasound speckle reduction using adaptive wavelet thresholding. Multidimens Syst Signal Process 33(2):275–300
5. Liu S et al (2021) SAR speckle removal using hybrid frequency modulations. IEEE Trans Geosci Remote Sens 59(5):3956–3966
6. Zhang K, Zuo W, Chen Y, Meng D, Zhang L (2017) Beyond a Gaussian denoiser: residual learning of deep CNN for image denoising. IEEE Trans Image Process 26(7):3142–3155
7. Mishro PK, Agrawal S, Panda R, Abraham A (2022) A survey on state-of-the-art denoising techniques for brain magnetic resonance images. IEEE Rev Biomed Eng 15:184–199
8. Liu X, Lian W (2022) Non-convex high-order TV and $\ell 0$-norm wavelet frame-based speckle noise reduction. IEEE Trans Circuits Syst II Express Briefs 69(12):5174–5178
9. Fan L, Zhang F, Fan H et al (2019) Brief review of image denoising techniques. Vis Comput Ind Biomed Art 2:7
10. Goyal B, Dogra A, Agrawal S, Sohi BS, Sharma A (2020) Image denoising review: from classical to state-of-the-art approaches. Inf Fusion 55:220–244
11. Ding K, Ma K, Wang S, Simoncelli EP (2022) Image quality assessment: unifying structure and texture similarity. IEEE Trans Pattern Anal Mach Intell 44(5):2567–2581
12. Ganesan P, Sathish BS, Vasanth K, Vadivel M, Sivakumar VG, Thulasiprasad S (2020) Color image quality assessment based on full reference and blind image quality measures. In: Lecture notes in networks and systems, vol 107. Springer, Singapore

13. Kalist V, Ganesan P, Leo Joseph LMI, Sathish BS, Murugesan R (2020) Image quality analysis based on texture feature extraction using second-order statistical approach. In: Lecture notes in networks and systems, vol 107. Springer, Singapore
14. Ravichandran M, Alexander TJ, Sathish BS, Joseph LMIL, Ganesan P, Sajiv G (2023) An effectual methodology to objective quality assessment for noisy and compressed images using full and blind reference quality metrics. In: 2023 IEEE 3rd international conference on applied electromagnetics, signal processing, and communication (AESPC), Bhubaneswar, India, pp 1–6
15. Simi VR, Edla DR, Joseph J (2021) A no-reference metric to assess quality of denoising for magnetic resonance images. Biomed Signal Process Control 70
16. Stpień, Oszust M (2022) A brief survey on no-reference image quality assessment methods for magnetic resonance images. J Imaging 8(6):160
17. Mittal, Moorthy AK, Bovik AC (2012) No-reference image quality assessment in the spatial domain. IEEE Trans Image Process 21(12):4695–4708

Hybrid Anomaly Detection System for Unsupervised Video Surveillance Using Isolation Forest and Robust Covariance

Premanand Ghadekar, Aparna Mete, and Deepali Deshpande

Abstract Anomaly detection is a crucial task in video for numerous applications such as surveillance, security, and detecting anomalies in industrial processes. This research proposes a new approach for anomaly detection in surveillance data using an ensemble method of isolation forest and robust covariance. The proposed method involves three stages: feature extraction, key frame extraction, and anomaly detection. Our approach for detecting anomalies involves using robust covariance-based feature extraction and thresholding to identify key frames. These key frames can then be inputted into an isolation forest algorithm. The proposed approach is tested on a surveillance dataset that is publicly available to study the effectiveness, and its performance is compared to other methods that are currently considered the most advanced in the field. The outcomes show that the proposed approach surpasses alternative approaches in accurately identifying anomalies, providing precise detections, and recalling a higher number of true anomalies.

Keywords Isolation forest · Robust covariance · Key frames · Covariance matrix · Video analysis

1 Introduction

An anomaly, or outlier, in machine learning, refers to an observation or data point that considerably diverges from the anticipated or usual pattern of the data. Anomaly detection is an important task in machine learning, as it helps identify and flag unusual or unexpected data points that may indicate errors, anomalies, or rare events in the data.

There can be numerous reasons for anomalies such as errors in measurement, data corruption, system malfunctions, fraud, or unusual behavior.

P. Ghadekar (✉) · A. Mete · D. Deshpande
Department of Information Technology, Vishwakarma Institute of Technology, Pune 411037, India
e-mail: premanand.ghadekar@vit.edu

A. C. Frery et al. (eds.), *Computational Intelligence and Data Analytics*, Lecture Notes on Data Engineering and Communications Technologies 236,
https://doi.org/10.1007/978-981-96-0451-7_19

Anomaly detection in machine learning can be performed using various techniques, such as statistical methods, clustering, and classification algorithms. Statistical methods, such as z-score, quartile deviation, and interquartile range, find anomalies based on their deviation from the expected distribution of the data.

Clustering algorithms, such as k-means and DBSCAN, group similar data points together and identify anomalies as those that do not belong to any cluster. Algorithms, such as support vector machines and decision trees, learn to classify data points as normal or abnormal based on their features or attributes.

The main categories of anomaly detection techniques: machine learning-based methods, statistical methods, and hybrid methods that incorporate both statistical and machine learning approaches. Machine learning-based methods, such as Isolation Forest and Robust Covariance, use algorithms to learn patterns and relationships in the data and identify anomalies based on their deviation from the learned model. Hybrid methods combine both statistical and machine learning approaches to achieve better accuracy and robustness in detecting anomalies.

The isolation forest algorithm uses a tree-based approach to identify anomalies, while the robust covariance method estimates the covariance matrix of the data to detect outliers. Anomaly detection is an important task in various fields. Identifying anomalies can be difficult, particularly when the data is complex, noisy, or high-dimensional.

Moreover, the definition of anomalies may vary depending on the context and domain of the data, making it challenging to define a universal threshold for detecting anomalies.

The proposed system aims on detecting the anomaly using Isolation Forest and Robust Covariance which will detect any anomalous behavior present in the videos. Ultimately, the goal of this project is to provide a more accurate and efficient way to detect anomalies in various applications.

2 Literature Review

In the paper [1], a new technique is presented for detecting abnormalities in aircraft data by utilizing "Recurrent Neural Networks (RNNs)". The authors highlight the significance of anomaly detection in aircraft data, which can prevent accidents and improve maintenance procedures. They provide a brief overview of RNNs and explain how they can be utilized for sequence modeling tasks such as detecting anomalies. The authors present their proposed method, which involves preprocessing the aircraft data, training the RNN model, and then utilizing the trained model to identify anomalies in new data. They evaluate their approach on a publicly accessible dataset of aircraft engine sensor readings and compare their results with other techniques for the detection of anomalies.

A novel method for detecting anomalies is introduced in the paper [2] in large-scale and high-dimensional datasets by utilizing a linear one-class "Support Vector Machine (SVM)" in conjunction with deep learning. The article underscores the significance of anomaly detection and the difficulties associated with detecting anomalies in datasets that are both high-dimensional and large-scale.

The suggested approach employs a linear one-class SVM as the primary classifier and integrates deep learning methods for both feature extraction and dimensionality reduction. The article outlines the design of the deep learning architecture and thoroughly describes the training procedure. The technique is evaluated on multiple datasets, and the experimental outcomes show that it performs better than other advanced methods in both computational efficiency and accuracy of detection.

The article referenced as [3] discusses the difficulties and concerns related to utilizing anomaly detection based on SVM to enhance security. The paper highlights the importance of anomaly detection in enhancing security and provides an overview of SVM-based anomaly detection techniques. In the paper, the challenges of using SVM-based anomaly detection are examined, which include the difficulty of selecting appropriate parameters for the SVM algorithm and the absence of a precise definition of what constitutes an anomaly in a particular dataset. This work also shows the advantages of selecting appropriate features for training the SVM model and the need for efficient methods for training and testing the model.

In reference [4], an introduction to a fresh approach for identifying abnormalities in time-series data utilizing neural networks in "Long Short-Term Memory (LSTM)" was given. The paper highlights the importance of anomaly detection and the challenges associated with detecting anomalies in time-series data.

The method proposed in [4] utilizes LSTM neural networks and unsupervised learning techniques to identify anomalies in time-series data. It offers an in-depth account of the training procedure and describes the design of the LSTM network architecture. Multiple datasets are used to evaluate the technique, and the outcomes of the experiments indicate that it outperforms other advanced methods in both detection accuracy and computational efficiency. Additionally, the article presents a comprehensive analysis of the proposed approach, including an investigation into how different parameters affect the model's performance. The study concludes that the suggested approach is effective in detecting anomalies in time-series data and could be applied in numerous domains such as finance, healthcare, and cybersecurity.

An improved method for detecting anomalies in data using the Isolation Forest algorithm was presented in [5]. The paper highlights the importance of anomaly detection in various fields and the challenges associated with detecting anomalies in high-dimensional data. The paper proposes an improved version of the algorithm by introducing a pruning mechanism that removes the unnecessary subtrees and reduces the computational cost of the algorithm.

A new scheme for detecting anomalies in cloud computing is suggested by the authors of [6], which is designed to be flexible and adaptable. The paper highlights the importance of anomaly detection in cloud computing environments and the challenges associated with detecting anomalies in such environments. The proposed method utilizes the "Local Outlier Factor (LOF) algorithm", a technique which is a density-based anomaly detection that is capable of effectively detecting anomalies in high-dimensional data with high efficiency. The proposed paper presents an adaptive form of the LOF algorithm that incorporates a dynamic adjustment mechanism to accommodate fluctuations in the cloud computing environment, thereby enhancing the precision of anomaly detection. The effectiveness of the proposed method is assessed across multiple datasets, and the experimental outcomes demonstrate that the adaptive LOF-based approach performs better than other advanced methods in both computational efficiency and detection accuracy.

The authors of [7] highlight the importance of anomaly detection in wireless sensor networks, as anomalies can lead to malfunctions and reduce the accuracy of the collected data. They then introduce their proposed approach, which combines DBSCAN clustering and SVM classification algorithms to detect anomalies.

The DBSCAN algorithm is used to group the sensor readings into clusters, and then SVM classifies each cluster as normal or anomalous. The authors evaluate their approach on a publicly available dataset and compare it to other anomaly detection methods.

Based on the experiment, it is evident that the suggested approach outperforms other methods in both detection accuracy and false alarm rate. Furthermore, the authors carried out a sensitivity analysis to assess the impact of various parameters on the model's performance.

In reference [8], the authors put forward a technique to identify abnormalities in electrocardiogram (ECG) readings utilizing stacked LSTM networks. The authors emphasize the importance of detecting anomalies in ECG readings since they may suggest severe heart conditions. The proposed approach includes preprocessing the ECG data, training a stacked LSTM network on the data, and using the trained model for the detection of anomalies in new data. The authors examine the approach on a publicly available dataset of ECG readings and compare their results with other anomaly detection methods. Based on the experimental results, the proposed approach exhibits superior performance compared to other methods while considering false positive rate and detection accuracy. Furthermore, the authors performed a sensitivity analysis to evaluate the effect of various parameters on the model's execution.

The authors of [9] argue that anomaly detection is critical in network traffic management, as it allows for the identification of unexpected or suspicious behavior. They then describe their proposed approach, which involves preprocessing the network traffic data, applying K-means clustering to identify normal traffic patterns, and using the cluster centroids to identify anomalies.

The authors [9] evaluate their approach on two different datasets, one containing simulated network traffic and the other containing real-world network traffic. By comparing their findings with other anomaly detection methods, they prove that their suggested technique performs better than the other methods in detecting anomalies, regarding both false positive rate and detection accuracy.

The objective of the study presented in research paper [10] is to create a robust model for detecting anomalies in cloud computing networks, utilizing autoencoders. The authors utilize the autoencoder algorithm to produce a fundamental portrayal of data and reduce the reconstruction error which is utilized as a metric for classifying to detect anomalies in the data, and the study investigates the ability to classify various types of anomalies. The authors also recommend a new approach, in which the reconstruction error is treated as a vector for each input feature, rather than a single value.

Research study [11] introduces anomaly detection systems for identifying anomalous situations, utilizing two different models: "Adversarial Autoencoder (AAE) and Deep Convolutional Generative Adversarial Networks (DCGAN)". Cumulative and reverse cumulative distribution is utilized by the authors to determine the most advantageous decision threshold for the DCGAN model. The optimal number of latent variables for the autoencoder network is determined using the reconstruction error ratio function.

The impact of image complexity on the performance of anomaly detection is explored in the study, where the detectors trained on less complex datasets demonstrated superior results. Specifically, when trained on one of the simplest databases available, MNIST, both detectors achieved a low anomaly detection error of 0.08% (AAE) and 1.89% (DCGAN) at an anomaly expectation probability of 0.5.

The task of anomaly detection in data analysis, which can be employed to identify network intrusions, is explored in detail in paper [12]. The paper provides an extensive analysis of key categories namely clustering, statistical, classification, and information theory. Additionally, the paper addresses the challenges associated with utilizing network intrusion detection datasets.

The aim of paper [13] is to provide an accurate overview of research on deep anomaly detection, presenting a thorough taxonomy that encompasses three main categories and eleven subcategories of methods. The paper delves into the fundamental concepts, objective functions, underlying assumptions, strengths, and limitations of each approach, evaluating how they tackle the challenges in the field. Additionally, the paper explores potential future opportunities and new perspectives for addressing these challenges.

The study in [14] suggests a comprehensive method that applies neural networks for detecting anomalies in video surveillance. This method uses frame-level feature extraction along with a one-class classifier for anomaly detection. There are two phases involved in the method, namely feature extraction and anomaly detection. In the feature extraction phase, a pre-trained convolutional neural network is used for the extraction of features from every frame of the video, which are then aggregated over a period. The one-class classifier employed is a variant of the "support vector data description (SVDD) algorithm", which creates a boundary around normal video data, and any data points outside of this boundary are identified as anomalies. The proposed technique has been compared with various baseline methods on two publicly accessible datasets, and it has been proven to be efficient in detecting anomalies in real-world surveillance videos. The combination of CNN-based feature extraction and one-class classification has been found to be effective in detecting anomalies in videos.

Table 1 Comparison of the proposed system with existing methods

Method	Pros	Cons
Isolation forest	Fast and scalable for high-dimensional data	Sensitive to outliers and may suffer from the masking effect
Robust covariance	Robust to outliers and can handle low-dimensional data	Computationally expensive and may be sensitive to noise
K-nearest neighbors	Simple and easy to implement	Sensitive to the number of neighbors and distance metric used
One-class SVM	Can handle non-linear data and has a well-defined decision boundary	Computationally expensive and sensitive to kernel parameters
Density-based clustering	Can handle arbitrary shapes and can adapt to varying densities	Sensitive to hyperparameters and may suffer from noise and outliers
Proposed approach	Can detect both point and contextual anomalies with high accuracy	More robust to noise and masking effects than individual techniques

3 Table of Comparison

See Table 1.

4 Methodology

Algorithms

1. **Isolation Forest**: Isolation Forest is an unsupervised machine learning technique designed to identify anomalies or outliers within a dataset. It builds a set of binary trees, called isolation trees, based on the concept of random forests. The algorithm randomly selects a feature and value range from the dataset to split the data recursively until each data point is isolated in a terminal node. The depth of each tree is limited to prevent overfitting, and the splitting process stops when a tree reaches a specified depth or when all data points are isolated.

After constructing the isolation trees, the algorithm calculates an anomaly score for each data point by determining the average depth of the tree where the point is isolated. The rationale behind this is that anomalies are usually separated from the rest of the data with fewer splits, resulting in a shorter path to the terminal node. Consequently, data points that have a low average depth across all trees are identified as anomalies, whereas those with a high average depth are regarded as normal.

The process of calculating the anomaly score for each data point is executed as follows:

(1) For each isolation tree, traverse the tree from the root to the terminal node where the data point is isolated, and record the depth of the node.

(2) Calculate the average depth of the terminal nodes across all trees.
(3) The calculation of the anomaly score of a data point is given as the reciprocal of the average depth, normalized by the maximum feasible depth of a tree (i.e., log2(n), where n is the number of data points).

The algorithm is computationally efficient and can handle high-dimensional datasets with good speed (Table 2), as the splitting process only requires comparing a random feature of the data with a randomly chosen threshold value. The method is also robust to irrelevant or redundant features in the data, thereby making it suitable for various use cases such as intrusion detection, fraud detection, and analysis of network traffic.

Isolation forest can be given by

$$s(x, n) = 2^{-\frac{E(h(x))}{c(n)}} \tag{1}$$

2. **Robust Covariance**: Robust covariance is a statistical technique that is utilized in machine learning to compute the covariance matrix of a dataset in a manner that is less susceptible to outliers or errors. Several machine learning algorithms, including "Principal Component Analysis (PCA), Linear Discriminant Analysis (LDA), and Gaussian Mixture Models (GMMs)", use the covariance matrix as a fundamental element. However, in datasets that contain anomalies or errors, the covariance matrix can be heavily influenced by these data points and may not provide an accurate representation of the underlying relationships between variables.

Table 2 Comparison of isolation forest, random forest, and other algorithms for anomaly detection and their sensitivity to outliers

Algorithm	Purpose	Labeled/ unlabeled	Speed/ scalability	Sensitivity to outliers	Customizability
Isolation forest	Anomaly detection	Unlabeled	Fast/ scalable	Less sensitive	Customizable
Random forest	Classification/ regression	Labeled/ unlabeled	Slower/less scalable	Sensitive	Customizable
One-class SVM	Anomaly detection	Labeled	Slow/ scalable	Sensitive	Customizable
Local outlier factor	Anomaly detection	Unlabeled	Medium/ scalable	Less sensitive	Customizable
Robust PCA	Dimensionality reduction/ anomaly detection	Unlabeled	Slow/not scalable	Less sensitive	Customizable
Autoencoders	Dimensionality reduction/ anomaly detection	Labeled/ unlabeled	Slow/not scalable	Less sensitive	Customizable

Robust covariance methods address this issue by reducing the impact of outliers or errors when estimating the covariance matrix. There are several robust covariance methods that can be used in machine learning, including:

(1) The "Minimum Covariance Determinant (MCD)" is a resilient covariance estimator that identifies a set of data that has the least determinant of its covariance matrix, which is a smaller subset of the original data. This subset is considered to have most of the non-anomalous data points, and the covariance matrix is calculated based on this subset. This method is less susceptible to outliers compared to conventional covariance estimation techniques and can produce more accurate covariance estimates in datasets having anomalies or errors.
(2) "Distance Covariance Matrix (DCM)": The Distance Covariance Matrix is a non-parametric technique that calculates the covariance matrix by measuring the distance between data points. This method is less prone to outliers than conventional covariance estimation methods and can provide more precise covariance estimates in datasets where the relationships between variables are non-linear.
(3) "Orthogonalized Gnanadesikan-Kettenring (OGK)" estimator: OGK is a robust covariance estimator that removes the effect of outliers by projecting the data onto a subspace orthogonal to the direction of the outliers. This method can produce more accurate covariance estimates in datasets with high-dimensional data and a small number of outliers.

Robust covariance techniques are employed as a preprocessing step to enhance the precision and robustness of machine learning algorithms that depend on the covariance matrix. For instance, in the context of detecting anomalies, robust covariance methods are employed to estimate the covariance matrix of the given data and are then utilized to identify anomalies based on their deviation from the expected behavior. This approach improves the accuracy and dependability of anomaly detection in datasets containing outliers or errors.

In general, robust covariance techniques are useful in improving the precision and reliability of machine learning algorithms that depend on the covariance matrix, especially in datasets with anomalies or mistakes. The selection of a robust covariance approach is determined by the specific demands of the job, the dataset's features, and the resources available.

Covariance Matrix Σ, also known as Mahalanobis Distance is given by

$$D_i = \sqrt{(x_i - \mu)^T \Sigma^{-1} (x_i - \mu)} \tag{2}$$

3. **Thresholding**: Anomaly detection typically involves identifying data points that exhibit significant deviations from the expected behavior. In this regard, thresholding is a commonly used technique for categorizing data points as normal or anomalous based on a predefined threshold value. For instance, in a Gaussian distribution model, data points that fall beyond a certain number of standard deviations from the mean can be identified as anomalous using thresholding. However, determining the appropriate threshold value in anomaly detection depends on several factors, including the

specific task requirements, dataset properties, and available resources. This value can be established using prior knowledge, empirical analysis, or optimization methods.

One major limitation of thresholding in anomaly detection is that it presupposes a clear distinction between normal and anomalous data points, which may not always be the case in real-world datasets. Furthermore, the sensitivity of thresholding to the chosen threshold value may result in false positives or negatives. Hence, thresholding is often used alongside other anomaly detection methods, like density-based or distance-based methods, to improve the dependability and accuracy of anomaly detection.

Project Flow Diagram

Algorithm of the Model Deployed

Step 1: Import necessary libraries: OpenCV, NumPy, scikit-learn covariance and ensemble modules, and OS.

Step 2: Set the new height and new width for resizing the frames.

Step 3: Load the video using the OpenCV VideoCapture object.

Step 4: Set the folder path for saving the resized frames as JPEG files.

Step 5: Initialize empty lists to store the resized frames and feature vectors.

Step 6: Loop through each frame in the video:

6a: Set the time position of the video to the current frame.

6b: Read the current frame.

6c: Resize the frame to the new height and new width.

6d: Save the resized frame as a JPEG file in the specified folder path.

6e: Convert the resized frame to a grayscale image.

6f: Flatten the grayscale image to a 1D array.

6g: Append the feature vector to the list of feature vectors.

Step 7: Convert the list of feature vectors to a NumPy array.

Step 8: Fit a robust covariance estimator to the feature vectors to model the distribution.

Step 9: Save the fitted robust covariance model using the pickle module.

Step 10: Calculate the Mahalanobis distances for all feature vectors.

Step 11: Set a threshold for the Mahalanobis distance using a percentile value (e.g., 90th percentile).

Step 12: Identify key frames with Mahalanobis distance above the threshold as potential anomalies.

Step 13: Convert the list of potential anomaly frames to a NumPy array and flatten them.

Step 14: Create an isolation forest model with specified parameters (e.g., number of trees, contamination rate, random state).

Step 15: Fit the isolation forest model to the flattened key frames.

Step 16: Predict outliers using the isolation forest model.

Step 17: Extract the indices of predicted outliers as anomaly indices.

Step 18: Print the anomaly indices.

5 Experimentation

(a) Data Collection and Preprocessing

The study used a dataset comprising a 15 s video captured at a frame rate of 60 frames per second (fps). From this video, 90 frames were extracted at a rate of one frame per second, yielding a collection of frames for further examination. The video's original frame size was 1080 * 720 pixels, which were shrunk to 160 * 90 pixels while keeping the aspect ratio intact for uniformity in feature extraction and future processing. The frames were resized to decrease processing costs while keeping critical information for anomaly detection.

(b) Finding Key Frames using Covariance Matrix and Thresholding

On the scaled frames, we first calculated the covariance matrix using a robust covariance estimation approach using Elliptic Envelope. The Mahalanobis distance was then computed for each frame using the estimated covariance matrix and the frame feature vectors. To determine critical frames, we established a threshold for the Mahalanobis distance. Key frames or potentially anomalous frames were defined as frames having a Mahalanobis distance larger than the threshold.

(c) Finding Anomalous Frames using Key Frames

We fed the previously discovered crucial frames into an Isolation Forest algorithm. The key frames were then passed for training the Isolation Forest model, which was then used to forecast the anomalous frames in the movie. Anomalous frames predicted by the Isolation Forest model were defined as frames in the video that are likely to include anomalies or uncommon occurrences.

Some other anomaly detection techniques we experimented with are as follows.

Local Outlier Factor: The Local Outlier Factor technique is a method of detecting anomalies by considering the density of data points. It is an unsupervised method but its training requires data with no outliers. It evaluates whether a point is an outlier or not by calculating an anomaly score, which measures its level of isolation compared to the neighboring data points. The anomaly score is determined based on both the local and global density of the data points.

The local Outlier Factor (LOF) is computed by

$$\mathrm{LOF}(x_i) = \frac{\Sigma x_j \varepsilon N(x_i) lrd(x_j)}{|N(x_i)|} \times \frac{1}{lrd(x_i)} \tag{3}$$

One-Class SVM: One-Class Support Vector Machines (OCSVM) is a machine learning algorithm used for anomaly detection in datasets where only one class of data (i.e., normal data) is available for training. OCSVM learns to describe the normal data distribution and identify deviations from it as potential anomalies.

In OCSVM, the algorithm learns a decision boundary that separates the normal data and the rest of the data given in the feature space. The decision boundary is chosen to have the largest possible margin such that it contains a specified fraction of the normal data points. Once the decision boundary is learned, the OCSVM can be used for predicting whether a new data point is normal or an anomaly by checking on which side of the decision boundary it falls.

If a new data point is on the same side as the majority of the normal data points, it is considered normal. If it falls on the other side of the decision boundary, it is considered an anomaly.

The performance of different algorithms for anomaly detection, such as Local Outlier Factor (LOF), Isolation Forest, and Support Vector Machine (SVM), was evaluated to identify fraudulent cases (as seen in Table 3). In this evaluation, the Isolation Forest algorithm detected 81 errors, indicating that it identified 81 instances as anomalous, while they were not. On the other hand, the LOF algorithm detected 97 errors, and the SVM algorithm detected 85.16 errors. Therefore, the Isolation Forest algorithm performed better than both LOF and SVM in terms of false positives.

Moreover, the accuracy of Isolation Forest was found to be 99.74%, which is higher than that of LOF at 99.65% and SVM at 70.09%. This means that Isolation Forest was able to identify 99.74% of the fraudulent cases accurately, which is a significant improvement compared to LOF and SVM.

The improvement in identifying fraudulent cases by Isolation Forest was around 30% compared to SVM. We tried to implement SVM as it works on unsupervised learning. But it showed very inaccurate results (predicting some non-anomalous frames as anomalous). In conclusion, the Isolation Forest algorithm showed better performance in detecting fraudulent cases and had a higher accuracy rate than LOF

Table 3 Accuracy of different algorithms

Sl. No.	Algorithm	Detected errors	Accuracy (%)
1	Local outlier factor	97	0.9965
2	Isolation forest proposed model	81	0.9971
3	Support vector machine	85.16	0.7009

and SVM. Therefore, Isolation Forest can be a useful tool in identifying anomalies and fraud cases in various domains.

6 Results

The results of our proposed approach showed promising performance in detecting both key frames and anomalous frames in the video. The utilization of robust covariance estimation contributed to a more accurate estimation of the covariance matrix, while the thresholding technique based on Mahalanobis distance effectively identified potential key frames. Furthermore, the Isolation Forest algorithm further enhanced the detection performance by incorporating the key frames as input for anomaly detection.

In Fig. 1 we can see the covariance matrix has returned key frame indices, across the entire dataset.

Among all the indices returned only frame number 43 had an anomaly.

Figure 2 displays what the frames at indices 3, 5, 8, 10, 17, 18, 19 and 20 contain.

Figure 3 shows the anomalous frames that were present at index 43, 49, 50, 51, 52, 53, 55, 57, 59, 68, 71, 72, and 77 among the key frames returned after thresholding.

The anomaly index produced after applying isolation forest on the crucial frames is shown in Fig. 4. "8, 9, 14, 15, 16, 17" indicates that frame number 53, 55, 57, 59 (as seen in the list in Fig. 1) are anomalous frames in the list keyframe indices at index number 14, 15, 16 and 17; whereas anomaly index 8 and 9 were not anomalous. After physically inspecting frames, they were discovered to be an anomalous frame. Practical challenges include computational constraints due to high data volumes, real-time processing needs, and deployment complexities in integrating with existing systems. Ethical considerations involve privacy concerns and data misuse, requiring stringent protection measures. Addressing these issues is crucial for successful real-world implementation in surveillance systems (Fig. 5).

7 Conclusion

In conclusion, Anomaly Detection Using Isolation Forest and Robust Covariance is a hybrid system that combines two techniques for detecting anomalies in data. Isolation Forest is a tree-based method that isolates anomalies by randomly partitioning the data into smaller subsets, while Robust Covariance is a matrix-based method that detects anomalies based on their deviation from the expected distribution of the data. The hybrid system can detect both types of anomalies with high accuracy and overcome the limitations of each technique, such as the masking effect and computational efficiency. The hybrid approach can be used in fraud detection and network intrusion detection.

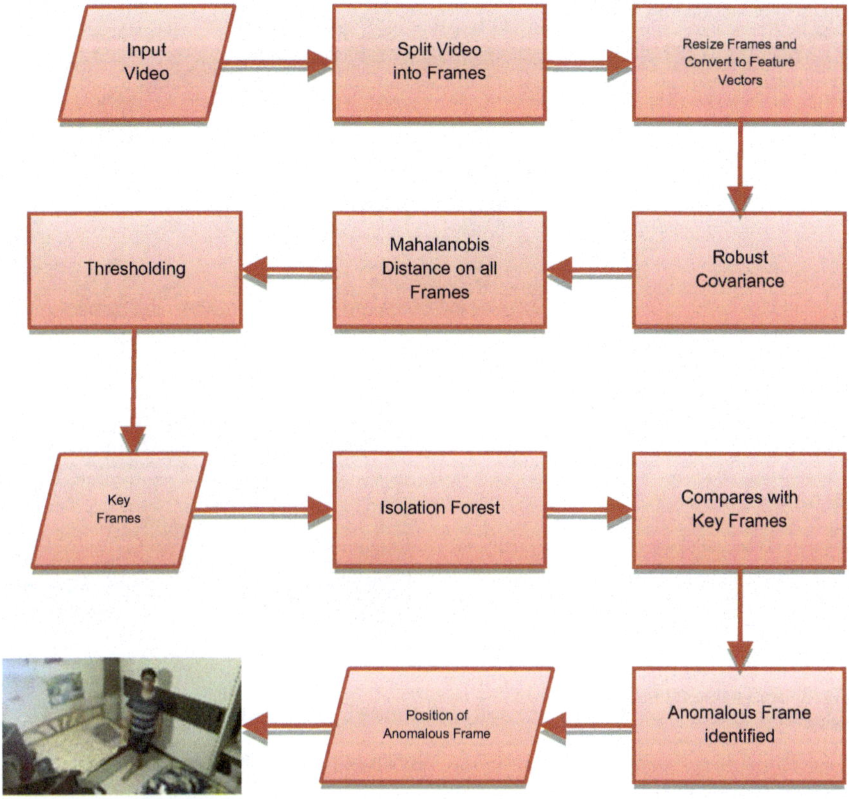

Fig. 1 Project flow diagram of hybrid anomaly detection system for unsupervised video surveillance using isolation forest and robust covariance

```
key_frames_indices

array([ 3,  5,  8, 10, 17, 18, 19, 20, 43, 49, 50, 51, 52, 53, 55, 57, 59,
        68, 71, 72, 77])
```

Fig. 2 List consisting of the key frame number

8 Future Scope

The proposed approach does have a scope for improvement and optimization in the future.

Fig. 3 Non-anomalous frame

Fig. 4 Anomalous frame contains a man walking in the room

Fig. 5 Anomalous frame index

```
anomaly_indices

array([ 8,  9, 14, 15, 16, 17])
```

1. Deployment in real-world systems: The hybrid system can be integrated into real-world systems, such as fraud detection systems, intrusion detection systems, and predictive maintenance systems, to detect anomalies in real-time and provide early warning signals for potential problems.
2. Optimization and parameter tuning: The hybrid system can be optimized and tuned to achieve better performance for specific datasets and applications. This can involve exploring different parameter settings, testing different combinations of techniques, and evaluating the performance of the hybrid system using different metrics.

References

1. Arun S, Xu L, Jain AK (2019) Anomaly detection in aircraft data using recurrent neural networks. IEEE Aerosp Electron Syst Mag 34(10):30–39
2. Erfani SM, Rajasegarar S, Karunasekera S, Leckie C (2016) High-dimensional and large-scale anomaly detection using a linear one-class SVM with deep learning. Pattern Recogn 58:121–134

3. Hosseinzadeh M (2017) Improving security using SVM-based anomaly detection: issues and challenges. J Ambient Intell Humaniz Comput 8(6):835–845
4. Eren T (2018) Unsupervised anomaly detection with LSTM neural networks. Expert Syst Appl 114:536–545
5. Xu D (2020) An improved data anomaly detection method based on isolation forest. IEEE Access 8:1514–1524
6. Huang T, Liu L, Tan W (2016) An LOF-based adaptive anomaly detection scheme for cloud computing. Futur Gener Comput Syst 54:12–21
7. Saeedi Emadi H, Mazinani SM (2018) A novel anomaly detection algorithm using DBSCAN and SVM in wireless sensor networks. Wirel Pers Commun 98:2025–2035
8. Thill M et al (2019) Anomaly detection in electrocardiogram readings with stacked LSTM networks. In: Proceedings of the 19th conference information technologies-applications and theory (ITAT 2019). CEUR-WS
9. Münz G, Li S, Carle G (2007) Traffic anomaly detection using k-means clustering. In: GI/ITG workshop MMBnet, vol 7, No 9
10. Torabi H, Mirtaheri SL, Greco S (2023) Practical autoencoder based anomaly detection by using vector reconstruction error. Cybersecurity 6:1. https://doi.org/10.1186/s42400-022-00134-9
11. Kopčan J, Škvarek O, Klimo M (2021) Anomaly detection using autoencoders and deep convolution generative adversarial networks. Transp Res Procedia 55:1296–1303
12. Ahmed M, Mahmood AN, Hu J (2016) A survey of network anomaly detection techniques. J Netw Comput Appl 60:19–31
13. Pang G, Shen C, Cao L, Van Den Hengel A (2021) Deep learning for anomaly detection: a review. ACM Comput Surv 54:2, Article 38, 38 pages. https://doi.org/10.1145/3439950
14. Franklin RJ, Dabbagol V (2020) Anomaly detection in videos for video surveillance applications using neural networks. In: 2020 Fourth international conference on inventive systems and control (ICISC). IEEE

Improving Air Quality Prediction: A Study on Data-Driven Techniques and Advanced Sensing Technologies

Yeshwanth Reddy and C. Sireesha

Abstract Air Quality Forecasting: This study explores data-driven techniques and advanced sensing technologies to improve air quality prediction. Focusing on deep learning models like LSTM, GRU, and DeepAR, the research evaluates their effectiveness in capturing dependencies among pollutants for accurate forecasts. The paper presents a comparative analysis of these models, highlighting the bi-directional GRU's superior performance in time series data processing. Future work suggests expanding datasets, exploring CNN techniques, and integrating external factors to enhance predictive accuracy.

Keywords Data-driven techniques · Advanced sensing technologies · Air quality prediction · LSTM and GRU models

1 Introduction

The growth of industries has resulted in increased air pollution, highlighting the need for accurate air quality prediction models. Predicting and assessing ambient air quality has become increasingly important worldwide. Traditional methods of estimating air pollutant concentrations are computationally intensive and limited in their ability to leverage available data. To address this, we propose a deep learning approach using LSTM [2] and GRU, which are specialized recurrent neural network architectures, to capture dependencies among pollutants and make accurate air quality predictions.

Y. Reddy (✉)
Stevens Institute of Technology, Hoboken, NJ 07307, USA
e-mail: ypeddama@stevens.edu

C. Sireesha
Vasavi college of Engineering, Hyderabad 500050, Telangana, India
e-mail: sireesha@staff.vce.ac.in

A. C. Frery et al. (eds.), *Computational Intelligence and Data Analytics*, Lecture Notes on Data Engineering and Communications Technologies 236,
https://doi.org/10.1007/978-981-96-0451-7_20

2 Literature Survey

This study [1] compared the performance of several deep learning algorithms, including CNN [7], LSTM [5], Bi-LSTM, GRU [3], Bi-GRU, and a hybrid CNN-LSTM model. According to experimental data, their solution, called "hybrid CNN-LSTM multivariate," outperforms all the stated classical models in terms of predictive performance and allows for more accurate predictions.

This research developed a prediction method based on integrated dual LSTM (Long Short-Term Memory) model [9], aiming to reduce the complexity of air quality prediction. First, a single-factor prediction model is created using the Sequence-to-Sequence (Seq2Seq) technology, which can determine the expected value of each component in the air quality data separately. In the forecasting process, every aspect of air quality is treated as a time series of data.

Our investigation focuses on three specific models: Long Short-Term Memory (LSTM), Gated Recurrent Unit (GRU), and DeepAR. Each of these models is carefully examined and succinctly described to highlight their distinctive characteristics and potential for effective air quality prediction [4].

2.1 LSTM (Long Short-Term Memory)

LSTM (Long Short-Term Memory) is an improved version of traditional Recurrent Neural Networks (RNNs) that addresses the problem of vanishing gradients. It introduces a memory block with memory cells and multiplicative gates. The memory block operates independently without external signal inputs. The core architecture of LSTM includes an input gate, a forget gate, and an output gate, which function within the range of 0 to 1 and are controlled by the sigmoid function. These gates regulate the flow of information and enable LSTM to capture long-term dependencies effectively (Fig. 1).

The LSTM architecture estimates an output $ot = (ot_1, ot_2, ot_3, ..., ot_{T-1}, ot_T)$ by updating three multiplicative units (input I output (op), and forget gate (fr)) on the memory cell with continuous write, read, and reset operations on the memory cell (mc) from time t = 1 to T. LSTM formulas are listed in Fig. 2. Since LSTMs are frequently used for sequential analysis, they can be trained to forecast air quality index levels for the upcoming hour or even the upcoming month using the historical data gathered by sensors at different weather stations. A lot of research has happened in the recent years and many researchers including [9–12] have implemented LSTM for predicting air pollution.

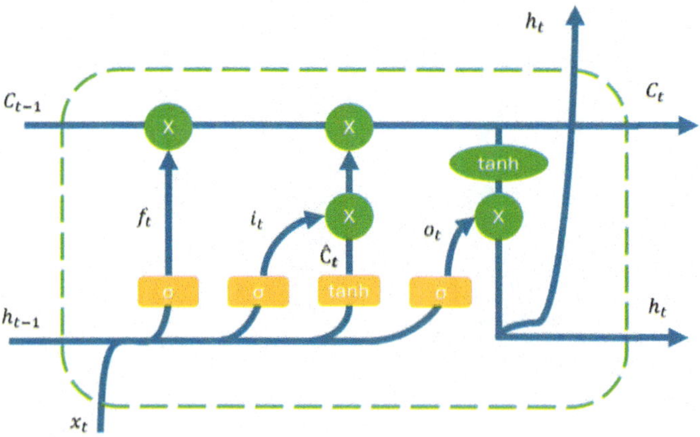

Fig. 1 Above shows these three sigmoid gates. The cell state is represented by a horizontal line that passes through it

Fig. 2 Where og and oh are activation functions; Wf, Wo, Wc are linear transformation matrix; bf, bt, bo, bc are corresponding bias. The final output is represented by the last hidden state

- **Forget Gate:** $f_n = \sigma_g(W_f \cdot [h_{n-1}, x_n] + b_f)$
- **Memory Gate::** $i_n = \sigma_g(W_i \cdot [h_{n-1}, x_n] + b_i)$
- **Temporary cell state:** $\tilde{C}_n = \sigma_h(W_C \cdot [h_{n-1}, x_n] + b_C)$
- **Current Cell State:** $C_n = f_n * C_{n-1} + i_n * \tilde{C}_n$
- **Output Gate:** $o_t = \sigma_g(W_o \cdot [h_{n-1}, x_n] + b_o)$
- **Hidden State:** $h_n = o_t * \sigma_h(C_n)$

2.2 *GRU (Gated Recurrent Unit)*

In addition to LSTM, the Gated Recurrent Unit (GRU) [13] is another recurrent neural network architecture used in air quality prediction. GRU incorporates update and reset gates to control information flow within the unit. It is comparable to LSTM and designed to handle time series data. The update gate determines the relevance of candidate activation for updating the cell state, while the reset gate resets the previous state between activations (Fig. 3).

The update gate is used to regulate how much data from hidden states in the past is carried over into the current state. More information about past states is brought in as the update gate value increases. The amount to which the information from earlier stages is discarded is controlled by the reset gate, and the smaller the value of the reset gate, the more it is ignored. As a result, long-term dependencies are accompanied by the activation of update gates, but short-term relationships are typically captured with frequent activation of reset gates. Though these architectures are quite simple, they can be very helpful and cost efficient for some predictive use cases. The paper [13] has used GRU for predicting the air pollution concentration.

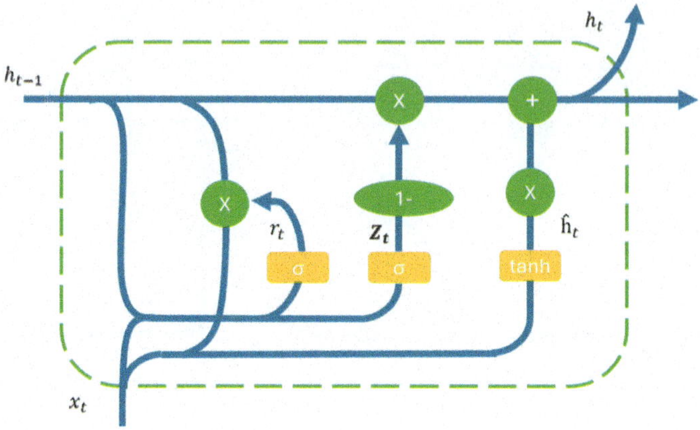

Fig. 3 Architecture of GRU cell

2.3 DeepAR

DeepAR [8] is a specialized forecasting algorithm developed by the Amazon research group. It is designed to predict scalar one-dimensional time series data. Unlike traditional methods like ARIMA [6] and ETS, DeepAR can learn from multiple related time series and utilize their combined knowledge for accurate forecasting. It outperforms traditional algorithms when applied to datasets with correlated time series. The model incorporates tunable hyper-parameters such as context_length, prediction_length, learning_rate, dropout_rate, embedding_dimension, num_cells, and num_layers. Notably, context_length determines the visibility of past records, while prediction_length defines the number of future records the model can accurately predict (Fig. 4).

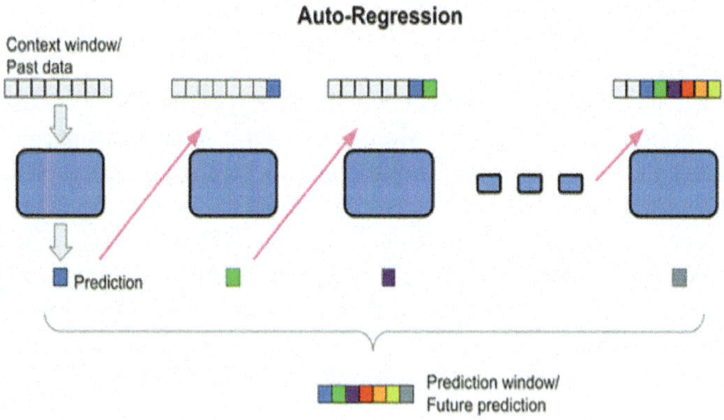

Fig. 4 Working of auto-regression

2.4 Auto-regression

Auto-regression is a predictive method that utilizes a linear combination of past values of a variable for forecasting. It involves constructing an auto-regressive model of order "p," denoted as AR(p), where the prediction at time step "t" is a function of the past values and a white noise term. The auto-regressive model equation is expressed as $y_t = c + \phi_1 y_{t-1} + \phi_2 y_{t-3} + \ldots + \phi_p y_{t-p} + \epsilon_t$ where y_t is the predicted value at time step "t"; ϕ is the intercept term; $\phi_1, \phi_2, ..., \phi_p$ are the parameters multiplied by the lagged values; and ϵ_t represents the white noise term. When building an auto-regressive model, several factors should be considered, including the length of the context window and the length of the prediction window. The context window determines the number of previous predictions ($y_{t-1}, y_{t-2}, ..., y_{t-p}$) used in the linear combination for making new predictions (y_t). The prediction window, on the other hand, defines the number of time steps the auto-regressive model can be invoked to forecast the future. For instance, an auto-regression with a prediction window of 10 can make predictions up to 10 days ahead, with each time step representing a single day. Thus, each new prediction relies on the previous predictions within the defined context window.

3 Implementation

3.1 Data

The data for this study was sourced from the Air Quality Data in India dataset available on Kaggle. This dataset contains air quality measurements from various monitoring stations across India. It includes data collected over several years on PM2.5 at hourly intervals.

Dataset Description:

- Source: The data was collected from official government monitoring stations and compiled into a comprehensive dataset by Kaggle user fedesoriano.
- Time Period: The dataset spans from 2017 to 2022, providing a wide temporal range for analysis.
- Geographic Coverage: The data covers major cities in India, including Delhi, Mumbai, Kolkata, Chennai, and Bangalore, among others.
- Pollutants Measured: The key pollutant measured is PM2.5 (particulate matter with a diameter less than 2.5 micrometers).

Data Preprocessing:

- Missing Values: The dataset contains some missing values, which were handled using interpolation techniques to ensure the continuity of time series data.

- Normalization: The data was normalized to a common scale to facilitate the comparison between different pollutants and cities.
- Aggregation: For certain analyses, the data was aggregated from an hourly basis to a daily basis to identify broader trends and patterns.

Usage in Study:

- Training and Testing: The dataset was divided into training and testing sets. The training set was used to develop the predictive models, while the testing set was used to evaluate their performance.
- Model Input: The dataset served as the input for various deep learning models, including LSTM, GRU, and DeepAR, enabling the prediction of air quality indices for different time horizons.
- By utilizing this extensive and detailed dataset, the study aims to provide accurate and reliable air quality predictions, contributing to better environmental management and public health strategies in India.

3.2 *Visualizing of Time Series*

Rolling analyses [14] are commonly employed to assess the stability of time series models over time. When analyzing financial time series data, it is crucial to assume that the model parameters remain constant. In order to visualize the anticipated air quality plot, the matplotlib library is utilized. The provided graphs exhibit the air quality data for India spanning the period of 2017 to 2022, as well as the air quality data for a specific year (2017–2018). To ensure consistency, the data has been aggregated from an hourly basis to a daily basis. This visualization aids in understanding the evolving patterns in the data across different time periods (Figs. 5 and 6).

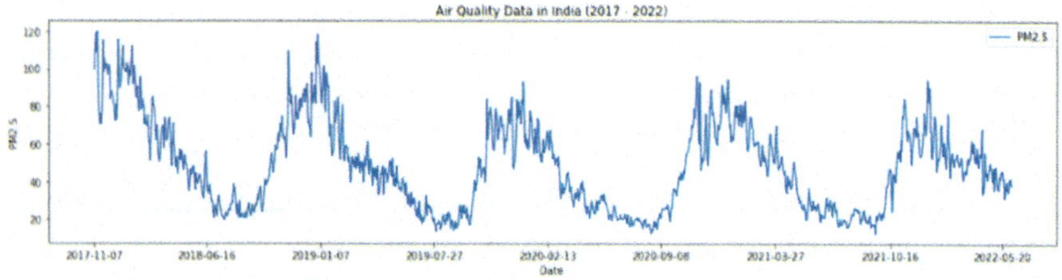

Fig. 5 Plot of data over time

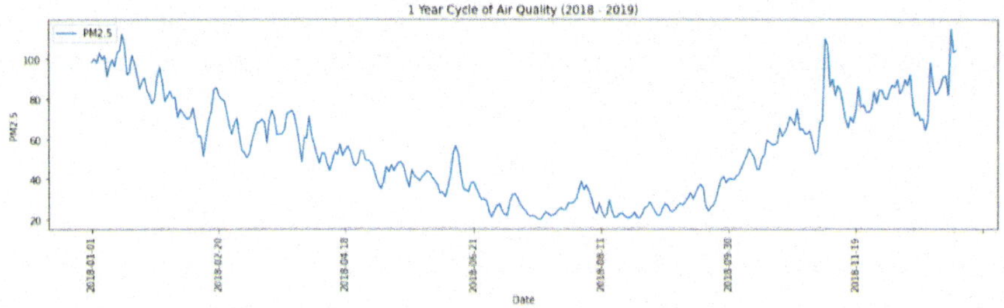

Fig. 6 Plot representing a subsection of data over time (1 month)

3.3 Prediction Using LSTM Model

The proposed model follows a sequential architecture, consisting of a linear stack of layers. The inclusion of the dense layer allows for dimensionality transformation of the output vector during backpropagation. Subsequently, a regression model is implemented using a recurrent neural network (RNN). Sequential data is input into the model, which is then incorporated into the regressor for prediction. The neural network undergoes training for prediction, utilizing randomly assigned biases and weights. The LSTM model comprises a sequential input layer, three LSTM layers, and a dense output layer with a linear activation function. The output generated by the RNN's output layer is compared to the desired value. The backpropagation algorithm is employed to minimize the error, which represents the discrepancy between the desired output and the actual output produced by the model (Figs. 7 and 8).

Output Value - The testing scores obtained for the LSTM model are as follows: - Test Score: 4.82 RMSE. - Test Score: 0.93414 R2.

We also employ the auto-regression technique over the existing LSTM model with the length of the context window as 100 and the length of the prediction window to be 365. This model takes 100 days of data as input and can predict up to 365 days in the future which is a full 1-year prediction based on the past 3 months' data. The following graph shows the output of the auto-regression model trying to predict 365 days into the future using past data.

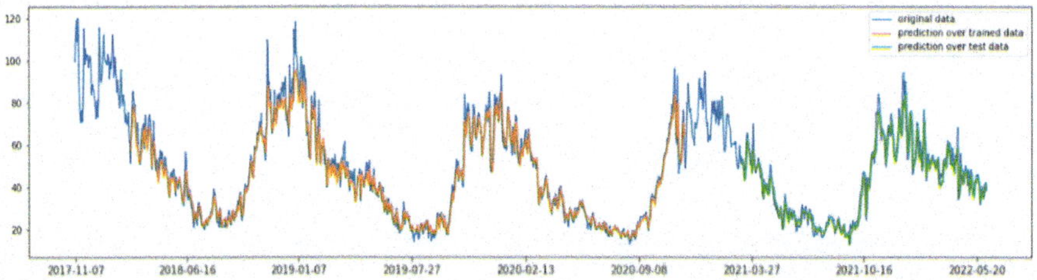

Fig. 7 The graph above is plotted against model predictions over training data, testing data, and actual data

As the context window is 100, the model's predictions are reliable up to 100 days into the future, i.e., a prediction window with 100 lengths is reliable. However, further predictions are made based on the previous predictions which will result in higher deviation from the original data.

3.4 Prediction Using Bi-Directional LSTM

Similarly, we experimented using a bi-directional LSTM model with 1 layer as well as an add and the outputs turned out to be as follows (Fig. 9 and 10):

The testing scores and root mean squared errors obtained for this model are Test Score: 4.80 RMSE. Test Score: 0.93461 R2.

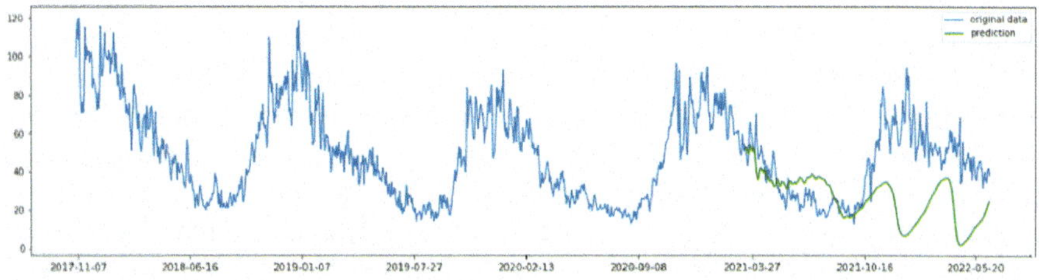

Fig. 8 Auto-regressive LSTM output

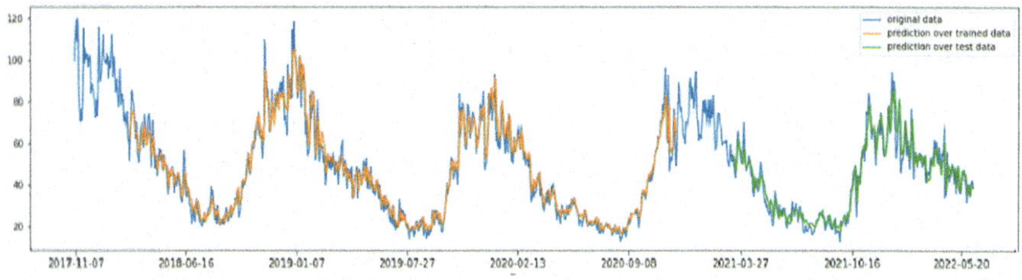

Fig. 9 The graph obtained below is plotted against predictions over training and testing data overlapped over original values

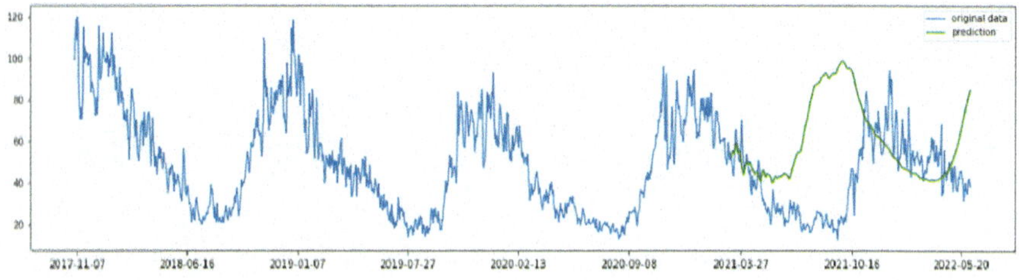

Fig. 10 Auto-regressive Bi-LSTM output

Similarly, we also employed an auto-regressive approach to the existing Bi-LSTM model and the following graph represents the prediction of 365 days based on the past data.

3.5 Prediction Using GRU Model

Our experiment consists of the usage of GRU for prediction. The test scores obtained for GRU are as follows (Figs. 11 and 12):

Test Score: 4.80 RMSE. Test Score: 0.93461 R2.

3.6 Prediction Using Bi-Directional GRU Model

Predictions from bi-directional GRU are as follows. The test scores are (Figs. 13 and 14: Test Score: 4.80 RMSE. Test Score: 0.93461 R2.

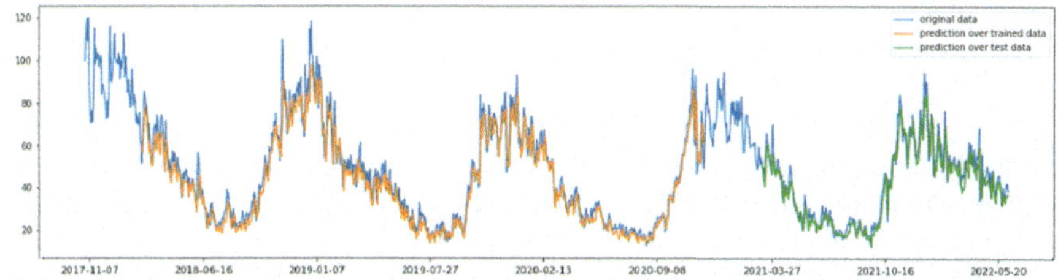

Fig. 11 The graph obtained above is plotted against predictions over training and testing data overlapped over original values

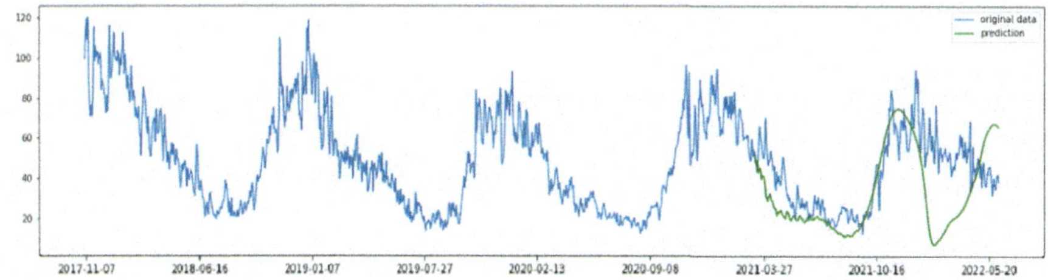

Fig. 12 Auto-regression output for GRU

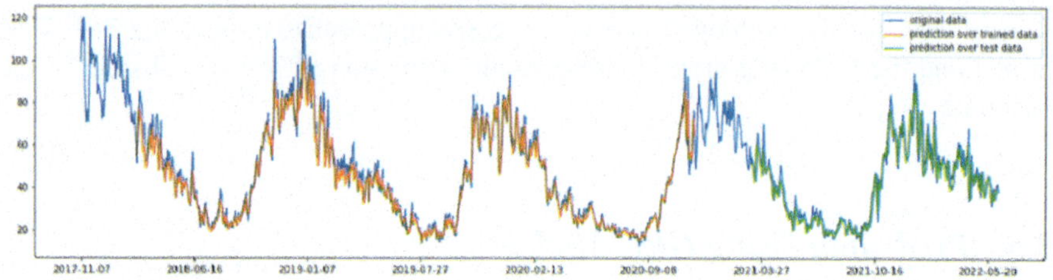

Fig. 13 The graph obtained below is plotted against predictions over training and testing data overlapped over original values

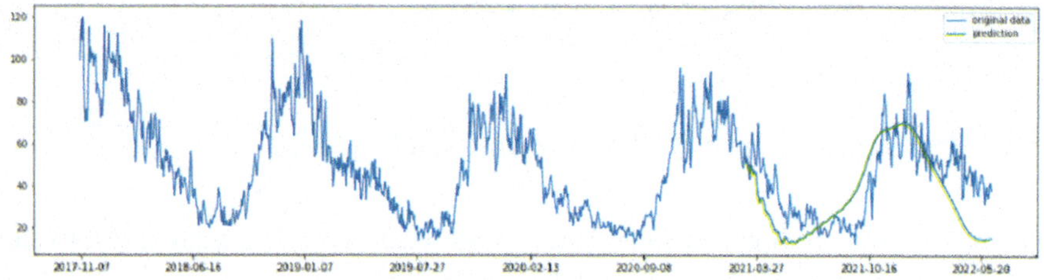

Fig. 14 Auto-regression output for GRU is as follows

3.7 Prediction Using DeepAR

DeepAR algorithm has also been implemented over the data. The experiment was to understand and compare the performance of various models versus the state-of-the-art DeepAR model. The model performed well over the data but the training time was relatively huge and was increasing exponentially with an increase in prediction_window. By using a prediction window of 100 days, the model was trained for 10 epochs and 3 minutes per epoch (Fig. 15).

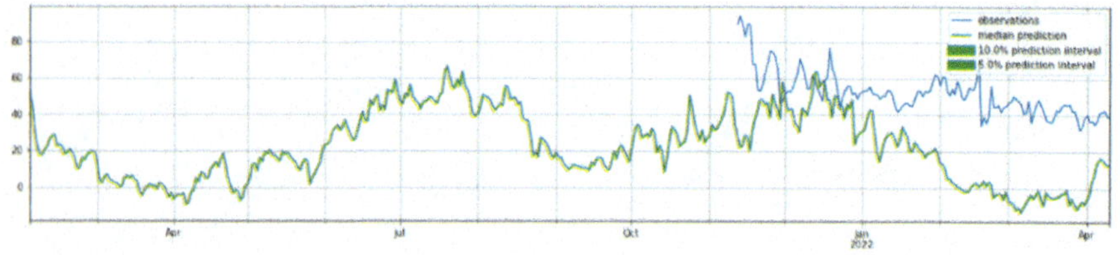

Fig. 15 The results of the DeepAR model are provided in the graph above

4 Conclusion

The R-squared (R2) values of all the models fall within the range of 0.90 to 0.92. The bi-directional GRU model achieved the highest score with an R2 value of 0.9175. Based on the findings, the Bi-GRU network demonstrates slightly superior performance compared to the LSTM network. Moreover, the bi-directional GRU model exhibits a lower error value compared to other models, indicating that its utilization can enhance the accuracy of predictions. This improvement can be attributed to the bi-directional GRU's ability to process time series data both chronologically and anti-chronologically. By capturing patterns that unidirectional GRUs may overlook, the bi-directional GRU enhances the feature learning capability of time series models.

5 Future Work

In future work, the model can be improved by using a larger dataset for evaluation and extending the prediction capabilities to other pollutants. The application of Convolutional Neural Network (CNN) techniques can be explored to detect subtle variations in air pollution data. Furthermore, analyzing the relationships between different characteristics can uncover hidden parameters that may influence the performance of distinct features. Considering the promising results of LSTM and GRU models, investigating the potential synergies of combining these models could lead to improved prediction accuracy. Additionally, future research can focus on incorporating external factors such as weather conditions and industrial activities to enhance the predictive capabilities. Exploring ensemble techniques and model interpretability methods can further strengthen the robustness and interpretability of the air quality prediction models.

References

1. Bekkar A, Hssina B, Douzi S, Douzi K (2021) Air-pollution prediction in smart city, deep learning approach. J Big Data 8:1–21
2. Chen H, Guan M, Li H (2021) Air quality prediction based on integrated dual LSTM model. IEEE Access 9:93285–93297
3. Dey R, Salem FM (2017) Gate-variants of gated recurrent unit (GRU) neural networks. In: 2017 IEEE 60th international midwest symposium on circuits and systems (MWSCAS). IEEE, pp 1597–1600
4. Du S, Li T, Yang Y, Horng S-J (2019) Deep air quality forecasting using hybrid deep learning framework. IEEE Trans Knowl Data Eng 33(6):2412–2424
5. Graves A, Graves A (2012) Long short-term memory. Supervised sequence labelling with recurrent neural networks, pp 37–45
6. Nelson BK (1998) Time series analysis using autoregressive integrated moving average (ARIMA) models. Acad Emerg Med 5(7):739–744
7. O'shea K, Nash R (2015) An introduction to convolutional neural networks. arXiv:1511.08458

8. Salinas D, Flunkert V, Gasthaus J, Januschowski T (2020) Deepar: probabilistic forecasting with autoregressive recurrent networks. Int J Forecast 36(3):1181–1191
9. Song X, Huang J, Song D (2019) Air quality prediction based on lstm-kalman model. In: 2019 IEEE 8th joint international information technology and artificial intelligence conference (ITAIC). IEEE, pp 695–699
10. Tiwari A, Gupta R, Chandra R (2021) Delhi air quality prediction using LSTM deep learning models with a focus on covid-19 lockdown. arXiv:2102.10551
11. Wang J, Li J, Wang X, Wang J, Huang M (2021) Air quality prediction using CT-LSTM. Neural Comput Appl 33:4779–4792
12. Zhang L, Liu P, Zhao L, Wang G, Zhang W, Liu J (2021) Air quality predictions with a semi-supervised bidirectional LSTM neural network. Atmos Pollut Res 12(1):328–339
13. Zhou X, Xu J, Zeng P, Meng X (2019) Air pollutant concentration prediction based on GRU method. J Phys: Conf Ser 1168:032058. IOP Publishing
14. Zivot E, Wang J, Zivot E, Wang J (2023) Rolling analysis of time series. Modeling financial time series with S-Plus®, pp 299–346

Insider Threat Detection in User Activity Data Using Optimized LSTM-AE

M. Krishna Kumar, Saloni Kumari, M. Bharathi, P. Lavanya, H. Anila Glory, and V. S. Shankar Sriram

Abstract Insider threat is exploiting the user's authorized access, which poses significant risks to the organization. Insider attacks are a distinct and sophisticated form of security hazards within the organization. Insiders can be employees of an organization, business partners, and vendors. In order to identify and detect insider threats, enormous traditional security actions such as firewalls and antivirus software exist. However, these actions focus on detecting external attacks rather than insider threats. This insufficiency encourages cybersecurity researchers to develop and implement a more refined insider threat detection technique. To overcome the abovementioned issues, this research concentrates on an optimized sequential neural network approach to detect insider threats with a high detection rate. It is evaluated using the benchmark dataset of CMU CERT r5.2 consisting of heterogeneous and highly imbalanced user activity logs. The proposed approach includes three modules to rectify the challenges in the dataset: (i) Obtaining the selective samples based on user activity for the curse of dimensionality, (ii) Implementing clustered sampling to cluster scenario-based samples that solve class imbalance problem which has the majority of benign samples and minority of insiders, and (iii) Deploying optimized Long Short-Term Memory (LSTM)-Autoencoder (AE) to detect the potential insiders with high detection rate 0.99 and less false negative rate 0.03 within an organization. The performance of the proposed approach is ensured by comparing it with state-of-the-art methods.

Keywords Insider threat · Insider threat detection · User activity · Content · Selective sample · Class imbalance · Long short-term memory-autoencoder

M. K. Kumar · S. Kumari · M. Bharathi · P. Lavanya · H. A. Glory (✉) · V. S. S. Sriram
School of Computing, Centre for Information Super Highway (CISH), SASTRA Deemed University, Thanjavur 613402, Tamil Nadu, India
e-mail: anilaglory@cse.sastra.edu

A. C. Frery et al. (eds.), *Computational Intelligence and Data Analytics*, Lecture Notes on Data Engineering and Communications Technologies 236,
https://doi.org/10.1007/978-981-96-0451-7_21

1 Introduction

The deployment of advanced technology improves the profitability and effectiveness of organizations affected by cyber attacks. Compared to external cyber attacks, organizations' network infrastructure is mainly criticized by internal attacks, known as insiders [1]. Recently, the report stated that important actors of insider threats are employees, contractors, and authorized users, with the insider threat vectors being privilege breaches and diversified mistakes that affect 71% of organizations. Some insider threat events are credential stuffing of authorized users to inject the attack and unauthorized access of confidential data, leading to data leakage, thus raising the essentiality of insider threat [2].

The detection and prevention of insider threats are highly complex these days due to the abnormal activities of users with legitimate access to organizations' network infrastructure. The software-as-a-service of the cloud has enormous data leakage and uses more external devices to access corporate resources. Traditional insider threat detection (ITD) approaches need to be revised to detect and identify the essential insiders within the organizations [3]. However, to condense the security harm caused by insider threats, cybersecurity researchers are still working on constructing ITD mechanisms. Therefore, developing a neural network model offers a great chance to detect insider threats with better performance. However, collecting real-time insider threat data is still challenging. Hence, the Computer Emergency Response Team (CERT) benchmark dataset from Carnegie Mellon University (CMU) is heterogeneous and possesses a skewed class distribution [4]. The issues of the dataset affect the performance of the ITD model. A sequential neural network detects insider threats with a high detection rate. This research work aims to solve these issues. The significant contributions of the research work are as follows:

- User activity-based behavioral sample selection is carried out for five files of CMU CERT releases r5.2, including logon, device, file, email, and HTTP. It includes content-based sample selection to notify the users of behaviors in files, email, and HTTP.
- Clustered sampling balances the identified class imbalance problem in the dataset. Depending on time-inclined user activity, classify the user and then form the random cluster using the scenarios of insiders. Each cluster should contain users from a specific scenario.
- Optimized Long Short-Term Memory-Autoencoder (LSTM-AE) is a Sequential Neural Network model using an RMS prop optimizer for detecting insider threats with a high detection rate. Comparing the different optimizers such as Adam, Stochastic Gradient Descent (SGD), Adadelta, AdaGrad, Adamax, and Gradient Descent with Momentum ensures the Root Mean Square propagation (RMSprop) works better than all the optimizers.

The rest of the research article is structured as follows: Sect. 2 is a detailed study of related works of ITD using machine learning, deep learning, and neural networks. Section 3 presents a description of the proposed model's three modules.

Section 4 discusses experimentation and results comparing different optimizers and state-of-the-art methods; the final section is the conclusion.

2 Related Works

The importance of identifying and detecting insider threats within an organization is explored by relevant studies addressing the issues of ITD using machine learning, deep learning, and neural networks. Xiao et al. [5] proposed a Convolutional Attention combined with a Transformer Encoder (CATE) and statistically integrated sequential analysis for ITD. The proposed model is evaluated using CMU CERT, which extracts time-based important features. However, the advanced transformer must ensure that the identification of time-based user behavior has satisfied all the necessary insider scenarios. Kan et al. [6] suggested a data-adjusting approach and optimized the XGBoost algorithm to detect insider threats. It is implemented in CMU CERT r4.2, and it underwent user behavior modeling by employing the data-adjusting strategy to eliminate meaningless data and the probability density curve to identify insiders among product line workers. However, the scalability and generalizability of the proposed model must be discussed to ensure that the model applies to other applications. Randive and Ramasundaram [7] had a novel approach, "MWcapsNet: A Multilevel Wavelet Capsule Network for insider threat detection" using the image. Converting the CSV file of the CMU CERT dataset into image representation using the IGTD framework, which collects the local and global features. The multilevel wavelet augments the model's ability to experience the relationship between multi-scale-based spectral and spatial features from the image. It proves that the model's performance was increased compared to other existing ITD models. Nevertheless, the computational complexity of the proposed model has not been described. d'Ambrosio et al. [8] had developed a Bayesian Threat Graph Network to manage the risk factors of insider threats. The proposed model includes an "exploit dependency graph" to address the scalability challenges in a traditional enumeration graph structure and explains the importance of bearing the risk management of insider threats owing to the ambiguity of different external factors. The Bayesian Belief networks mainly concentrate on temporal and spatial complexity feature data. However, the proposed model offers high computational complexity. Racherache et al. [9] had implemented an ITD model for user profiling and cyber-persona identification, which creates persona profiles depending on users' network traffic patterns. The proposed supervised machine learning model segregates the user-generated data from the device to achieve high accuracy, AUC, and f1-score. Nevertheless, the user's network traffic patterns rely solely on the context-dependent indicators of insiders. Overall, the complications of ITD are high dimensional, heterogenous, and imbalanced data are addressed by employing the appropriate sample selection technique and sampling technique for solving class imbalance problem and the presence of a neural network model for detecting the insiders with high detection rate and false negative rate.

3 Proposed Method

The proposed ITD comprises three modules: Module 1 is the user activity-based behavioral sample selection to achieve dimensionality reduction. The clustered sampling technique is deployed to solve the class imbalance problem in module 2, and module 3 is the optimized LSTM-AE of sequential Neural Network for ITD (see Fig. 1).

3.1 *Data Collection*

The insider threat data is collected from different IT organizations, which are user activity logs. Usually, organizations produce a massive volume of data, organized as log data and time-stamped with user action details. The log data are aggregated and scrutinized to derive the structure of the data within organizations [10]. The Computer Emergency Response Team (CERT) from Carnegie Mellon University established the most common and benchmark dataset of insider threats [4].

3.2 *Data Preprocessing and Feature Extraction*

It is necessary to preprocess the log dataset of insider threats, which are heterogeneous and high dimensional. The dataset comprises categorical and ordinal values, which do not convey any information for further process. The label encoding method is used to encode the original data, and based on the user's activity, the data are preprocessed to remove the noisy and redundant values. Then, the preprocessed data is forwarded to reduce the dimensionality by selecting the samples depending on time-based user activity and content-based user activity, which are combined and named as user

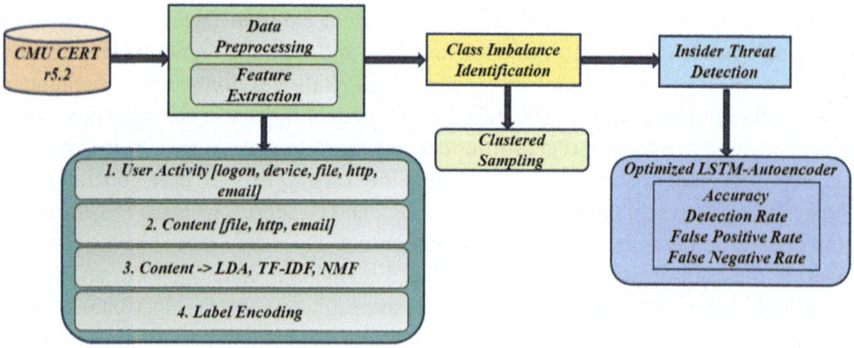

Fig. 1 Architecture of insider threat detection

activity-based behavioral sample selection. In the case of content-based user activity, the methods of Term Frequency-Inverse Document frequency (TF-IDF) and Non-negative matrix Factorization (NMF), and Latent Dirichlet allocation (LDA) are used for selecting the essential samples from the description of the available folders. The time-based user activity is deployed to select the sample of user actions with time from user activity concentrated folders.

TF-IDF is a statistical method primarily used for text analysis tasks such as document classification, clustering, and information retrieval. It is used for retrieving information and mining text to measure the importance of a word in a document. In term frequency in Eq. (1), it computes the recurrent occurrence of a specific term present in a document. The importance of TF-IDF is to identify the terms that are essential to the document [11].

$$tf\,(t, d) = (t_n)_d \Big/ (t_d) \tag{1}$$

where $(t_n)_d$ is count of specific term in the document and (t_d) is the total number of terms present in document. In IDF in Eq. (2), the weight of the frequent occurrence of the exact term in the corpus is reduced because those terms are not essential to the document. It measures the importance of the particular term across the document.

$$idf\,(t, D_c) = \log \frac{TN\,(D)_C}{N_D} \tag{2}$$

where $TN\,(D)_C$ is the total number of documents containing in Corpus and the N_D is the specific term present in the document. Then, the TF-IDF in Eq. (3) can be represented in the form

$$tf - idf\,(t, d, D_c) = (t_n)_d \Big/ (t_d) \,*\, \log \frac{TN\,(D)_C}{N_D} \tag{3}$$

The NMF unsupervised algorithm reduces the dimensionality of input data without disturbing its underlying structure. It decomposes the original matrix into two lower-rank non-negative matrices. It is used to obtain essential and interpretable samples from the content and find a more concise original structure of the data. Depending on the iterative procedure, the initial values are modified to obtain the matrix, and it might conclude by completing the particular number of iterations or converging at error occurrence [12].

LDA is a generative statistical model under Natural Language Processing (NLP) that defines dataset generation. It is used for topic modeling to catch essential topics from the corpus. In LDA, each corpus document combines different topics, and each topic is a group of words that tend to appear together. In general, the model assumes that the document has information about multiple topics and that each word relates to one. It analyzes the co-occurrence pattern of words and then automatically discovers the underlying topic present in the document. The objective is to find topics in a way

Fig. 2 Flowchart of clustered sampling

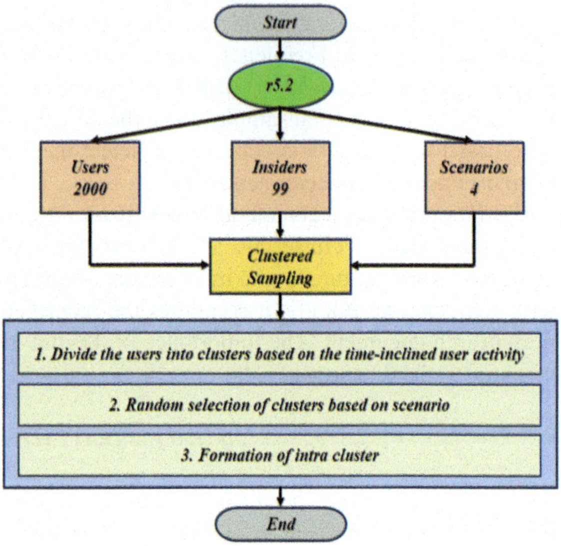

that helps us understand what the document contains. It helps find the document's primary context without any human input [13].

3.3 Sampling the Imbalance Data

The selected sample of user activities based on time and content shows that insider threat data contains a majority of normal samples and a minority of malicious samples, and it concludes that the preprocessed data is purely class-imbalanced and heterogeneous. To address the class imbalance problem, one of the random sampling methods, known as clustered sampling, is deployed to balance the samples [14]. The Cluster sampling divides the samples into a more significant number of subgroups at the initial stage, and then, depending on the criteria of insider threat data, the clustered are selected randomly (see Fig. 2).

3.4 User Activity-Based Insider Threat Detection Using Optimized LSTM-AE

The sampled trained and tested insider threat data were taken for ITD using the sequence neural network of optimized long-short-term memory—autoencoder. Different types of optimizers, such as Adam, Stochastic Gradient Descent (SGD), degrade, adamax, rmsprop, and Gradient Descent with Momentum (GDWM), are used in LSTM-AE [15] to notify the detection of insider threat and classify the attack. Among the optimizers, the rmsprop optimizer in LSTM-AE has achieved high DR with less FNR while detecting the insiders. The structure of LSTM-AE includes the LSTM encoder in input layers and the LSTM decoder in output layers. Combining LSTM with AE learns compressed representations of sequential data. LSTM's ability to capture and retain long-term dependencies makes it well-suited for encoding sequential data. The sequence form of insider data fed into the LSTM network is encoded in compressed representations.

The latent space of LSTM captures the most important features of the sequential data from the compressed representations. The obtained compressed representation is then forwarded to another LSTM network known as a decoder. The work of the LSTM decoder is to reconstruct the original input sequence data from the compressed representation. The output of the LSTM decoder is compared with the original input sequence data; then, the autoencoder is trained to minimize the reconstruction error. It encourages the autoencoder to learn a compressed representation that preserves the essential characteristics of the input data. The dense layers of LSTM-AE convert the encoded data into the concerned output format and learn the complex pattern in the data. The dense layer usually has a size corresponding to the dimensionality of the input data. The repeat vector is one more layer present in the LSTM-AE, which is used to repeat the encoded representation of the input sequence before passing it through the decoder LSTM layers. It ensures the decoder receives the encoded representation as input at each time step. It can effectively absorb the representations of sequences while handling long-term dependencies and is more essential than traditional AE.

4 Experimentation and Results

4.1 Experimental Setup

The experimentation was carried out on a PC with an INTEL Core i7-1255U, 1.70 GHz CPU, and 16 GB RAM running Windows 11 OS. Different optimizers were implemented in LSTM-AE using Python 3.10 with essential packages and libraries [16].

4.2 Dataset Description

The CMU CERT insider threat test dataset is used to experiment with the different optimizers-based LSTM-AE. It has released from r1 to r6.2, the releases from r3 to r6 have two versions, r3.1, r3.2, r4.1, r4.2 and r5.1, r5.2 and the latest versions are r6.1, r6.2 with the activity log records of 17 months based on five different scenarios from Dec 2009 to May 2011 of Lightweight Directory Access Protocol (LDAP). Release r5.2, with its scenarios, is implemented using five folders: logon, device, file, email,

Table 1 Secarios of insider threat data of r5.2

Scenarios	Description
1	Employee behavioral changes are logon after office timing, thumb drives, organization data upload to external sources
2	Employee violations are job search in rival companies and transfer of high-rate data using thumb drives
3	System administrator's abuse the downloads a keylogger and uses a thumb drive to transfer it to his supervisor's machine and using key logs, he is logging in as the supervisor and sending alarming emails to the employees
4	Another machine accesses the employee of an organization access another employee system and transfer all important files and emails to his/her system

HTTP, and psychometric [4]. The common attributes of these files are ID, date and time, user, and pc. The r5.2 has four scenarios are listed in Table 1.

4.3 Results and Discussion

The real insider threat data has user activity log files of logon, device, file, email, and HTTP, which are formatted into a single homogenous file with LDAP. The single homogeneous files are high dimensional, heterogenous, and have skew class distribution. The file is preprocessed by computing the user activity per day of the week based on the count of PC usage occurrences. Only the most frequent PC for a day is selected for all records. Once the single homogeneous files are preprocessed, the sample selection is carried out to reduce the dimensionality of the insider threat data. The time-based user activity samples are extracted for logon and device records, and the content-based user activity samples are extracted from the file, email, and HTTP. Depending on the office hours, which start from morning eight to five, the evening is projected for a day with its corresponding user actions in the login and device. The TF-IDF, NMF, and LDA were implemented to extract necessary samples depending on the content of the file, email, and HTTP. The attributes of the file's content are file_tree, email is content, and HTTP is website links. In the primary step, applying TF-IDF, all the three-content computes the recurrent occurrence of a specific term present in a record by term frequency, and then the weight of the frequent occurrence of the exact term in the record is reduced because those terms are not essential to the document that measures the importance of the particular term across the file, email and HTTP by inverse document frequency. By observing the important term from the record, the NMF is applied to reduce the dimensionality of the data by converting the low-rank non-negative matrices by an iterative procedure. The LDA combines the different topics, and each topic is a group of words that tend to appear together. When the feature extraction is completed and obtained, the dimensionality reduces samples by merging all the records into a single file, splitting the data into training and testing data, and identifying the data with a class imbalance problem.

The classified 70% of training and 30% of testing data have defined the majority of normal samples and the minority of malicious samples, which leads to a class imbalance problem. The imbalanced insider threat data are solved using random sampling. Among the various random sampling methods, cluster sampling is used to diversify the sample into subgroups by time-inclined user activities. Depending on the insider threat scenarios, the subgroups are then randomly generated with scenarios. The cluster sampled data are forwarded to the detection model of LSTM-AE, which is trained with 30 epochs, batch size = 10,000, learning rate = 0.0001, and optimizer is rmsprop. The input and output dimensions are similar, and the loss function is computed by comparing the input and output data sequence using Mean Squared Error (MSE). In the model training stage, the input is passed through the dense and repeat vector, then it reaches the output layer, and depending on the malicious or non-malicious sample, the model is trained. During the testing phase, the model occurred with extensive reconstruction errors demonstrating abnormal behavior. In order to differentiate the malicious and non-malicious samples, a threshold value is specified. If the threshold value with reconstruction error is greater, the sample is declared malicious or non-malicious. The performance of the optimized LSTM-AE is evaluated in terms of accuracy, precision, detection rate (DR), false alarm rates of false positive rate (FPR), and false negative rate in the Eqs. (4), (5), (6), and (7).

$$\text{Accuracy} = \frac{\text{TP} + \text{TN}}{(\text{TP} + \text{TN} + \text{FP} + \text{FN})} \tag{4}$$

$$\text{Detection Rate (DR)} = \frac{\text{TP}}{(\text{TP} + \text{FN})} \tag{5}$$

$$\text{False Positive Rate (FPR)} = \frac{\text{FP}}{(\text{TN} + \text{FP})} \tag{6}$$

$$\text{False Negative Rate (FNR)} = \frac{\text{FN}}{(\text{FN} + \text{TP})} \tag{7}$$

The optimized LSTM-AE is compared with the existing models, such as bidirectional LSTM-CNN and normal LSTM-CNN, in terms of accuracy, DR, FNR, and loss. The working of LSTM is similar for all three BiLSTM-CNN, LSTM-CNN, and optimized LSTM-AE. However, the hyperparameter tuning and the add-on models combined to produce better performance. The optimized LSTM-AE has higher DR 0.9904 and 0.0095 as FNR than other models listed in Table 2. Then, the optimized LSTM-AE is compared with various optimizers such as SGD [19], Adam [17], and Adamax [20]. Adadelta [21], GDWM [22], Adagrad [18], and RMS prop in terms of accuracy (ACC), Detection Rate (DR), False Positive Rate (FPR), False Negative Rate (FNR), and region of characteristics (ROC) are listed in Table 3.

Among the optimizers, the RMSprop has performed better due to preventing oscillations. Its different behavior leads to increased stability convergence, and it adapts

Table 2 Optimized LSTM-AE compared with existing models

ITD detection model	ACC	DR	FNR	Loss
BiLSTM-CNN	0.519	0.534	0.012	0.069
LSTM-CNN	0.437	0.456	0.013	0.250
LSTM-AE	**0.943**	**0.990**	**0.009**	**0.057**

Table 3 Comparison of various optimizers in LSTM-AE

Optimizer	ACC	DR	FPR	FNR	ROC
SGD	0.923	0.987	0.775	0.012	0.93
GDWM	0.925	0.990	0.680	0.009	0.95
Adam	0.939	0.987	0.600	0.012	0.94
Adagrad	0.921	0.982	0.797	0.017	0.91
Adadelta	0.607	0.989	0.802	0.010	0.90
Adamax	0.921	0.986	0.606	0.013	0.95
RMSprop	**0.943**	**0.990**	**0.621**	**0.009**	**0.95**

by modifying the learning rate depending on the gradient scales, which are advantages over SGD and GDWM. In Adam and Adadelta's case, the rmsprop initially sets the simple learning rate that normalizes the hyperparameter tuning and makes rmsprop more robust. Compared to Adam and Adamax, the rmsprop is computationally efficient. Overall, the rmsprop is more efficient regarding momentum than another optimizer. Therefore, in LSTM-AE, the rmsprop has achieved a high detection rate (DR) of 0.994 and a false alarm rate FNR of 0.009 in Table 3 and the optimizers-based LSTM-AE model for epoch 30 and the accuracy of each model (see Fig. 3) and the confusion matrix of non-malicious and malicious (see Fig. 4).

5 Conclusions

Insider threats pose a substantial hazard to organizations, which makes it challenging to provide the necessary detection and mitigation approaches. The identification and detection of insider threats is a difficult task in cybersecurity. Cybersecurity researchers in insider threat detection face challenges in the available insider threat data. Recently, neural network models have played an important role in detecting and mitigating insider threats. In this research work, the sequence neural network of LSTM-AE is used to detect the insider threat. The user activity-based behavioral insiders are detected by selecting the samples based on user activity and the content of the user's document. Then, selected samples are clustered based on insider scenarios. The optimized LSTM-AE of ITD is identified by comparing the experimental results of various optimizers with those of other neural network models. Thus, the optimized

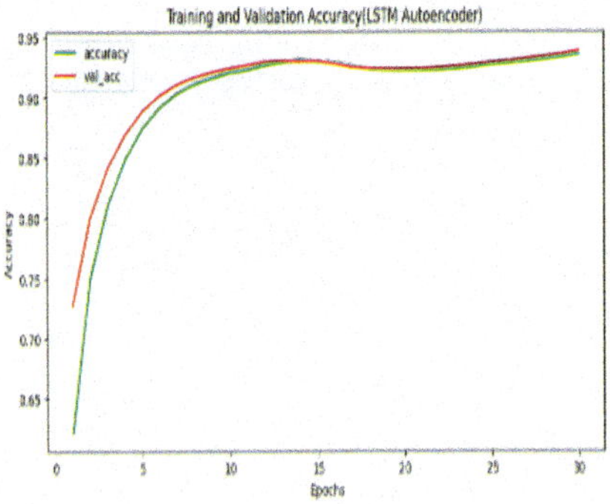

(a)

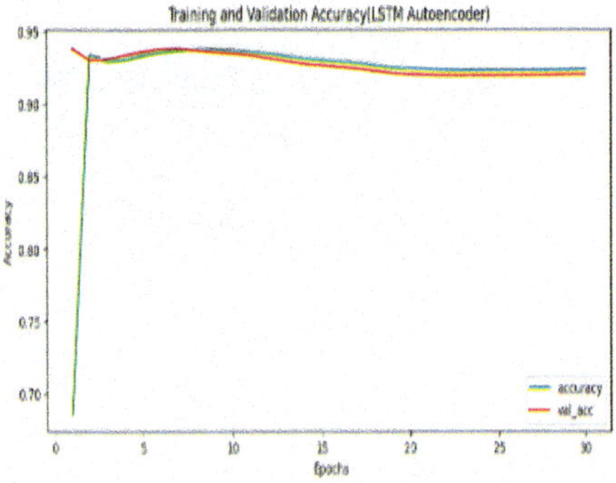

(b)

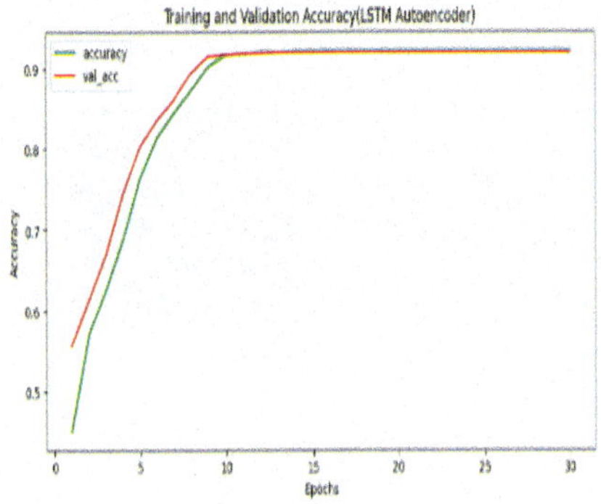

(c)

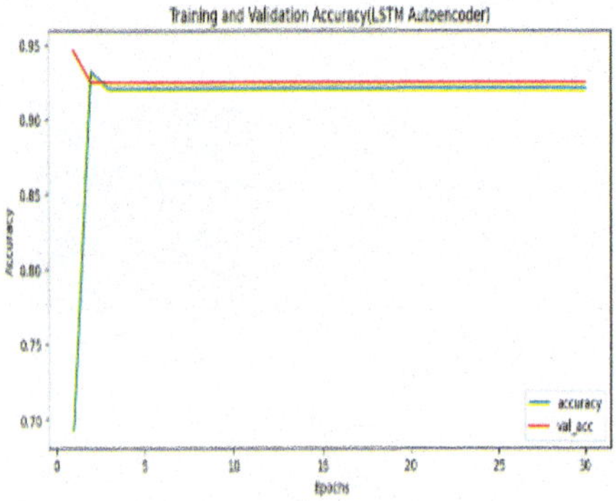

(d)

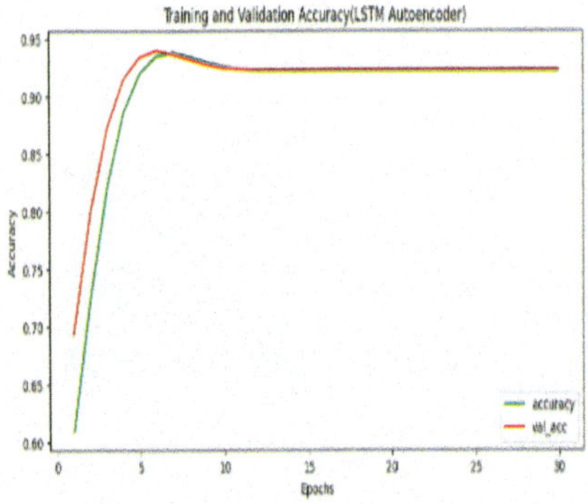

(e)

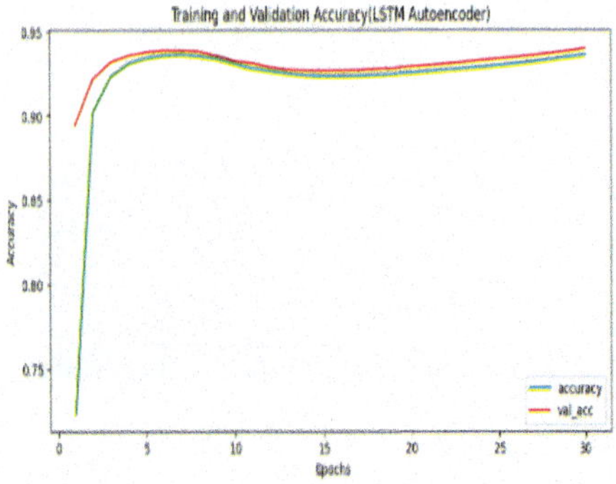

(f)

Fig. 3 Comparing the results of accuracy versus epoch for training and validation data of **a** Adam, **b** Adamax, **c** Adagrad, **d** GDWM, **e** SGD, **f** RMSprop of LSTM-AE

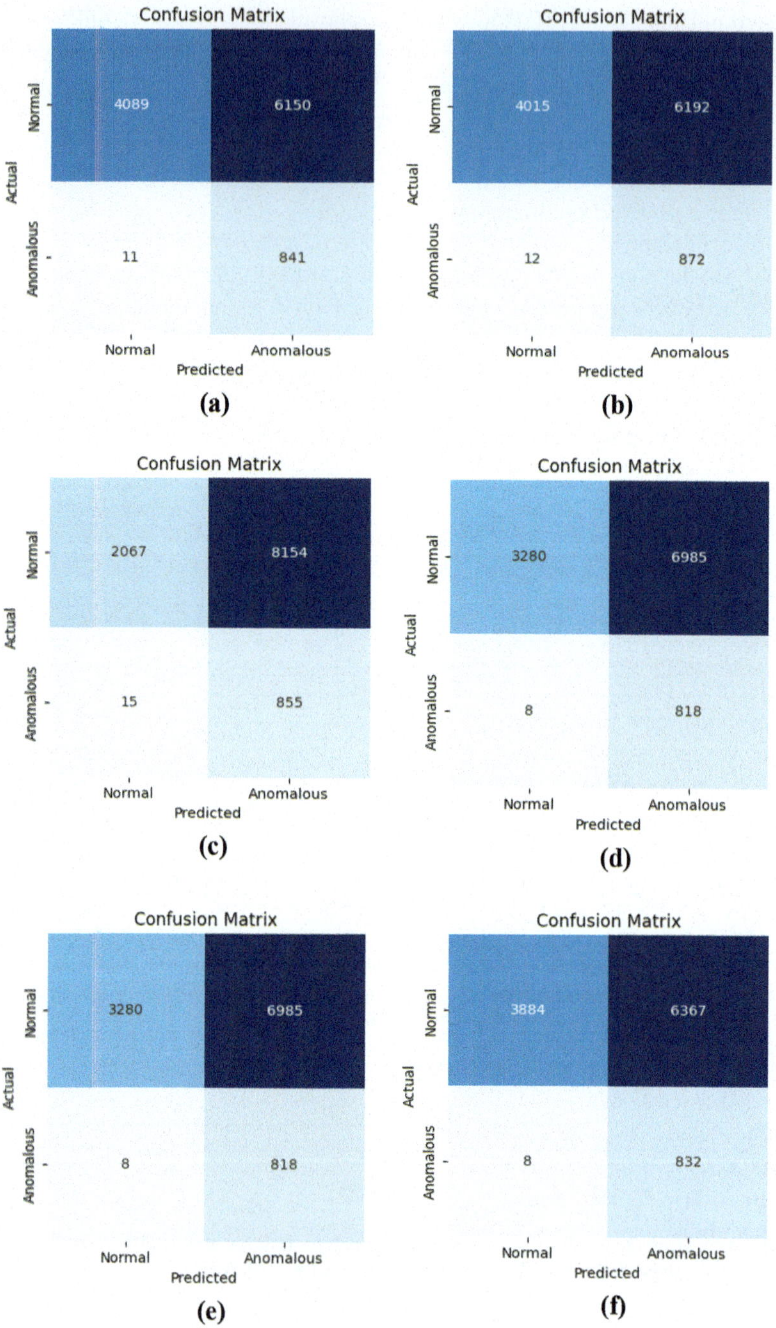

Fig. 4 Comparing the results of confusion matrix for insider threat data of **a** Adam, **b** Adamax, **c** Adagrad, **d** GDWM, **e** SGD, **f** RMSprop of LSTM-AE

LSTM-AE has achieved a high detection rate and minimized false negative rate. However, in the future, the framed ITD model will be deployed in real-time insider threat data for detection and mitigation.

References

1. Yuan S, Wu X (2021) Deep learning for insider threat detection: Review, challenges and opportunities. Comput Secur 104:102221
2. https://www.ekransystem.com/en/blog/insider-threat-statistics-facts-and-figures#id-main-reasons-for-insider-threat-incidents. Accessed 28 March 2024
3. Bin Sarhan B, Altwaijry N (2022) Insider threat detection using machine learning approach. Appl Sci 13(1):259
4. https://kilthub.cmu.edu/articles/dataset/Insider_Threat_Test_Dataset/12841247. Accessed 13 Feb 2024
5. Xiao H, Zhu Y, Zhang B, Lu Z, Du D, Liu Y (2024) Unveiling shadows: a comprehensive framework for insider threat detection based on statistical and sequential analysis. Comput Secur 138:103665
6. Kan X, Fan Y, Zheng J, Chi CH, Song W, Kudreyko A (2023) Data adjusting strategy and optimized XGBoost algorithm for novel insider threat detection model. J Franklin Inst 360(16):11414–11443
7. Randive KD, Ramasundaram M (2023) MWCapsNet: a novel multi-level wavelet capsule network for insider threat detection using image representations. Neurocomputing 553:126588
8. d'Ambrosio N, Perrone G, Romano SP (2023) Including insider threats into risk management through Bayesian threat graph networks. Comput Secur 133:103410
9. Racherache B, Shirani P, Soeanu A, Debbabi M (2023) CPID: Insider threat detection using profiling and cyber-persona identification. Comput Secur 132:103350
10. Pal P, Chattopadhyay P, Swarnkar M (2023) Temporal feature aggregation with attention for insider threat detection from activity logs. Expert Syst Appl 224:119925
11. Janjua F, Masood A, Abbas H, Rashid I, Khan MM (2021) Textual analysis of traitor-based dataset through semi supervised machine learning. Futur Gener Comput Syst 125:652–660
12. Kan X, Fan Y, Zheng J, Kudreyko A, Chi CH, Song W, Tregubova A (2023) User-level malicious behavior analysis model based on the NMF-GMM algorithm and ensemble strategy. Nonlinear Dyn 111(22):21391–21408
13. Mittal A, Garg U (2023) Prediction and detection of insider threat detection using emails: a comparison. In: 2023 Second international conference on electrical, electronics, information and communication technologies (ICEEICT). IEEE, pp 01–06
14. Wang J, Sun Q, Zhou C (2023) Insider threat detection based on deep clustering of multi-source behavioral events. Appl Sci 13(24):13021
15. Sivakrishna AM, Mohan R, Randive K (2023) AUBIT: an adaptive user behaviour based insider threat detection technique using LSTM-autoencoder. In: Recent trends in computational intelligence and its application. CRC Press, pp 267–274
16. https://www.python.org/downloads/release/python-3100
17. Han Z, Tian H, Han X, Wu J, Zhang W, Li C, Qiu L, Duan X, Tian W (2024) A respiratory motion prediction method based on LSTM-AE with attention mechanism for spine surgery. Cyborg Bionic Syst 5:0063
18. Kim T, Lee S (2024) Deep learning integrated Bayesian health indicator for cross-machine health monitoring. Struct Health Monit 14759217241227599
19. Mavikumbure HS, Cobilean V, Wickramasinghe CS, Varghese BJ, Carlson B, Rieger C, Pennington T, Manic M (2024) Cy-Phy ADS: cyber physical anomaly detection framework for EV charging systems. IEEE Trans Transp Electrification

20. Dabaghi-Zarandi F, Behjat V, Gauvin M, Picher P, Ezzaidi H, Fofana I (2024) Using deep learning to detect anomalies in on-load tap changer based on Vibro-Acoustic signal features. Energies 17(7):1665
21. Jogunola O, Ajagun AS, Adebisi B, Aibinu AM, Ojo JA (2021) Comparative analysis of hybrid deep learning frameworks for energy forecasting. In: The 5th international conference on future networks and distributed systems, pp 214–219
22. Liu YT, Chen JJ, Tseng YC, Li FY (2022) An auto-encoder multitask LSTM model for boundary localization. IEEE Sens J 22(11):10940–10953

Learning Emotions : A Comparative Study of EEG-Based Emotion Recognition with Machine Learning and Deep Learning Techniques

Bhargav Burman and Anupam Agrawal

Abstract EEG signals can be used as an accurate metric for human emotions. Using machine learning, and subsequently deep learning, we can accurately classify EEG signals and predict emotions. Several deep learning methodologies have been experimented with for this purpose. Thus, we have reviewed notable published research on the field and compared their results. EEG is a complex data format that holds further potential for analysis. Therefore, we have also compared and discussed the contribution of different feature extraction techniques in EEG data. We have tested traditional machine learning models (SVM, kNN, MLP and ELM) and deep learning models ViT and RNN. The results show deep learning models with attention mechanisms give the best accuracy in emotion recognition.

Keywords Electroencephalogram · Emotion detection · Self-attention · Multi-channel · Feature extraction · Transformer

1 Introduction

Even with the recent advancements in technology, we know very little about emotions and the working of the human mind for a specific emotional state. Thus, a longstanding field of research has been recognition and identification of the emotional state of a person. This research has taken many forms. With HCI rising as a discipline, the focus has been on using machine and deep learning models for the purpose of emotion classification. Also, emotional data can be recorded in many forms. Facial data, text data, audio data are all common forms of emotional data representation [1, 2].

B. Burman (✉) · A. Agrawal
Department of IT, Indian Institute of Information Technology, Allahabad, Devghat, Jhalwa 211012, Prayagraj, India
e-mail: MHC2022011@iiita.ac.in

A. Agrawal
e-mail: anupam@iiita.ac.in

A. C. Frery et al. (eds.), *Computational Intelligence and Data Analytics*, Lecture Notes on Data Engineering and Communications Technologies 236,
https://doi.org/10.1007/978-981-96-0451-7_22

Electroencephalogram (EEG) signal is a representation of the electrical signals generated by brain activity [3]. Thus, information distilled from these signals can be used as an accurate gauge of the emotional state of mind in response to stimuli. Compared to external manifestations of emotion like facial expressions or gestures, extracting information from EEG signals is more direct and reliable. In addition, it is more subject-independent [4].

Human brain activity is mainly reflected in certain bands of frequencies of EEG signals. Thus, using these bands, several deep learning methodologies have been implemented to try and classify emotions.

Pusarla et al. [5] used Extreme Machine Learning (ELM) for classification EEG signals combined with wavelet transform. Before the advent of such specialised neural networks, general neural networks were being applied to emotion detection problems using EEG. Khubani et al. [6] used an ANN model optimised for better performance on EEG data. Traditional models also are still being researched with optimisations to try and achieve performance on par with deep learning techniques. Nivedha et al. [7] used the Support Vector Machine (SVM) [7] model along with Particle Swarm Optimisation [8] to achieve promising results in emotion detection. Anushka Pradhan et al. [9] have used such an optimisation technique on the ELM mode, termed the "Hierarchical extreme puzzle learning machine". Blaiech et al. [10] have also modified the ELM model to create their own "OS-ELM" model. Bhatt et al. [11] have used the modified LibSVM model on the emotion detection task.

The process later achieved a breakthrough with the introduction of attention to deep learning techniques. Attention is a fundamental mechanism in deep learning that allows models to focus on relevant information while ignoring irrelevant information. Attention can be added in various ways.

Mohsen et al. [12] used the LSTM [12] model for EEG emotion detection. The intuition behind the application being that EEG is a long-time series data and requires suitable processing techniques. Pratiwi et al. [13] used LSTM [12](Long Short Term Memory) and Bi-directional LSTM to classify emotions between happy and sad. Tao et al. [14] used an attention-based Recurrent Neural Network (ACRNN) [14] for the classification of images based on arousal and valence. The method uses an additional convolutional neural network (CNN) for attention on specific channels. Arjun et al. [15] used the powerful Vision Transformer (ViT) [16] model to classify emotions. The Transformer model is of special note as it has intrinsic attention capabilities. It itself consists of the attention layers stacked with feed-forward layers [17]. RNNs have the ability to handle time series data with long-term dependencies but are not parallelisable [18]. Transformers do not suffer from this shortcoming of RNNs, as a natural consequence of its architecture.

In this paper, traditional machine learning models and deep learning models have been compared, on their performance on EEG emotion detection. We have also discussed and contrasted the feature extraction techniques and their advantages. There is also a discussion on the trend of attention mechanisms—both current and emerging, that may hold further scope for improvement in emotion detection.

2 Background

For emotion recognition, we need a digital representation of the emotions. The data recorded by EEG have high precision time measurements that can be closely studied. With modern advances in the technology, the electrodes are only ever in direct contact with the scalp, so it is a safe and non-invasive procedure for the subject as well. For these reasons listed above, EEG data have been used extensively for emotion detection through machine learning methods (Fig. 1 shows an example of such data with 32 channels).

The EEG signals examine brain electrical activity that relates to thinking and emotions as discovered in medical studies [20]. The amplitude varies across the range 2–10μV, with frequencies in range 1–100 Hz. These are divided into five sub-bands named delta(<4 Hz), thete(4–7 Hz), beta(14–30 Hz) and gamma (31–100 Hz). In this, the delta waves are only observed in deep sleep [20].

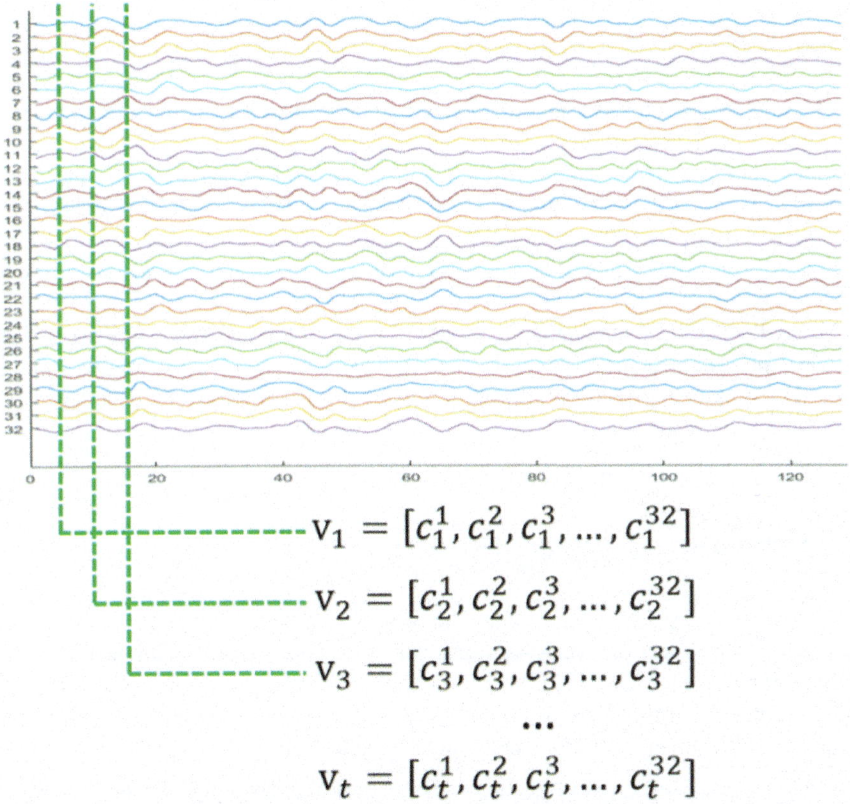

Fig. 1 32 channel EEG data [19]

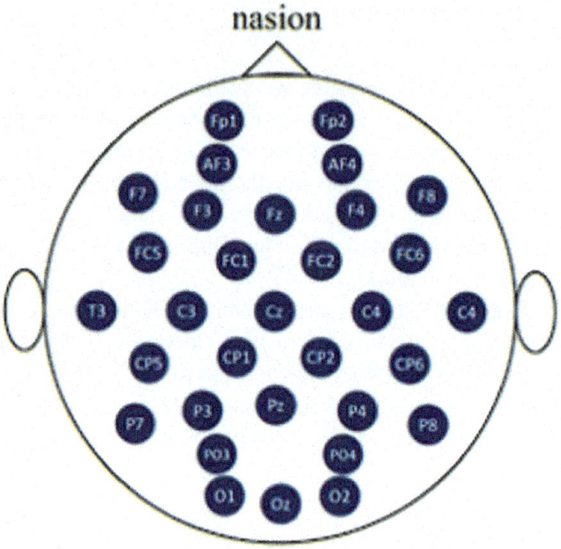

Fig. 2 Example of 10/20 system of electrode [19]

The electrodes, placed over the scalp, record brain waves using the internationally recognised 10/20 system [21]. This signifies that the adjacent electrodes are placed at intervals either 10% or 20% of the skull's front-back or right-left distance [21] (as shown in Fig. 2). Also standardised is the nomenclature of EEG electrodes: Frontal(F), Temporal(T), Occipital(O), parietal(P). This is followed by a number, which is even on the right side of the brain (F8, C4, etc.) and odd for the left side (F7, T5, etc.).

Russell et al. [22] suggested the modern format of arousal valence plane format of emotion representation, which is more prevalent in recent years. In this format, the EEG signals are rated on arousal and valence and emotions are designated (as shown in Figs. 3 and 4).

3 Methods

The methods to be detailed for understanding the comparative work have been separated into three sections corresponding with the three integral components of the study. Firstly, we have detailed the traditional machine learning methods. Following this, the deep learning methods used in the study have been detailed. Lastly, the multi-resolution feature extraction methods used have been explained.

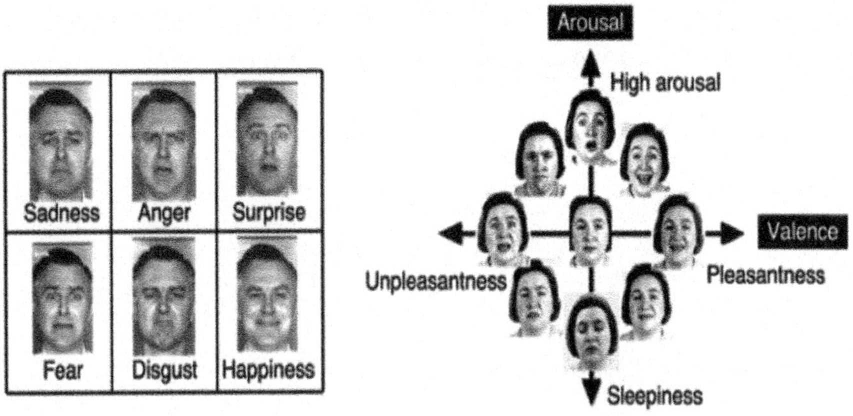

Fig. 3 The two common emotional representations: categorical(left) and 2-D arousal-valence plane(right) [23]

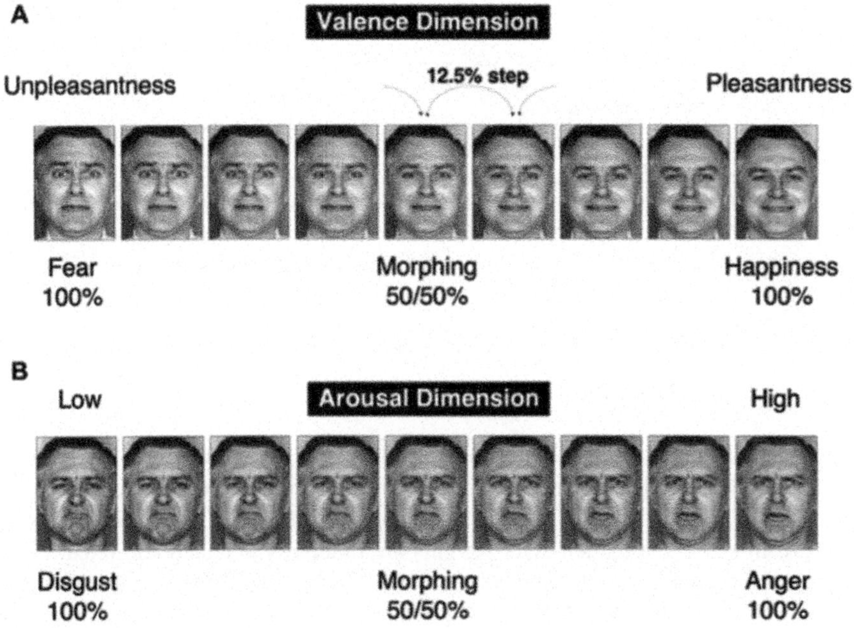

Fig. 4 Emotional position in arousal-valence plane and representative emotion [23]

3.1 Traditional Machine Learning Methods

We have compared the task of EEG emotion recognition of some traditional machine learning models using shifting window feature extraction of traditionally prevalent Short-time Fourier Transform using Welch's methods. These are ELM, SVM, MLP and kNN. These provide a baseline performance to be expected from the use of traditional machine learning techniques in EEG detection.

3.1.1 Support Vector Machine

Support Vector Machine (SVM) is an acclaimed machine learning algorithm, first developed for supervised classification tasks. It functions by determining the best hyperplane that separates data points into different classes [24].

3.1.2 Multi-Layer Perceptron

Multi-Layer Perceptron (MLP) is a simple neural network that can perform classification and regression. MLP consists of multiple layers of interconnected neurons, termed nodes. It has backpropagation for training, adjusting weights to minimise prediction errors, which enables it to learn complex non-linear relationships [24].

3.1.3 K-Nearest Neighbour

k-Nearest Neighbour (kNN) is a traditional machine learning algorithm used for classification and regression. It classifies data points based on the majority class of their nearest neighbours. The number of such neighbours is denoted as k, and it is a user-defined parameter. It can find very complex patterns in the data. It has no explicit training phase; predictions are made at runtime [24].

3.1.4 Extreme Learning Machine

Extreme Learning Machine [25] (ELM) is a machine learning paradigm that belongs to the family of single-layer feedforward neural networks (architecture given in Fig. 5). It was introduced as a novel learning algorithm by Guang-Bin Huang et al. [25]. In ELM, the hidden layer weights are randomly assigned and fixed, eliminating the need for iterative optimisation during training. This contributes to the fast learning speed of ELM. ELM also consists of a single hidden layer between the input and output layers. This simplicity reduces computational complexity and training time. ELM has been shown to be a universal approximator, meaning that it can approximate any continuous function with arbitrary accuracy given a sufficient number of hidden nodes.

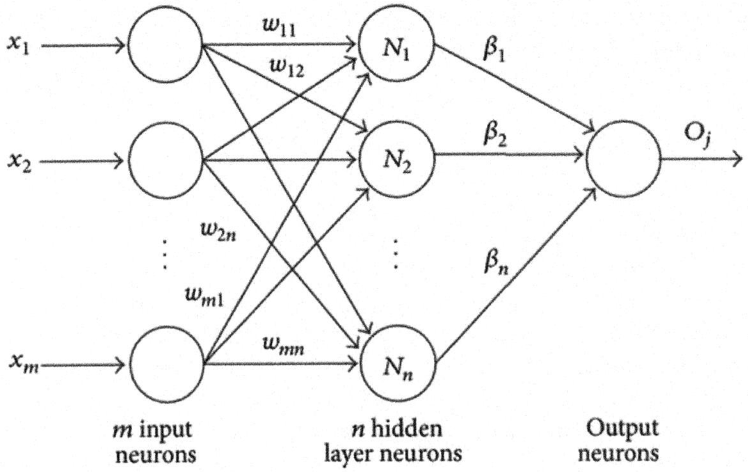

Fig. 5 Extreme learning machine architecture [26]

3.2 *Deep Learning Methods*

We have focused on the state-of-the-art models, which have been deep learning-based. Here, we describe the attention mechanism of such models, which the study found were integral to their performance. Following that is the description of two such recent deep learning architectures we have used for comparison, along with their relevant feature extraction techniques.

3.2.1 Attention in Deep Learning

The success of the state-of-the-art deep learning models in the field of EEG emotion recognition can be partly attributed to their inclusion of attention mechanisms. These allow the models to properly process the long time-series data of EEG.

Attention mechanisms augment deep learning models by allowing selective focus on particular input elements, which are deemed important as context, improving accuracy in prediction [17]. They provide proper emphasis on relevant parts of the data, highlighting them to enhance overall model performance. The attention mechanism mimics this selective concentration by assigning weights to different input elements, which are a measure of the relative importance or relevance of each element, as perceived by the model. By incorporating attention, models can properly process long-range dependencies and improve accuracy [17]. Attention mechanisms comprise two main steps:

Computing attention scores: Calculating relevance between a query element and all other elements in the input.

Weighted summation: Creating a context vector to emphasise important information based on these attention scores.

Attention can be implemented in various ways. Earlier RNN models had a separate attention layer. LSTM models have attention as part of their modules. Transformers were envisioned as having stacked attention layers [17].

3.2.2 Attention-Based Convolutional Neural Network

ACRNN uses a CNN with integrated mechanism for channel-specific attention, to explore spatial information. This gives importance to different channels which is beneficial in co-dependant multichannel signals like EEG [14]. The extended self-attention mechanism also takes into account the temporal information in the EEG signals, which uses the LSTM method (as shown in Fig. 6).

3.2.3 Vision Transformer

The Vision Transformer, or ViT [16], is a deep learning architecture introduced for computer vision tasks.

ViT employs the transformer architecture, which was originally developed for sequence-to-sequence tasks in NLP. This architecture is based on self-attention mechanisms, enabling the model to find long-range dependencies in the input data. Unlike CNNs that operate on grid-like structures of pixels, ViT divides an input image into fixed-size non-overlapping patches, treating each patch as a token. These patches are

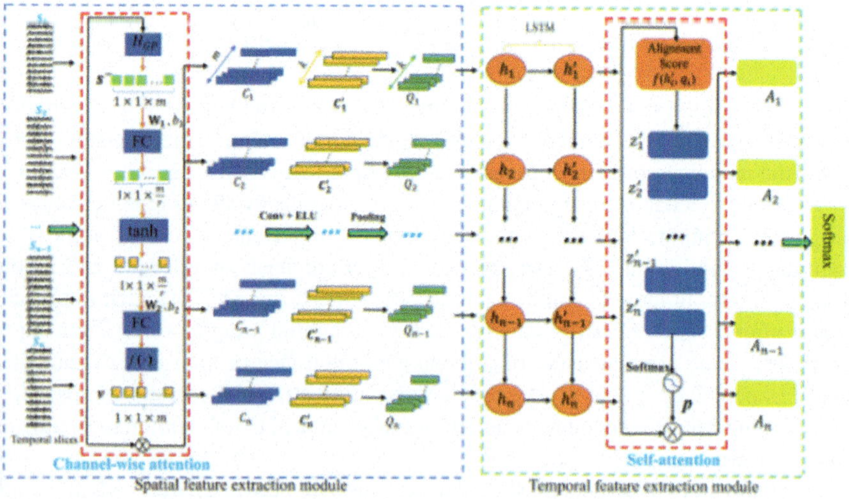

Fig. 6 ACRNN structure with channel-wise attention [14]

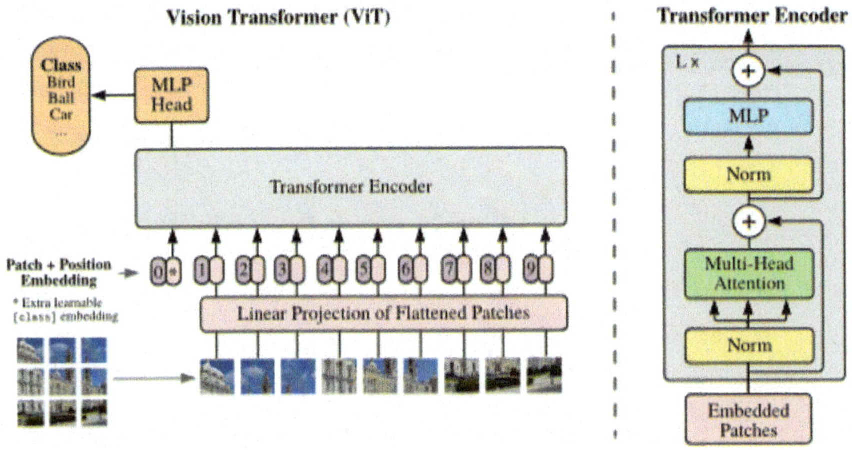

Fig. 7 Vision transformer architecture [16]

then linearly embedded and processed by the transformer. To retain spatial information, positional embeddings are added to the patch embeddings (process represented in Fig. 7) . The final output of the transformer is typically fed into a classification head to predict the class labels of the input image.

Its ability to capture global dependencies without relying on spatial hierarchies has made it a significant development in the field of deep learning for computer vision.

3.3 *Multi-Resolution Features and Their Necessity*

Multi-resolution feature extraction is valuable for non-stationary signals due to the dynamic nature of these signals over time. Non-stationary signals exhibit variations in their statistical properties, frequency content and other characteristics across different time intervals. Traditional Fourier analysis has a fixed time and frequency resolution, which may not be suitable for non-stationary signals. Multi-resolution analysis, like in wavelet transforms, can decompose the signal into various scales, providing localised information about different frequency bands and time intervals. It also allows for the adaptation of the analysis window to capture changes in the signal's characteristics over time [27] (example shown in Fig. 8).

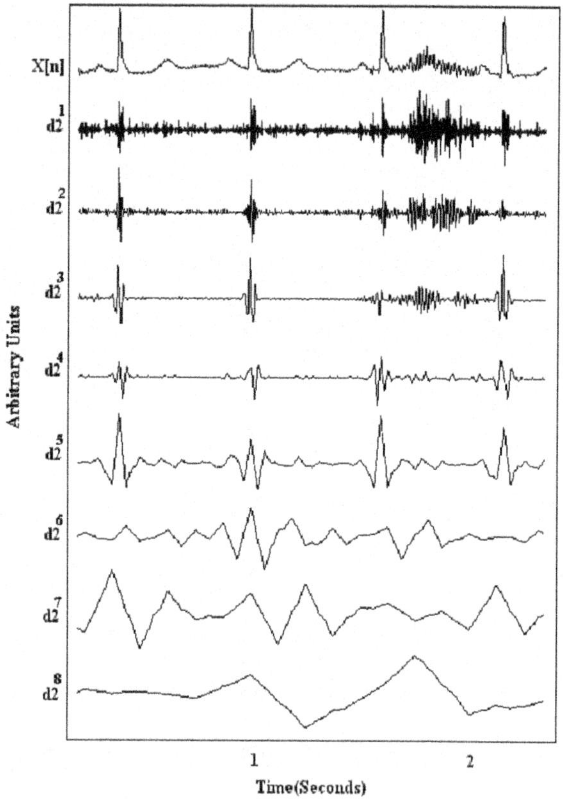

Fig. 8 Multi-resolution feature extraction on EEG [29]

3.3.1 Welch's Method

Welch's method is a signal processing technique used for calculating the power spectral density (PSD) of a given signal. Welch's method has the following steps—divide a signal into overlapping segments, apply a window function to each segment, compute the Fourier transform of each segment and then average the squared magnitudes of the resulting spectra to calculate the PSD [28] (example shown in Fig. 9). This method is particularly helpful when dealing with non-stationary signals, where the characteristics of the signal change over time.

3.3.2 Continuous Wavelet Transform

The Continuous Wavelet Transform (CWT) provides a way to decompose a signal into its constituent frequency components as they evolve over time. The CWT is calculated by convolving the signal with version of a wavelet function that has

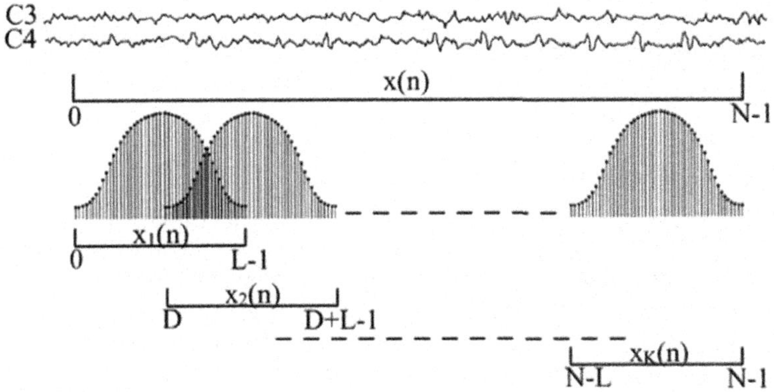

Fig. 9 Welch's method diagram [30]

been scaled and translated to coincide [31]. The formula for the continuous wavelet transform is given by:

$$CWT(a,b) = \int_{-\infty}^{\infty} x(t) \cdot \frac{1}{\sqrt{a}} \psi^* \left(\frac{t-b}{a} \right) dt$$

3.3.3 Db4 Wavelet

The Daubechies Wavelet, often denoted as dbN, is a family of orthogonal wavelets. The "db4" wavelet is a specific member of this family, and it is widely used in signal processing and image compression. The "4" in "db4" refers to the length of the filter, which is 4 [31].

The Daubechies wavelets have several desirable properties, making them popular in various applications. The Daubechies wavelets form an orthonormal basis for signal representation. This means that the wavelet coefficients provide a complete and non-redundant representation of the signal. The wavelet functions have compact support, meaning they are non-zero only over a finite interval. This property is useful in applications where localised information in both time and frequency domains is essential.

The mathematical expression for the "db4" wavelet function is defined in terms of its filter coefficients. For "db4," the filter coefficients are:

$$h = \frac{1+\sqrt{3}}{4\sqrt{2}}, g = \frac{3+\sqrt{3}}{4\sqrt{2}}, h' = \frac{3-\sqrt{3}}{4\sqrt{2}}, g' = \frac{1-\sqrt{3}}{4\sqrt{2}}$$

Here, h and g, respectively, represent low-pass and high-pass filter coefficients for analysis (decomposition), and h' and g' are the corresponding coefficients for synthesis (reconstruction).

3.3.4 Discrete Wavelet Transform

The Discrete Wavelet Transform (DWT) is another technique for analysis and processing of signals and images. DWT decomposes a signal into approximation coefficients (representing low-frequency components) and detail coefficients (representing high-frequency components). This is achieved through a series of filtering and downsampling operations. The decomposition is typically carried out using a filter bank structure. The signal is convolved with low-pass and high-pass filters, and the results are downsampled to produce the approximation and detail coefficients at a coarser scale [31].

Approximation Coefficients (low-pass filter):

$$a_{j+1}[n] = \sum_k h[k] \cdot a_j[2n - k]$$

Detail Coefficients (high-pass filter)

$$d_{j+1}[n] = \sum_k g[k] \cdot a_j[2n - k],$$

The coefficients obtained from the DWT represent the contribution of different frequency components at different scales. The original signal can be reconstructed by combining the approximation and detail coefficients through a process of upsampling and filtering. This allows for signal reconstruction with varying levels of detail.

It's important to distinguish the Discrete Wavelet Transform from the Continuous Wavelet Transform (CWT). DWT provides variable time and frequency resolution through its multi-resolution analysis. However, resolution is constrained to dyadic scales. It is more effective for sampled time series like a single-channel EEG signal CWT offers continuous and flexible time and frequency resolution, allowing analysis at any scale. Provides a continuous time-frequency representation. This makes it better suited for handling the combination of the EEG channels.

4 Experiments and Results

4.1 Dataset Description

The open source dataset DEAP [32] was used to evaluate the models. DEAP dataset is made up of EEG data recorded from 32 participants, with equal number of subject

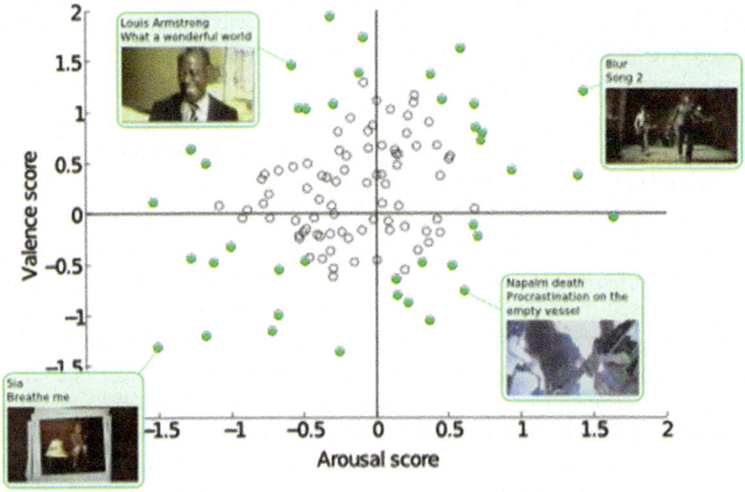

Fig. 10 DEAP dataset of video in arousal-valence plane [32]

of each gender. The participants were 40 videos to watch, each 1 minute long and EEG signals from the participants were recorded during the process. After watching, each participant rated the experience in terms of Valence, Arousal, Dominance and Liking (plot shown in Fig. 10). The collected raw EEG data were then subjected to extensive pre-processing. The steps taken for pre-processing are as follows:

1. The data were down-sampled to 128 Hz.
2. Electrooculogram (EOG) artefacts were separated out.
3. A band-pass filter with a range from 4.0 to 45.0 Hz was applied, to only get relevant signals.
4. The data were averaged to the common reference.
5. The data were divided into 60 s trial segments and the 3 s pre-trial baseline was removed.

4.2 Experimental Setting

The ELM, SVM, kNN and MLP were compared on arousal and valence scores. The traditional models have used Welch's method for power spectral density calculation. This has been used as the feature for classification. SVM was set the Radial Basis Function (RBF) kernel. The k-NN was using $k = 4$ corresponding with the four states of high valence-low arousal (HVLA), high valence-high arousal (HVHA), low valence-high arousal (LVHA), low valence-low arousal (LVLA) with uniform weightage. MLP was using the ReLU activation with ADAM solver and alpha value of 3, operation for maximum of 500 iterations. ELM was being tested with tanh and bounded ReLU activations with 1500 hidden nodes.

Deep learning-based models were also tested. The ViT model has intrinsic attention compared to the hybrid ACRNN model. But both used raw EEG data, wholly or partially, for their classification. EEG data can give unreliable results if used without processing. It also increases the training and testing time of the model. Thus, a separate ViT model was also developed as an experiment, which would use only CWT and DWT features.

Raw EEG is avoided as CWT features that can be more robust to noise and artefacts compared to raw EEG signals. By focusing on specific frequency components and time intervals, CWT can help mitigate the influence of background noise or non-brain-related artefacts, enhancing the signal-to-noise ratio and improving the discriminative power of the extracted features. CWT and DWT can effectively reduce the dimensionality of EEG data while preserving relevant information. By selecting a subset of informative wavelet coefficients or frequency bands, CWT features can capture the most discriminative aspects of the EEG signal, reducing the computational complexity of subsequent classification algorithms and improving generalisation performance.

The Vision Transformer model works by sectioning image into patches with their own embedding, reshaping each patch as a vector and providing the vectors as input. The EEG sequence can be sectioned into such patches and input into the transformer. The channel-specific attention mechanism works for patches of different channels, finding their dependencies.

4.3 Results and Discussion

In the traditional machine learning models, we see that ELM outperforms the traditional models by a small margin. It reaches a maximum accuracy of 62% in subject and band independent evaluation. This can be attributed to the good generalisation capabilities of ELM. It should also be noted that ELM also takes far less time (shown in Table 1).

Table 1 Traditional machine learning method comparison

Model	Accuracy (%)	F1 score
SVM	57	0.36
kNN	60	0.34
MLP	61	0.627
ELM	62	0.622

Table 2 Deep learning method comparison

Method	Model	Features	Accuracy (%)
Arjun et al. [15]	ViT	CWT and raw EEG	94
Tao et al. [14]	ACRNN	Raw EEG	92.4
Feature ViT(Experimental)	ViT	CWT and DWT	91.6

The deep learning models all outperform traditional models by a significant margin. The ViT model reports an accuracy of 94% on the DEAP dataset. The similarly attention-based ACRNN reported a maximum accuracy of 92.4% on the same dataset. The results show that the experimental ViT model, using only CWT and DWT features, in place of raw EEG, also performs reasonably well. In the experiments performed, the model reaches a peak accuracy of 91.6% (shown in Table 2). This is close to state-of-the-art performance without the complexity of processing raw EEG signals and the pitfalls that come with it. It can be seen that using the CWT and DWT approximation is still a viable option.

5 Conclusion

A thorough comparative analysis of various methods pertaining to EEG emotion recognition was conducted. Various traditional machine learning models were compared based on accuracy of arousal and valence. Following this, two of the most recent deep learning models were compared. An experimental model was developed using wavelet transform feature extraction techniques. Using both CWT and DWT for global and channel-specific attention, a vision transformer model was tested. On the DEAP dataset, the model gave near state-of-the-art results without the complexity of processing raw EEG data.

A few observations can be made from the study. We see that deep learning models with attention mechanisms outperform traditional methods by a significant margin. This can be attributed in part to attention mechanisms that allow them to perform well on long-series data. Transformers present a particularly fruitful proposition with intrinsic attention mechanisms. The current state-of-the-art focuses on channel-specific attention as well as overall attention for EEG. The study presented an approximation of channel-specific attention to prove its viability. Future work can focus on implementing more sophisticated architectures that prioritise channel-specific data. The current methods for processing hyper-spectral images show such suitable techniques.

References

1. Zhang X, Wang J, Cheng N, Xiao J (2022) Metasid: singer identification with domain adaptation for metaverse. In: International joint conference on neural networks. IEEE, pp 1–7
2. Duan W, Yi Yu, Zhang X, Tang S, Li W, Oyama K (2023) Melody generation from lyrics with local interpretability. ACM Trans Multimedia Comput Commun Appl 19(3):1–21
3. Pooja SKP, Veer K (2022) Recent approaches on classification and feature extraction of EEG signal: a review. Robotica 40(1):77–101
4. Zhao G, Ge Y, Shen B, Wei X, Wang H (2018) Emotion analysis for personality inference from EEG signals. IEEE Trans Affect Comput 9(3):362–371
5. Pusarla N, Singh A, Tripathi S (2022) Automatic EEG based emotion recognition using extreme learning machine. In: 2022 IEEE 9th Uttar Pradesh section international conference on electrical, electronics and computer engineering (UPCON), Prayagraj, India, pp 1–6
6. Khubani J, Kulkarni S (2022) Optimized artifical neural network For EEG-Based emotion recognition. In: 2022 6th international conference on computing methodologies and communication (ICCMC), Erode, India, pp 1624–1633
7. Nivedha R, Brinda M, Vasanth D, Anvitha M, Suma KV (2017) EEG based emotion recognition using SVM and PSO. In: 2017 international conference on intelligent computing, instrumentation and control technologies (ICICICT), Kerala, India, pp 1597–1600
8. Kennedy J, Eberhart R (1995) Particle swarm optimization. In: Proceedings of ICNN'95-international conference on neural networks, Perth, WA, Australia, pp 1942–1948
9. Pradhan A, Srivastava S (2023) Hierarchical extreme puzzle learning machine-based emotion recognition using multimodal physiological signals. Biomed Signal Process Control 83:1746–8094
10. Blaiech H, Liouane N (2020) Recursive identification based on OS-ELM for emotion recognition and prediction of difficulties in video game. Stud Inf Control 29:337–351
11. Bhatt A, Bhatt AR (2019) EEG based Emotion Recognition using SVM and LibSVM. Int J Comput Appl
12. Mohsen S, Alharbi AG (2021) EEG-Based human emotion prediction using an LSTM model. In: IEEE international midwest symposium on circuits and systems (MWSCAS), Lansing, MI, USA, pp 458–461
13. Pratiwi M, Wibawa AD, Purnomo MH (2021) EEG-based happy and sad emotions classification using LSTM and bidirectional LSTM. In: 2021 3rd international conference on electronics representation and algorithm (ICERA), Yogyakarta, Indonesia, pp 89–94
14. Tao W et al, EEG-based emotion recognition via channel-wise attention and self attention. IEEE Trans Affect Comput
15. Arjun A, Rajpoot AS, Raveendranatha Panicker M (2021) Introducing Attention Mechanism for EEG signals: emotion recognition with vision transformers. In: 43rd annual international conference of the IEEE engineering in medicine and biology society (EMBC), Mexico, pp 5723–5726. https://doi.org/10.1109/EMBC46164.2021.9629837
16. Alexey D, Lucas B, Alexander K, Dirk W, Xiaohua Z, Thomas U, Mostafa D, Matthias M, Georg H, Sylvain G, Jakob U, Neil H (2021) An Image is Worth 16 × 16 Words: transformers for image recognition at scale. In: International conference on learning representations
17. Vaswani A, Shazeer NM, Parmar N, Uszkoreit J, Jones L, Gomez AN, Kaiser L, Polosukhin I (2017) Attention is all you need. Neural Inf Process Syst
18. Agrawal S, Sharma DK (2022) Feature extraction and selection techniques for time series data classification: a comparative analysis. In: 2022 9th international conference on computing for sustainable global development (INDIACom), New Delhi, India, pp 860–865
19. Cho, Jungchan and Hyoseok Hwang. "Spatio-Temporal Representation of an Electoencephalogram for Emotion Recognition Using a Three-Dimensional Convolutional Neural Network." Sensors (Basel, Switzerland) 20 (2020)
20. Kolb B, Whishaw I (2009) Fundamentals of human neuropsychology. Macmillan, pp 73–75
21. Haas LF (2003) Berger Hans, Richard Caton, and electroencephalography. J Neurol Neurosurg Psychiatr 74:9

22. James R (1980) A circumplex model of affect. J Pers Soc Psychol 39:1161–1178
23. Yoshi-Taka M, Tomomi F, Kentaro K, Masato O, Kenichi U, Kang C, Kazuo O (2013) The implicit processing of categorical and dimensional strategies: an fMRI study of facial emotion perception. Front Hum Neurosci 7:551
24. Jafar A, Nayyar A, Kumar A, Machine learning from theory to algorithms. An overview. J. Phys.: Conf. Ser. **1142**, 012012. https://doi.org/10.1088/1742-6596/1142/1/012012
25. Huang G-B, Zhu Q-Y, Siew C-K (2004) Extreme learning machine: a new learning scheme of feedforward neural networks. In: 2004 IEEE international joint conference on neural networks (IEEE Cat No 04CH37541), Budapest, Hungary, pp 985–990
26. You Z, Li S, Gao X, Luo X, Ji Z (2014) Large-Scale protein-protein interactions detection by integrating big biosensing data with computational model. BioMed Res Int
27. Yamaguchi T, Fujio M, Inoue K (2008) Feature extraction of EEG waves based on morphological multiresolution analysis. In: 2008 3rd international conference on innovative computing information and control, Dalian, China, pp 60–60
28. Naderi MA, Mahdavi-Nasab H (2010) Analysis and classification of EEG signals using spectral analysis and recurrent neural networks. In: 17th Iranian conference of biomedical engineering (ICBME), Isfahan, Iran, pp 1–4
29. Mahmoodabadi SZ, Ahmadian A, Abolhasani M, Eslami M, Bidgoli JH (2005) ECG feature extraction based on multiresolution wavelet transform. In: 2005 IEEE engineering in medicine and biology 27th annual conference, pp 3902–3905
30. Trigui O, Zouch W, Messaoud MAB (2017) Hilbert-Huang transform and welch's method for motor imagery based brain computer interface. Int J Cogn Informatics Nat Intell 11:47–68
31. Prochazka A, Kukal J, Vysata O (2008) Wavelet transform use for feature extraction and EEG signal segments classification. In: 3rd international symposium on communications. Control and signal processing, Saint Julian's, Malta, pp 719–722
32. Koelstra S, Muehl C, Soleymani M, Lee J-S, Yazdani A, Ebrahimi T, Pun T, Nijholt A, Patras I (2012) DEAP: a database for emotion analysis using physiological signals (PDF). IEEE Trans Affect Comput 3(1):18–31

Modified GANs Based on GNN for Anomaly Detection in Graphs

Premanand Ghadekar, Ruchita Chaudhari, Kshitij Bisen, Ashish Biradar, Chetan Chaudhari, and Aditya Bhosale

Abstract Recent years have seen a significant increase in the research work regarding graph-based datasets and GNN. Research in node classification in a graph, link prediction between nodes, anomaly detection in graphs, etc. have become significant in various applications. Graph neural networks are used to perform various tasks on graphs one such task is anomaly detection which is identifying anomalous nodes in the graph. This paper proposes a novel GNN-based architecture that uses GANs for identifying anomalies in the graph. The model outperforms other MLP, RNN-based architecture. GNN is used because a graph is a complex data structure difficult to represent in a latent shape due to variable shapes, etc. GNN helps us to create proper graph embedding which helps in incorporating maximum information from the graph. The model consists of a discriminator network, a generator network, and an encoder made up of graph convolutional layers as a base neural network. The proposed model especially focuses on a single graph-based dataset where the whole graph is passed as input and anomalies are identified based on the anomaly score calculated by combining different reconstruction losses calculated from the encoder and generator. A higher anomaly score is observed for the anomalous nodes. The model proposed outperforms all the traditional outlier detection models when compared based on ROC metrics.

Keywords Graph neural networks · Deep learning · Anomaly detection · GAN

1 Introduction

The proposed model focuses on identifying anomalies in graph datasets using the unsupervised technique where labels regarding anomalies are not used to train the model. The proposed model is a modified architecture of a generative adversarial

P. Ghadekar · R. Chaudhari · K. Bisen (✉) · A. Biradar · C. Chaudhari · A. Bhosale
Vishwakarma Institute of Technology, Pune, Maharashtra, India
e-mail: kshitij.bisen20@vit.edu

A. C. Frery et al. (eds.), *Computational Intelligence and Data Analytics*, Lecture Notes on Data Engineering and Communications Technologies 236,
https://doi.org/10.1007/978-981-96-0451-7_23

network model consisting of a generator network, a discriminator network, and an encoder.

Anomalies in graphs are outliers or irregularities in nodes or subgraphs. These are unusual or rare node occurrences that differentiate from the normal nodes very significantly. These anomalous nodes may differ from normal nodes either structurally like in a sparsely separated graph, or a group of densely connected nodes may be anomalous. Either these nodes may vary in a contextual manner that is attributes or node features significantly deviate from the attributes of neighboring normal nodes.

Anomalies are identified on the basis of an anomaly score calculated by finding out various reconstruction losses while generating data based on input noise. For anomalies, the anomaly score is higher, and for normal nodes, it is lower.

The proposed model is all about building a robust model to identify these anomalous nodes in the given graph data. This model could be applied to find frauds or irregularities in financial graph data, social media graphs, network graphs, and much more.

2 Literature Review

Graph anomaly detection is a crucial task in various applications, including social network analysis, fraud detection, and cyber security. Recent advancements in graph anomaly detection have been made using graph neural networks (GNNs).

GLAM, a graph-level anomaly detection model, is proposed in [1] for unsupervised model selection and a new pooling layer called MMD pooling for detecting distributed anomalies. Graph GAN, proposed in [2], is a graph representation learning framework based on a game-theoretical minimax game.

The authors in [3] propose an approach for non-overlapping anomaly-type graph construction using unsupervised learning and graph features. In [4], two graph-based techniques for anomaly detection using subdue are introduced. The authors in [5] propose a novel graph convolutional network (GCN) architecture, BWGNN, for anomaly detection in graph-structured data.

AGNN, proposed in [6], learns the graph representation by encoding the structural data of the graph. The authors in [7] present an in-depth analysis of graph anomaly detection using GNNs and highlight the difficulties of graph anomaly detection. The Graph AnoGAN framework is proposed in [8] for identifying anomalous snapshots in applications such as social network analysis, cyber security, and fraud detection.

The authors in [9] develop a related encoder training method and quick anomaly detection using GANs. The Beta Wavelet Graph Neural Network (BWGNN) is proposed in [10] for more effective anomaly information capture on graphs. The study in [11] identifies, evaluates, and synthesizes several GBAD techniques used in data mining-disseminated fraud detection studies.

The authors in [12] provide a thorough explanation of each model, highlighting its unique features, benefits, and drawbacks. The paper in [13] discusses the use of pre-trained CNNs to improve the efficiency of the classification model and the potential benefits of transfer learning. The paper in [14] proposes using a convolution neural network for large graph data. The GraphSAGE framework, proposed in [15], creates node embeddings from the feature matrix of previously learned nodes from the graph.

3 Methodology

3.1 Dataset

For the objective of anomaly detection Cora dataset from Pytorch, the geometric planetoid is used which is a citation network dataset of scientific publications. Cora is an Attributed graph dataset that consists of an attribute matrix also called node features and an Adjacency matrix defining the connection between different nodes.

As shown in Fig. 1 graph consists of a citation network graph with different classes. It is a single graph consisting of 2708 total nodes belonging to 7 different classes. For the task of Anomaly Detection impurities or anomalies are injected by generating outliers using Pygod library.

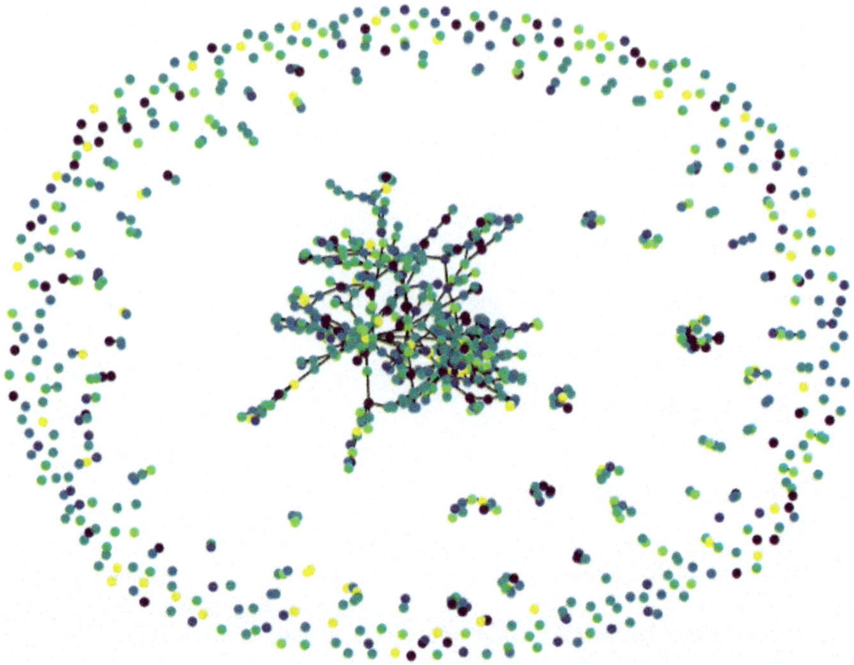

Fig. 1 Cora dataset visualization

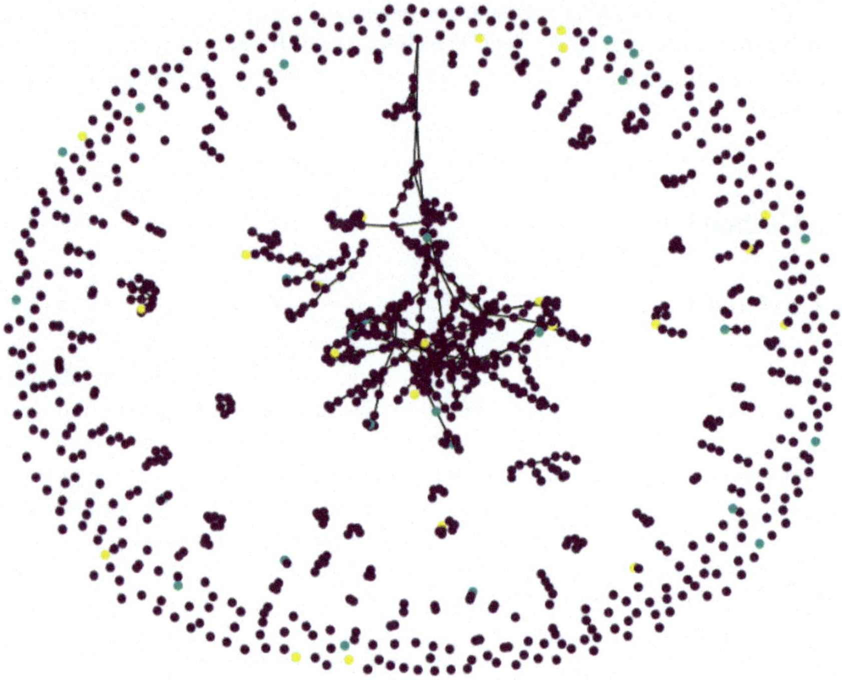

Fig. 2 Cora dataset visualization with injected outliers

As shown in Fig. 2 graph consists of a citation network graph injected with outliers of two types

1. Structural Outliers (Yellow)—Based on structure
2. Contextual Outliers (Green)—Based on node attributes.

Purple color samples represent the in-liners or the normal nodes in the dataset.

3.2 Algorithm

Algorithm of anomaly detection:

1. Start
2. Take the input graph for anomaly detection and load the dataset in the environment.
3. Process the graph into a suitable input format. Convert the raw data object into a tuple of sub-data of objects for the model. This includes processing information like edge index list, attribute matrix, and adjacency matrix.
4. Train the Generator with input Gaussian noise and generate fake data.

5. Process Fake and real data from the encoder.
6. Using a discriminator to differentiate between real and fake data. Sigmoid is used for the same.

$$\hat{A} = \text{Sigmoid}\left(ZZ^{T}\right) \tag{1}$$

7. Optimization of GAN architecture by solving the optimization equation.

$$\min(G)\max(D) = f(D, E, G) \tag{2}$$

8. Calculating the anomaly score using the loss functions.
9. Detect Anomalies using the model predict function.
10. Evaluating the Results based on several evaluation metrics such as TPR, FPR, AUC Score, and precision.

3.3 *Technology*

The model proposed is a graph-based architecture defined as

G = (V, A, X) where G is the Generator network. V denotes N nodes of a graph. A denotes the adjacency matrix that helps in mapping the nodes to each other and X denotes the node feature matrix.

The Generator takes the Gaussian noise as input and then produces the attribute or node feature matrix for the node and then the real and fake nodes are encoded by using the encoder E mapped between the generator and discriminator. Later encoded vectors are passed through a discriminator to binary classify whether the node belongs to the graph or not.

The generator network G is a basic Graph Convolutional Network with leaky relu as the activation function and a dropout layer at the end for dimension-reducing purposes. Adam Optimizer is used in Generator Network.

The Encoder Network E mapped between the Generator and Discriminator is used to encode real graph nodes and fake nodes generated using a generator. GCN layers are used to build an encoder with layers connected using the leaky relu function and a dropout layer at the end is used. Adam optimizer is used. Encoder is used twice first to encode the real graph node and second to encode the generated graph node.

The Discriminator (D) Network is used to differentiate real graph nodes from generated graph nodes. The Sigmoid Function is used.

Optimization of GAN is done by solving the following equation where D is a discriminator, G is a generator and E is an Encoder.

$$\min(G)\max(D) = f(D, E, G) \tag{3}$$

$$f(D, E, G) = E_{x \sim pX}\left[E_{z \sim pE(|x)}\left[\log D(Z)\right]\right]$$

$$+ E_{x'\sim pG}\left[E_{z'\sim pE(|x')}\left[1 - \log \mathrm{D}(\mathrm{G}(Z'))\right]\right] \quad (4)$$

As shown in Fig. 3 Architecture of GAN is made up of a generator, an encoder, and a discriminator.

Anomaly is detected using an anomaly score. The anomaly score is higher for the Anomalous node. The score is calculated for each node using discriminator loss and reconstruction loss.

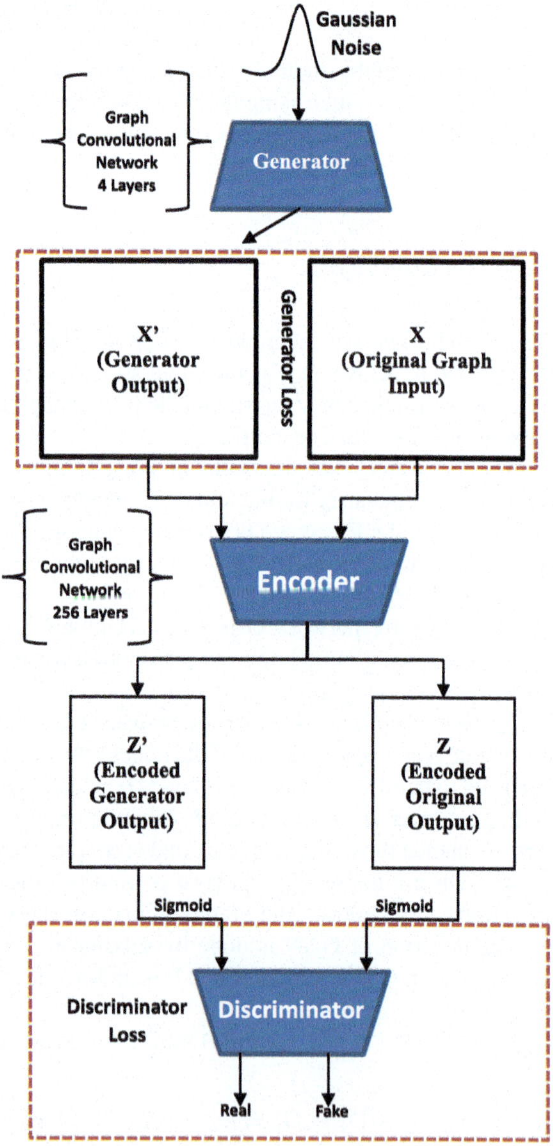

Fig. 3 Architecture of the proposed model of modified GAN architecture based on GNN for graph anomaly detection

$$\text{score(node)} = \alpha\text{L(G(node))} + (1 - \alpha)\text{L(D(node))} \tag{5}$$

α is cross-entropy loss. L(D(node)) represents the discriminator loss and L(G(node)) represents the Reconstruction loss of the generator.

3.4 Hyperparameter Tuning

Hyperparameters of the model are used to improve the performance of the base model and then create an efficient model. Hyperparameters such as layers of the encoder, layers of the generator, learning rate, and activation function can be adjusted to optimize performance.

From manual hyperparameter tuning changing no. of layers in encoder and generator stood most significant.

Figures 4 and 5 show the variation in AUC score with change in no. of Encoder layer for Generator layers 2 and 4 respectively.

Table 1 shows the AUC score and precision scores for different encoder layers and generator layers. On further increasing layers no good results are observed.

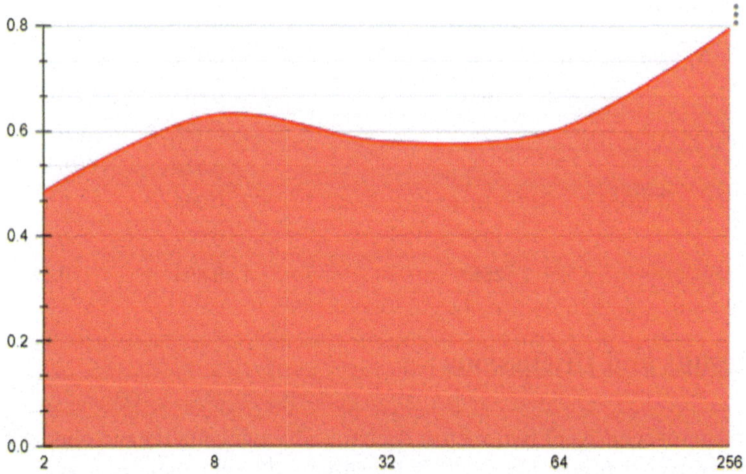

Fig. 4 AUC versus no. of encoder layer (Generator Layer = 2)

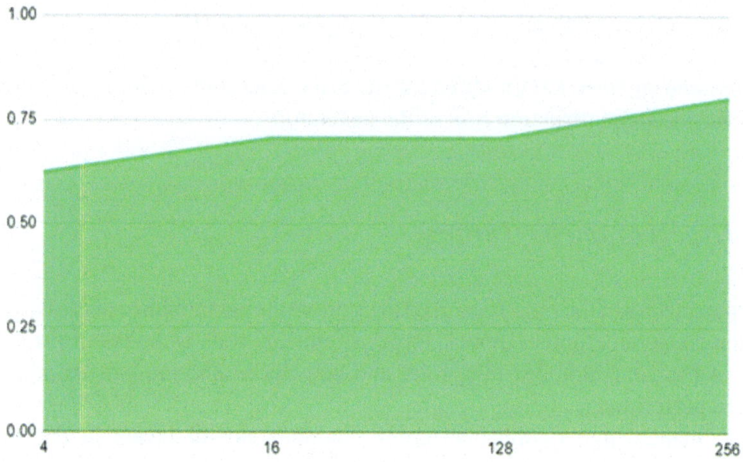

Fig. 5 AUC versus no. of encoder layer (Generator Layer = 4)

Table 1 Hyperparameter tuning report

Number of generator layer	Number of encoder layer	AUC score	Precision
2	2	0.485	0.048
2	8	0.63	0.09
2	32	0.579	0.09
2	64	0.603	0.096
2	256	0.795	0.183
4	4	0.624	0.07
4	16	0.706	0.135
4	128	0.707	0.13
4	**256**	**0.803**	**0.193**

4 Results and Discussions

Graph datasets consist of two vectors, one for adjacency matrix and one for feature matrix some models are only available for structural level anomaly prediction which makes them incapable of feature level anomaly detection.

From Table 2 we can infer the average anomaly scores for normal and anomalous nodes in the graph and from the results we can infer that for normal nodes we have lower AAS.

Table 3 presents the comparison of AUC (Area Under Curve) Scores of the models for geometrical anomaly detection.

Table 4 shows the comparison of different Autoencoder-based and graph embedding-based models with the proposed model on the attributed graph dataset. All the results are validated on the CORA dataset.

Table 2 Different evaluation metrics for model

Metrics	Value
Average Anomaly Score (AAS)	0.66
AAS for anomalous node	1.09
AAS for normal node	0.63
Average confidence	0.99

Table 3 Comparison of models for structural anomaly prediction based on AUC scores

Methods	AUC scores
IForest	53.09 ± 0.03
OCSVM	54.35 ± 0.02
PCA	62.17 ± 0.01
AE	62.17 ± 0.01

Table 4 Comparison of models for attributed graph anomaly prediction based on AUC Scores

Models	AUC score
Deep walk + AE	62.39 ± 0.02
GAAN	0.501 (MLP based)
Struc2Vec + Features + AE	62.83 ± 0.01
SDNE + Features + AE	62.30 ± 0.02
GAE	60.15 ± 0.18
DOM	64.50 ± 0.25
Proposed model	80.30

From Fig. 6 we can infer the graph of TPR versus FPR and from that, we can conclude the efficiency of the model. TPR overpowering FPR is observed which defines the accuracy of the model in correctly classifying the anomalies in the graph.

From Fig. 7 displaying the confusion matrix we can conclude the proportion of true labels and predicted labels dominating over false predictions. 2473 true label matches with predicted labels which is dominating over other parameters.

The proposed model is suitable for both structured and attributed graph datasets unlike traditional models only suitable for structural graph datasets without any attribute features making it a robust model for anomaly detection on graphs. From the results and comparisons with other models studied it is observed that based on AUC metrics the proposed model is outperforming other compared models.

5 Limitation

The model depicts an AUC score of nearly 0.8 which can be increased by building a more complex model.

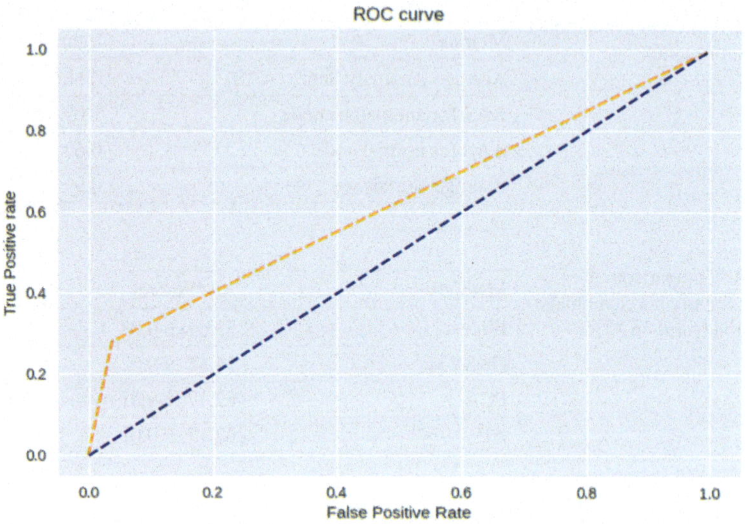

Fig. 6 ROC curve for anomalous node classification model performance

Fig. 7 Confusion matrix for anomalous node classification model performance

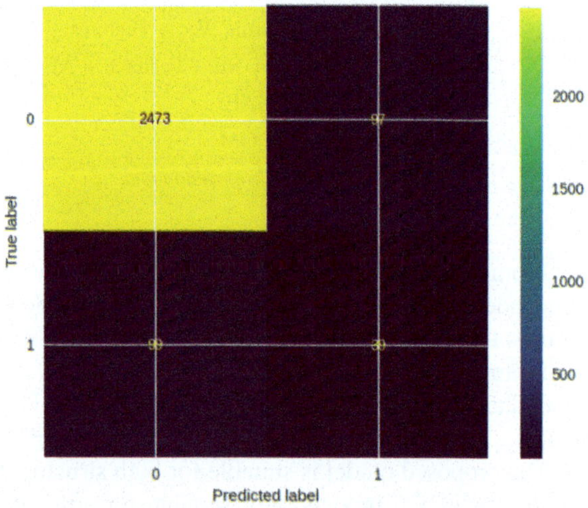

Model is applicable for static graphs for the graph with temporal features model is not applicable.

6 Future Scope

The AUC score of models could be increased by adding some complexity to the model and by using some advanced and modified activation and loss functions. More Robust Models should be created.

Models applicable for multiple graphs and also applicable to graphs with temporal features can be created by making some modifications to the model.

7 Conclusion

In this paper, a modified GAN model with a graph neural network is proposed for graph anomaly detection tasks. The model can be used in various domains such as in financial fraud detection, social media anomaly detection, transactional fraud detection, and various cyber security applications such as detecting attacks from some group of nodes in the network, etc. The model outperforms various traditional techniques by leveraging the strength of GNN which helps the model to understand the graph both structural-wise and attribute-wise. The model shows an AUC score of 80 on the Cora dataset with injected anomalies. As the research is still going on more advanced and robust techniques for the task of graph anomaly detection can be expected.

References

1. Bai Z, Zhang W, Zheng W, Huang Y (2021) Graph anomaly detection with unsupervised GNNs. In: Proceedings of the AAAI conference on artificial intelligence, vol 35, No 15, pp 13454–13462
2. Wang H, Wang J, Wang J, Zhao M, Zhang W, Zhang F, Xie X, Guo M. GraphGAN: graph representation learning with generative adversarial nets. In: Proceedings of the 2018 conference on empirical methods in natural language processing, pp 3617–3626
3. Vengertsev D, Thakkar H (2015) Anomaly detection in graph: unsupervised learning, graph-based features and deep architecture
4. Noble CC, Cook DJ (2003) Graph-based anomaly detection. Knowl Discov Data Mining
5. Tang J, Li J, Gao Z-C, Li J (2022) Rethinking graph neural networks for anomaly detection. In: International conference on machine learning
6. Chaudhary A, Mittal H, Arora A (2019)Anomaly detection using graph neural networks. In: 2019 international conference on machine learning, big data, cloud and parallel computing (COMITCon), Faridabad, India, pp 346–350. https://doi.org/10.1109/COMITCon.2019.8862186
7. Kim H, Lee BS, Shin W-Y, Lim S (2022) Graph anomaly detection with graph neural networks: current status and challenges. IEEE Access 10:111820–111829. https://doi.org/10.1109/ACCESS.2022.3211306
8. Abu-El-Haija S, Wang R, Feng Y, Song L (2021) GraphAnoGAN: detecting anomalous snapshots from attributed graphs. IEEE Trans Knowl Data Eng 33(3):1118–1132. https://doi.org/10.1109/TKDE.2019.2953141

9. Schlegl T, Seeböck P, Waldstein SM, Schmidt-Erfurth U, Langs G (2019) f-AnoGAN: fast unsupervised anomaly detection with generative adversarial networks. Med Image Anal 54:30–44
10. Zhao Y, Akoglu L, Tong H (2019) Rethinking graph neural networks for anomaly detection. In: Proceedings of the 2019 ACM SIGKDD international conference on knowledge discovery and data mining (KDD'19)
11. Barcelos RA, Parpinelli RS, Dutra LV (2018) Fraud detection: a systematic literature review of graph-based anomaly detection approaches. J Netw Comput Appl 117:43–67
12. Wu Z, Pan S, Chen F, Long G, Zhang C, Yu PS (2021) A comprehensive survey on graph neural networks. IEEE Trans Neural Netw Learn Syst 32(1):4–24. https://doi.org/10.1109/TNNLS.2020.2978386
13. Ding K, Li J, Bhanushali R, Liu H (2019) Deep anomaly detection on attributed networks. In: Proceedings of the SIAM international conference on data mining (SDM), May
14. Niepert M, Ahmed M, Kutzkov K (2016) Learning convolutional neural networks for graphs. In: Proceedings of the 33rd international conference on international conference on machine learning—Volume 48 (ICML'16). JMLR.org, 2014–2023
15. Hamilton WL, Ying R, Jure (2017) Inductive representation learning on graph. In: NIPS

Probabilistic Approach to Detect PCOS Using Medical Imagery with Automated Report Generation

Sneha Saravanan, R. Rachana, Tejas Kalluraya, P. Sai Deepika, and H. R. Mamatha

Abstract Polycystic ovary syndrome (PCOS) is a prevalent endocrine disorder affecting women of reproductive age. Early and accurate diagnosis is crucial for managing its associated health complications. This research proposes a novel probabilistic approach for PCOS detection using ovarian ultrasound images, integrating deep learning models with Bayesian inference and automatic report generation. Our approach leverages the strengths of two convolutional neural networks (CNNs): PCONet and MobileNet. PCONet, specifically designed for PCOS detection, captures subtle ultrasound features indicative of polycystic ovaries. MobileNet, known for its lightweight architecture, facilitates efficient deployment on resource-constrained platforms. We employ Bayesian CNN ensembles to learn from the combined knowledge of these models, generating probabilistic predictions for each image. Further, we incorporate XGBoost, a gradient boosting algorithm, to refine these probabilities and improve diagnostic accuracy. Through our study and work, we are successfully able to make a high-performing ensemble model, exhibiting 98.6% accuracy on validation data. To streamline the diagnostic workflow, we develop an automatic report generation module. This module extracts key findings from the ultrasound images, analyzes predictions from the probabilistic model, and generates a comprehensive report summarizing the PCOS diagnosis and potential clinical implications. This automation expedites decision-making and enhances healthcare provider efficiency. Our proposed framework, combining deep learning, Bayesian

S. Saravanan (✉) · R. Rachana · T. Kalluraya · P. Sai Deepika · H. R. Mamatha
PES University, Banashankari, Bengaluru, Karnataka 560085, India
e-mail: pes1ug20cs721@pesu.pes.edu

R. Rachana
e-mail: pes1ug20cs677@pesu.pes.edu

T. Kalluraya
e-mail: pes1ug20cs467@pesu.pes.edu

P. Sai Deepika
e-mail: pes1ug20cs718@pesu.pes.edu

H. R. Mamatha
e-mail: mamathahr@pes.edu

A. C. Frery et al. (eds.), *Computational Intelligence and Data Analytics*, Lecture Notes on Data Engineering and Communications Technologies 236,
https://doi.org/10.1007/978-981-96-0451-7_24

inference, and automatic report generation, offers a promising avenue for accurate and early PCOS detection, ultimately improving patient care and management of this prevalent condition.

Keywords Polycystic ovarian syndrome · CNN · Deep learning · Ultrasound images · Image classification

1 Introduction

Polycystic ovarian syndrome (PCOS) is a condition occurring in women of reproductive age when ovaries produce abnormal amounts of the male sex hormone, androgen which is usually present in small amounts. While the exact cause for PCOS is still unclear, according to an article from Johns Hopkins Medicine [1], women diagnosed with PCOS seem to have insulin resistance. This also may cause higher androgen levels. A few common symptoms observed in women with PCOS include irregular periods, weight gain, acne, thinning hair, and, infertility. It affects several women worldwide, affecting their overall well-being and reproductive health. Early diagnosis is critical as it helps to control the various symptoms and is beneficial to prevent the occurrence of serious long-term health conditions. PCOS is diagnosed over a course of three steps. The initial assessment consists of the doctor's oral evaluation of the patient's symptoms. This is followed by an ovarian ultrasound which shows the presence of numerous cysts or enlarged ovaries. The third step is an analysis using a blood test, which shows the androgen levels, glucose levels as well the levels of various other hormones. PCOS is a serious condition, which if not treated at the right time may lead to various other complications like Type II Diabetes and in some cases even ovarian cancer.

While using computation methods to detect PCOS is a hard task due to various reasons like the availability of dataset, and computation challenges to acquire results of all three tests, the prevalent methods provide an image and text-based binary classification over a seemingly small dataset. The aim of this paper is to present a probabilistic approach, that is provide the percentage of the presence of PCOS in every ovarian ultrasound image over a large dataset. This way, we do not aim to replace healthcare professionals, instead we help them in scenarios where mass testing takes place- like health camps, and fasten the process in areas that lack healthcare professionals. The sensitivity of providing patient details to public, and lack of cohesive test records for all three tests were a major problem at the initial phase of research. This paper uses publicly available ovarian ultrasound images with the help of augmentation to enhance the training of neural networks. The paper also involves the generation of a report with the ovarian ultrasound image, a histogram denoting the percentage of PCOS presence, and a numerical value for the same.

2 Literature Review

Machine learning algorithms have shown remarkable potential in analyzing clinical data for PCOS prediction. Traditional methods like Naive Bayes, KNN, and Decision Trees have laid the groundwork, with Decision Trees achieving high accuracy depicted in Huang et al. [2]. However, the real excitement lies in hybrid approaches like CatBoost and XGBRF, which combine these algorithms for even greater accuracy, reaching 95% in a recent study (e.g., [3]). Feature selection and sampling techniques like those employed in a modified XGBoost [4] can further improve performance, pushing accuracy up to 98%. The potential does not stop there—recent explorations of ensemble models and RFLR [5, 6] suggest even more avenues for accurate PCOS detection through machine learning.

Meanwhile, the world of image analysis has unveiled a powerful tool in Convolutional Neural Networks (CNNs) for analyzing ovarian ultrasound images. PCONet, a dedicated CNN architecture, stands as a testament to this potential, achieving an impressive 98.12% accuracy, outperforming even established models like Inception V3 [7]. Feature extraction techniques like area and perimeter can further enhance CNN performance for classification, while automated follicle identification methods offer even more exciting possibilities [8–10]. A recent study even combined ensemble ML models with CNNs to achieve the highest accuracy among image-based approaches [11]. Powerful techniques of image denoising like non-local means, which can significantly improve cyst detection accuracy when used in conjunction with machine learning [12] showed significant results. PCONet is a promising model that showed better accuracy, 98.12%, than fine-tuned InceptionV3—96.56%. PCONet had a total of 582,690 parameters, of which all were trainable. On the other hand, InceptionV3 had a total of 48,018,722 parameters, of which 34,432 were non-trainable and 47,984,290 were trainable. PCONet is a much lighter model than InceptionV3 [13]

In conclusion, the synergy of ensemble models and Convolutional Neural Networks in image analysis, exemplified by PCONet, showcases a promising frontier. As we explore feature extraction and image denoising techniques, the future holds exciting possibilities for even more precise and efficient PCOS detection through machine learning.

3 Data Preparation

The project required ultrasound images of the ovary. The initial data were acquired from a [14] publicly available repository in Kaggle. Since the dataset was split into train, test, and validation, the folders were reformed to normal—indicating that the ovaries have minimal chances of being affected with PCOS and infected—indicating that the ovaries are infected with PCOS. Figure 1 shows an example of a normal ovarian ultrasound image and a PCOS infected ovarian ultrasound image. A major problem identified through extensive literature review was the lack of availability of large amount of data for training purposes. Hence, the dataset was expanded using augmentation, from 2000 images in both categories to 14,955 images in both categories, with 7500 images in PCOS infected and 7455 images in the normal category. The images were converted to a standard.jpg format. The augmentation methods used in this process were mainly rotation, contrast enhancement, horizontal flipping, and the image enhancement from the library. The data were then split to train, test, and validate where 70% of the data from both the classes were present in the train folder, 15% in the test folder, and the rest 15% in the validation folder. The pre-processing was further enhanced by implementing the watershed algorithm to segment the images to compare the results obtained by training deep learning models with unsegmented pre-processed images and the segmented images. Figure 1 represents the ovarian ultrasound images just after augmentation but before the application of the watershed algorithm. Figure 2 depicts the images after the usage of watershed algorithm. The augmentation and other pre-processing techniques using watershed algorithm and ImageDataGenerator were employed by keeping in mind the sensitivity of medical images and their usability quotient. The test data were also verified from a doctor prior to usage.

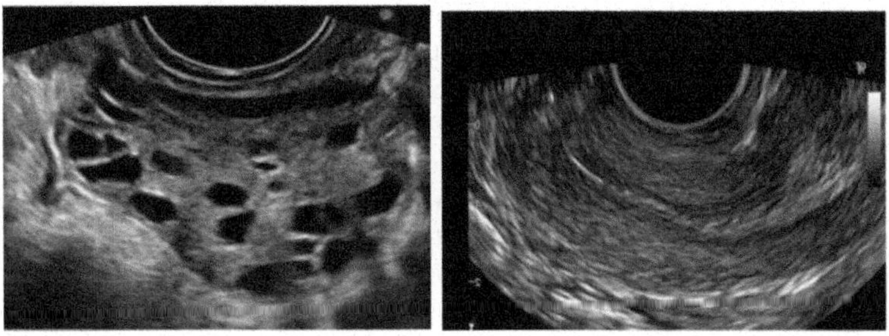

(a) (b)

Fig. 1 **a** PCOS infected ovary, **b** normal ovary

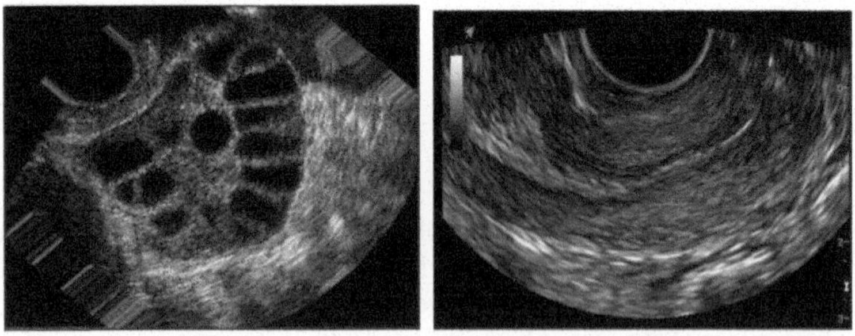

(a) (b)

Fig. 2 **a** PCOS infected, **b** normal ovary, after employing watershed algorithm

4 Method

Three convolutional neural network methods were employed in this study. Binary classification was utilized in most of the papers studied as a part of literature review, and probability-based classification is the proposed approach in this study. Hence, three models, modified MobileNetV2 with custom classification layers, a fine-tuned version of the previously developed [7] PCONet model, and a combination of the Bayesian CNN and XGBoost model were employed. The combination of Bayesian CNN and XGBoost is our proposed approach to probabilistic classification. Additionally, a report generation was included in the Bayesian CNN + XGBoost model using the reportLab library. The purpose of employing three models was to compare the results of a binary classification method Mobilenet, a previously developed model [7] PCONet, and our Bayesian CNN + XGBoost model. This was done to study the differences and the advantages our probabilistic model hold over prevalent binary classification methods.

4.1 Modified MobileNetV2 Model

MobileNetV2, a pre-trained convolutional neural network layer, was used as the base model for feature extraction. The weights pre-trained on the ImageNet dataset were used to obtain extensive image understanding capability. Custom classification layers were added on top of this base model. To condense spatial information, GlobalAveragePooling2D was added, a fully connected layer with 1024 neurons and ReLu activation function was used. A dropout layer with 0.5 rate to handle overfitting was also added. The output layer consisted of the sigmoid function for binary classification to classify images as either infected or normal category. To enhance performance, the pre-trained layers of MobileNetV2 were frozen to prevent the updating of weights during the training phase. Adam optimizer with a learning rate of 0.01 and binary cross-entropy loss function was used to compile and train the model (Fig. 3). Table 1 represents the model training details. The evaluation metrics consisted of accuracy, F1 score, precision, and recall. To tackle the problem of overfitting, early stopping and model checkpoint were implemented.

4.2 PCONet Model [7]

The model was built according to the description mentioned in [7] with minor fine-tuning in the values of the variables. The main aim behind deploying this model was to check its nature and performance with our dataset and as a comparison to the MobileNetV2 model. Five convolutional blocks, each containing a convolutional layer with ReLu activation, a 3×3 filter, and valid padding are employed. A max

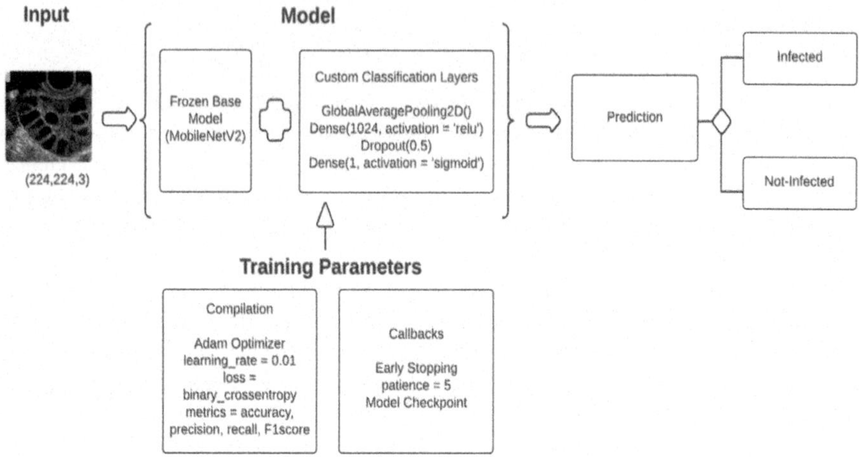

Fig. 3 Modified MobileNetV2 architecture

Table 1 Model name and the information regarding the number of parameters for training

Model name	Total parameters	Number of trainable parameters	Number of non-trainable parameters
Modified MobileNet	357,075	357,075	0
PCONet [7]	568,605	568,605	0
Bayesian CNN + XGBoost	497,268	497,268	0

pooling layer with 2 × 2 pooling and stride of 2 was used. The flattened output was fed into two dense layers (32 and 64 neurons) with ReLu activation. A dropout layer with a 0.65 was incorporated between the dense layers. The sigmoid function was used in the output layer. The model was compiled with Adam optimizer with a learning rate of 0.0001 and binary cross-entropy loss function (Fig. 4). Table 1 represents the model training details. Evaluation metrics remained the same as that of MobileNetV2.

4.3 *Bayesian CNN + XGBoost*

Bayesian CNN

A Bayesian convolutional neural network is used to estimate uncertainty. It is a weak learner on its own, hence, it is combined with a powerful model like XGBoost to make it robust. The input layer was of the format (224,224,3). The first convolutional block—Convolutional2DReparametrization layer consisted of eight filters, with a

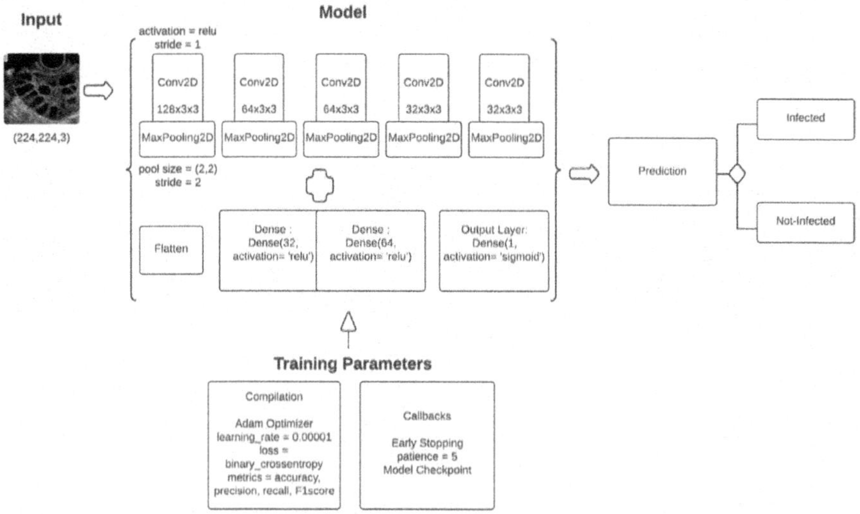

Fig. 4 PCONet architecture

kernel size of 16 × 16, Bayesian priors and a ReLu activation. The Conv2D layers consisted of filter numbers in the following order—128, 128, 64, 32, 32. All the four Conv2D layers had L2 regularization incorporated with the kernel_regularizer value of 0.01 and bias_regularizer value being 0.15. Each Conv2D layer was followed by a MaxPooling2D layer and a dropout layer. The first two layers of Conv2D had a dropout of 0.4 and the next two with 0.3. Following this was a flatten layer followed by a dense layer with 64 neurons, ReLu activation and L2 regularization as well as dropout of 0.3. All the values for dropouts and filters were derived after numerous iterations of trial and error with various values and patterns. The compilation of the model consisted of the Adam Optimizer with a learning rate of 0.0025, a custom log-likelihood function for Bayesian model fitting. The evaluation metrics were precision and accuracy. EarlyStopping was employed to tackle overfitting and ReduceLROn-Plateau to reduce the learning rate in the validation accuracy plateaus for 5 epochs (Fig. 5).

XGBoost

XGBoost classifier, a powerful gradient boosting model, was trained with the data. The model was trained on the training data with no automatic label encoding since the labels were previously numerically encoded and a logistic regression objective for binary classification tasks.

Bayesian CNN + XGBoost Classifier

The features were extracted first from the Bayesian CNN model. The XGBoost was created and trained on those features extracted. Following this, the individual models were made to make predictions and finally the predictions were combined using

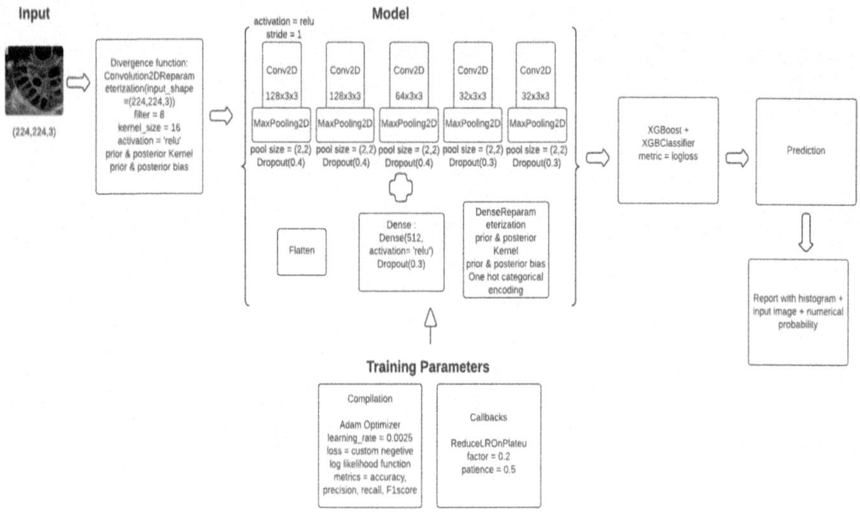

Fig. 5 Bayesian CNN + XGBoost architecture

weighted averaging. The output is a numerical probability indicating the chance of an ovary being affected by PCOS (Fig. 5).

Report Generation

Once the Bayesian CNN + XGBoost model is trained and tested, it is validated on the validation data. Post this, the reportLab library is used to present a report. reportLab is a Python library, possessing powerful API to generate documents in the PDF format. Its functionalities include dynamic document generation and incorporation of complex charts. The report is generated in the.pdf format with and comprises of ovarian ultrasound image provided as input, a histogram consisting of the probability value as well as the numerical probability. The input is manually given for this purpose. Figure 6 depicts the final report obtained.

5 Results and Analysis

5.1 Individual Binary Classification Models

MobileNet: MobileNet was trained with a batch size of 32 for 10 epochs. It exhibits a training accuracy of 95.7%, validation accuracy of 96%, indicating its robust performance in binary image classification. Table 2 depicts the values of accuracy, precision, recall, and F1 score for MobileNetV2.

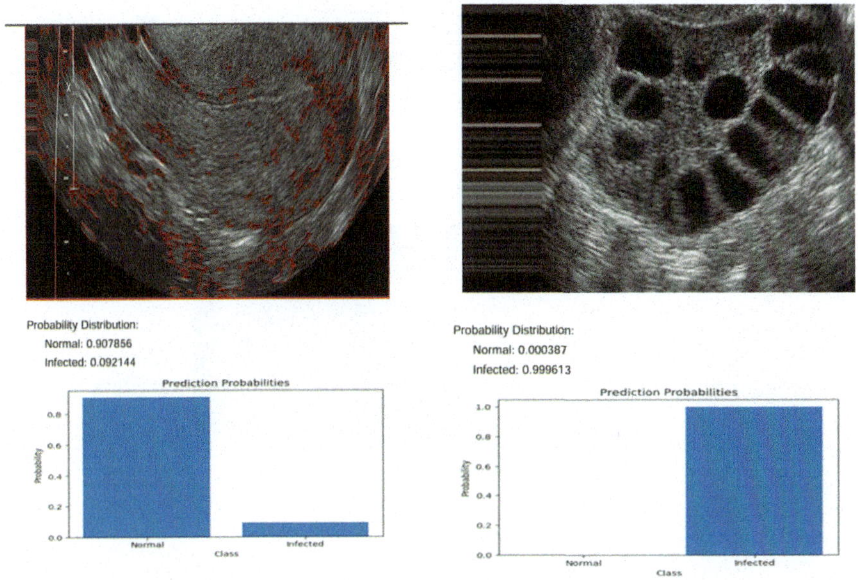

(a) (b)

Fig. 6 Final reports of **a** Normal ovary and **b** PCOS-infected ovary

Table 2 Evaluation metrics of the models on validation data

Model name	Accuracy	Precision	Recall	F1 score
Modified MobileNet	0.960	0.980	0.970	0.979
PCONet [7]	0.957	0.960	0.964	0.962
Bayesian CNN + XGBoost	0.986	0.990	0.988	0.991

PCONet [7]: PCONet is trained with a batch size of 16 for 10 epochs. It demonstrates a training accuracy of 94.9% and a commendable validation accuracy of 95.7% in the binary classification task, showcasing its efficacy in image classification. Table 2 depicts the values of accuracy, precision, recall, and F1 score for the model.

Probabilistic Approach: Bayesian CNN Ensemble with XGBoost

The novel model, an ensemble of Bayesian Convolutional Neural Network (CNN) and XGBoost, exhibits a training accuracy of 98.05% and achieves an impressive validation accuracy of 98.6%. This superior performance signifies the effectiveness of the proposed probabilistic approach in enhancing classification accuracy. Figure 7 shows the confusion matrix on the validation data. Additionally, Table 2 depicts the values of accuracy, precision, recall, and F1 score for the model.

Probabilistic Advantage: The probabilistic approach introduces a nuanced understanding of uncertainty in classification. By leveraging Bayesian CNN and XGBoost ensembling, the model not only provides accurate predictions but also offers insights

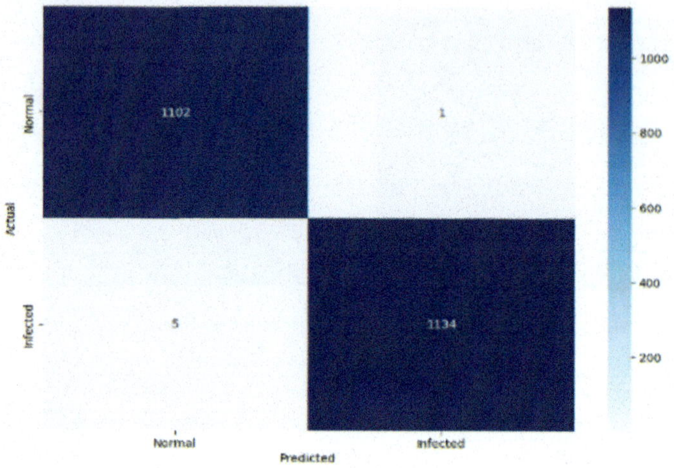

Fig. 7 Confusion matrix on the test data for Bayesian CNN + XGBoost

into the uncertainty associated with each classification, contributing to a more informed decision-making process.

Comparative Analysis: The probabilistic approach outperforms the individual binary classification models (PCONet and MobileNet) in terms of accuracy, highlighting its potential for more reliable and robust predictions.

6 Conclusion

The aim of this study is to develop a novel probabilistic approach to detect the presence of polycystic ovarian syndrome (PCOS) in all ovarian ultrasound images. Three models were leveraged, one modified version of a prevalent model PCONet [7], another model that best works for binary classification MobileNet, and the third, a novel probabilistic approach consisting of an ensemble of Bayesian CNN and XGBoost. A relatively high validation accuracy of 98.6% in the ensemble model followed by MobileNet exhibiting 96% and PCONet [7] with 95.7% on validation data was found. The aim of this study is to aid healthcare professionals, especially during times of mass testing and in rural areas with the lack of human resources and technology support. The future scope entails collaboration with hospitals and healthcare centers to amplify and automate the other two procedures during the detection of PCOS. We further hope to incorporate various other ovarian conditions and provide a probability of the presence of all ovarian conditions along with PCOS, for instance ovarian cyst, ovarian cancer, etc. The report generation is also aimed to be built as a tool with useful and appealing user interface with more details like name, age, and date along with the current image, histogram, and numerical value.

References

1. Polycystic Ovary Syndrome (PCOS) (2022), 28 Feb. www.hopkinsmedicine.org. https://www.hopkinsmedicine.org/health/conditions-and-diseases/polycystic-ovary-syndrome-pcos#:~:text=The%20name%20polycystic%20ovary%20syndrome
2. Huang P et al (2020) Comparative analysis of machine learning algorithms for prediction of PCOS. IEEE Xplore
3. Pal A et al (2021) Hybrid machine learning algorithms for polycystic ovary syndrome detection. IEEE Xplore
4. Wang P, Wang J, Li Y, Li L, Zhang H (2021) Improved sampling and feature selection to support extreme gradient boosting for PCOS diagnosis. IEEE Xplore
5. Nabi N, Islam S, Khushbu SA, Masum AKM (2021) Machine learning approach: detecting polycystic ovary syndrome & its impact on Bangladeshi women. In: Proceedings of the 12th international conference on computing communication and networking technologies (ICCCNT), Kharagpur, India, pp 1–7
6. Bharati S, Podder P, Hossain Mondal MR (2020) Diagnosis of polycystic ovary syndrome using machine learning algorithms. In: Proceedings of the 2020 IEEE region 10 symposium (TENSYMP), Dhaka, Bangladesh, pp 1486–1489
7. Kurnianingsih et al (2020) PCONet: a convolutional neural network architecture to detect PCOS from ovarian ultrasound images. IEEE Xplore
8. Manickam S, Perumal C, Prabha R, Srilatha K (2021) Study and detection of PCOS related diseases using CNN. In: IOP conference series: materials science and engineering, vol 1070. Article number 012062
9. Nazarudin AA, Zulkarnain N, Hussain A, Mokri SS, Nordin INAM (2020) Review on automated follicle identification for polycystic ovarian syndrome. Bull Electr Eng Inform 9(2): 588–593
10. Prapty AS, Shitu TT (2020) An efficient decision tree establishment and performance analysis with different machine learning approaches on polycystic ovary syndrome. In: Proceedings of the 2020 23rd international conference on computer and information technology (ICCIT), Dhaka, Bangladesh, pp 1–5
11. Bhargava S, Choubey A, Nandan D, Mahajan A (2021) Polycystic ovarian syndrome detection by using two-stage image denoising. Traitement du Signal 38:1217–1227
12. Suha SA, Islam MN (2022) An extended machine learning technique for polycystic ovary syndrome detection using ovary ultrasound image. Sci Reports 12. Article number 17123
13. Salman Hosain AKM, Mehedi MHK, Kabir IE (2022) PCONet: a convolutional neural network architecture to detect polycystic ovary syndrome (PCOS) from Ovarian ultrasound images. In: 2022 international conference on engineering and emerging technologies (ICEET), Kuala Lumpur, Malaysia, pp 1–6. https://doi.org/10.1109/ICEET56468.2022.10007353
14. Choudhari A, Korde A. PCOS detection using ultrasound images. www.kaggle.com/datasets/anaghachoudhari/pcos-detection-using-ultrasound-images

Quality Estimation for Dynamic Pricing of Perishable Products: A Deep Learning Approach

Ashish Kumar and Sunil Agrawal

Abstract The current world is facing tremendous challenges related to food security, which are likely to increase in the future due to rising population and environmental concerns. Perishable product management is highly complex due to perishability dynamics. The quality and price of the product are the main factors that govern the buying tendency for these products. Thus, the main objective of this work is to develop a deep learning-based image processing model that is both accurate and efficient in order to identify products, like pomegranates, depending on their quality. A unique Convolutional Neural Network (CNN) model is developed for this work. For the same reason, pre-trained models from the ResNet family, including ResNet18, ResNet34, ResNet50, ResNet101, and ResNet152, are also modified using transfer learning approaches. The results of every model are compared with respect to performance metrics, including f1-score, accuracy, precision, recall, and execution time. The base CNN model achieved the highest accuracy of 90.66% and performed best in all other performance criterion. With the help of new age technologies like machine learning, deep learning, etc., real-time non-destructive quality estimation of the perishable products can be done in real time to attach dynamic pricing based on their quality. In the long run, this will increase food security by reducing post-harvest losses through the ability to make supply chain management decisions more quickly and correctly.

Keywords Quality · Deep learning · Convolutional neural network · Dynamic pricing · Supply chain management

A. Kumar (✉) · S. Agrawal
ME Department, PDPM IIITDM, Jabalpur, India
e-mail: 1913603@iiitdmj.a.c.in

S. Agrawal
e-mail: sa@iiitdmj.ac.in

A. C. Frery et al. (eds.), *Computational Intelligence and Data Analytics*, Lecture Notes on Data Engineering and Communications Technologies 236,
https://doi.org/10.1007/978-981-96-0451-7_25

1 Introduction

Concerning food safety and security, the world is up against enormous obstacles. The primary cause is the approximately one-third of the world's food production that is lost each year as a result of various issues with perishable supply chain management [1]. With the rising population worldwide especially in the developing economies like India, it becomes of immense necessity that the post-harvest losses must be reduced along with the development of sustainable methods in agriculture to feed the upcoming population that is estimated to reach around 9.1 billion by 2050 [2].

Perishable items like fruits, vegetables, flowers milk, etc. normally have shelf life ranging from 1 day to up to 3 weeks forms a major part of the daily intake of all the population to meet their daily nutritional requirements. The management of the perishable supply chain becomes more complex and challenging when perishability dynamics are taken into account. Determining the product's quality at a given point in its life cycle is vital since perishability dynamics influence the rate of product deterioration. Quality assessment is a difficult and complex process that requires a great deal of work to correlate sensory and subjective judgments with traditional destructive methods. Consumers want the best possible quality produce in terms of appearance, safety, and nutritional content. Also, the relation between quality and price must be appropriately determined so that it increases the chances of being accepted by the consumer. These have prompted companies and researchers to provide quick, accurate, and affordable food analysis methods.

1.1 Research Background

The primary determinants of consumers' inclination to purchase perishable goods are price and quality. Fruit and vegetable quality is correlated with their level of maturity, whose combination of chemical and physical characteristics dictates consumer acceptability [3]. Fruit and vegetable quality can be assessed using five different criteria: texture, flavour (taste and aroma), safety, nutritional content, and visual quality or appearance (i.e., size, shape, colour, and flaws). While consumers can directly assess appearance, texture, and flavour using their senses, safety and nutritional values cannot be felt or ascertained at the point of purchase. Consumers allegedly rely on their initial assessments of fruit and vegetable quality exclusively on external attributes [4]. Next, internal quality parameters that relate to flavour and scent during eating such as acidity, sweetness, sugar to acid ratio, and texture indicate the extent to which customers are satisfied with their initial experience. Fruit and vegetables should therefore have an appealing look to encourage first sales, a flavour and texture that consumers will find acceptable, and appropriate nutritional contents to persuade them to make additional purchases. Therefore, determining the quality and attach appropriate price based on it is a highly complex process with needs innovation and implementation of new age technologies.

The quality of the perishable items can be determined by both destructive and non-destructive techniques. Conventional methods, i.e., destructive ones are costly, time-consuming, and damaging. Analytical and destructive procedures are still often utilized, although they are costly, time-consuming, and environmentally harmful. They might also call for specific tools and careful sample preparation. Ultimately, they are unsuitable for use in industrial lines where fast, trustworthy, non-destructive, less expensive, and less polluting methods are needed for fruit and vegetable grading, quality assessment, and flaw identification [5]. The second group includes the quick non-destructive testing approach. It encompasses the following methods: mechanical and acoustic; electromagnetic (e.g., nuclear magnetic resonance, electron spin resonance method); sensory (electronic tongue, biosensors, electronic nose, etc.); X-ray diffraction; and other detection methods (microbial indication indicators, enzyme indication indicators, etc.) [6]. When compared to traditional approaches, these techniques offer the advantages of non-destructive analyses, minimal sample contamination, and quick response times.

One of the emerging technologies under non-destructive testing is the computer vision/image processing for real-time quality estimation of the product based on its external features. Numerous external quality attributes, such as colour, shape, size, and texture, as well as flaws, can be measured and found by the image processing system [7]. Using appropriately trained classification models, it automatically extracts the most discriminative features associated with product quality from photos and grades the result. The new age technologies like machine learning, deep learning and cyber-physical systems can now be integrated with each other to develop a robust system for the afore-mentioned task [8, 9]. As a result, the various supply chain participants are able to make more informed decisions on supply chain management, including those pertaining to pricing, procurement, storage, and transportation.

2 Literature Review

The last 5 years have seen a significant amount of study on image processing with deep learning for fruit categorization and grading. The technique of identifying which fruit variety is visible in a photograph including one or more fruit kinds is known as classification. One of the most crucial tasks to maintain extra value in the marketplaces is fruit grading. Therefore, efforts to enhance the techniques for identifying fruit damage, diseases, freshness, and maturity level are ongoing. The three main uses of image processing in agriculture are tasks related to classification [10], grading [11] and detection [12].

Convolutional neural network (CNN) models were attempted to be created for classification tasks in the majority of previous works [13–17]. They also employed several pre-trained models that were already in use, in case the accuracy gained beat CNN models in the improbable occasion that it did. Sakib et al. [14] developed and analysed a number of CNN architectures for the classification of fruits using the Fruits-360 dataset. They used various hidden layer and epoch setups for various conditions. CNN models developed by Lu [15] also investigated the impact of data augmentation on the classification task accuracy of the model's classifier. It was discovered that incorporating data augmentation techniques increases the model's dependability and accuracy. Zeng [18] used the pre-trained VGG model that was already in use, which yielded higher accuracy than their CNN models.

2.1 Objectives

Developing an accurate and efficient image processing model that can categorize products based on their quality is the primary goal of the ongoing study. The product selected is pomegranate, which has three grades and under each grade there are four classes based on its quality. Various tasks completed in order to meet our primary objective are listed below:

- Development of an entirely new CNN (Convolutional Neural Network) model.
- Application of the concept of transfer learning for modification of pre-trained ResNet models, including ResNet18, ResNet34, ResNet50, ResNet101, and ResNet152.
- Comparison of developed and modified pre-trained models using model performance metrics such as accuracy, recall, F1-score, and precision/accuracy.

The remainder of the document is structured as follows: In Sect. 3, the research technique is provided. The evaluation methods, findings, and discussion of the model are outlined in Sect. 4. In Sect. 5, references are included after the presentation of conclusions and future work.

3 Research Methodology

Figure 1 illustrates the suggested approach for creating image processing models and all of its sub-steps.

3.1 Data Collection and Pre-processing

Two common issues with datasets used for image processing jobs are imbalance and insufficiency, which both lead to subpar performance. Also, the data labelling is a complex and tedious task. Therefore, in this research work, the authors have used a standard dataset collected and used in previous research [19]. The details of the dataset are mentioned below.

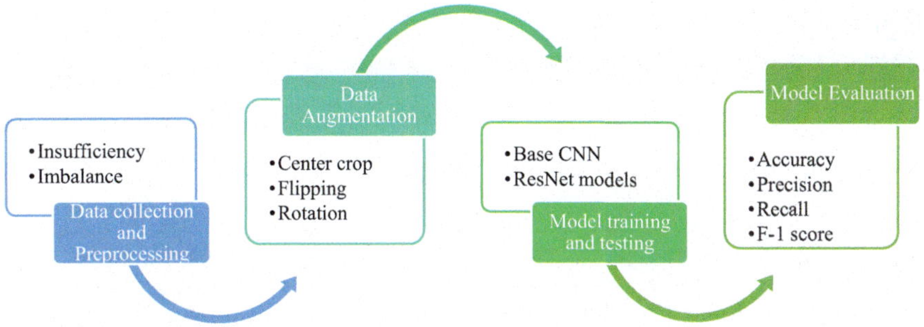

Fig. 1 Research methodology

Table 1 Details of collected dataset

Serial number	Particulars	Description
1	Fruit	Pomegranate *(Punica granatum)*
2	Cultivar	Ruby
3	Number of grades considered	3
4	Number of samples of each grade	30
5	Total number of samples	90
6	Geographical location	16°12′17.863″N 75°30′5.27″EAltitude: 452 msl
7	Temperature during harvest	26 °C

The fruits of the cultivar Ruby make up the collected dataset. There are three varieties of pomegranate-grade fruits collected. Table 1 provides information on the dataset gathering.

The description of each of the grades is presented in Table 2. A weighing machine is used to determine the weight of each fruit sample, and a Vernier calliper is used to determine its diameter. For every grade, pomegranate fruits are gathered and divided into four categories. To accomplish this, pictures were taken every other day for a total of eight days. With four qualities inside each grade, the entire procedure is repeated for a total of twelve classes of effective quality standards. To maintain track of every fruit, tagging is used. The designations of the class labels are mentioned in Table 3, and Fig. 2 shows the images captured for each class.

3.2 Data Augmentation

By producing altered versions of the photos in a dataset, a technique known as image data augmentation can be used to artificially create variations of the images in order to create a robust model. The fit models' capacity to apply what they have learnt to

Table 2 Description of the grades

Serial number	Grade	Descriptive characteristics					
		Aesthetics	Colour	Range of weight (gm/fruit)	Range of size (mm)	Outer surface	Tolerance (%)
1	G1	Superior quality look with fully ripe	Attractive red or rose pink	300–400	85.39–106.12	Smooth, slight superficial defects not affecting the look and quality	10
2	G2	Good look with fully ripe	Attractive red or rose pink. Improper colouring may be present	200–300	74.72–89.23	Slight defects like scar, scratch, scrape, blemish, etc. may be allowed, not affecting the look and quality	10
3	G3	Good look with fully ripe	Attractive red or rose pink Improper colouring may be present	100–200	64.85–78.15	Slight defects like scar, scratch, scrape, blemish, etc. may be allowed, not affecting the look and quality	10

Table 3 Class labels of pomegranate quality grades

Class	1	2	3	4	5	6	7	8	9	10	11	12
Effective quality	G1 Q1	G1 Q2	G1 Q3	G1 Q4	G2 Q1	G2 Q2	G2 Q3	G2 Q4	G3 Q1	G3 Q2	G3 Q3	G3 Q4

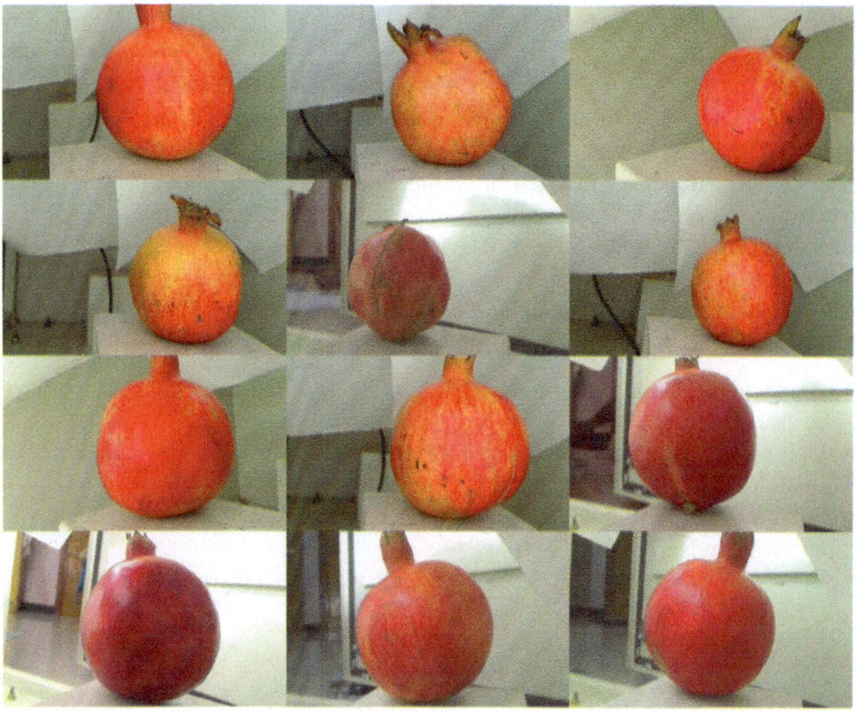

Fig. 2 Images of different pomegranate quality grades from G1_Q1 to G3_Q4

new images can be enhanced by the various augmentation techniques that produce alterations of the images. In this study, we performed data augmentation techniques like rotation, flipping along the horizontal and vertical axes and centre crop to training images, but we only applied centre crop to test images. Below is a description of the data augmentation approaches that have been executed:

- Centre crop: The photos were centre cropped to 224 × 224 pixels per inch, which is the minimum size required by all pre-trained models.
- Flipping: With a chance of 0.5, we randomly flipped both horizontally and vertically.
- Rotation: We applied a 20-degree rotation to the photos in our dataset.

3.3 Model Training and Testing

The 1080 images in the entire dataset are separated into training and testing folders. There are 12 classes in all. The training folders contain a total of 864 images with each class having around 72 images and test dataset has a total of 216 images with 18 or 19 images in different classes. Testing and training cannot be done until the architecture of our model is complete. Subsequently, the model's parameters were determined by conducting preliminary tests and consulting prior research conducted on comparable datasets. The effect of several parameters on the accuracy of the CNN base model was investigated in preliminary testing. Most of the parameters were shown to have impacts that were similar to those reported in earlier studies, including learning rate, optimizer function, and activation function. Our model worked best when the batch size was eight when it came to batch size selection. Furthermore, we obtained similar accuracy after just 50 epochs of training and testing. We determined the final model parameters for each model based on a combination of data from earlier studies and the preliminary experimental trials. Below are the details of the standard model parameters along with their chosen values:

- Activation function: A sigmoid activation function was used in the output layer of each model.
- Optimizer function: We employed stochastic gradient descent, or SGD, as the optimizer function in each model.
- Learning rate: We employed the OneCycleLR scheduler in all of the models, with a maximum learning rate of 0.01.
- Epochs: A total of 50 epochs were used to train and evaluate each of our models.

3.3.1 Base CNN Model

In the baseline CNN model, three convolutional layers were employed, with each layer performing three operations: convolution, batch normalization, ReLU activation, and max pooling. Each convolutional layer utilized three filters, repeated three times. Following the convolutional layers, a dropout layer was placed between two fully connected layers. The first fully connected layer has 25,088 input features and 600 output features, while the second fully connected layer consists of 600 input features and 12 output features. The architecture of the baseline CNN model is illustrated in Fig. 3.

3.3.2 Pre-trained Models

Using a trained model to perform a similar task is known as transfer learning. Stated differently, it is a research process for machine learning that retains the information gleaned from addressing a particular problem and applies it to a similar but unrelated topic. By recycling the knowledge gained from the previously completed activity, this boosts efficiency.

In this study, several of the popular pre-trained models are compared with the built CNN model's performance in terms of accuracy and execution time (training and testing). Since there is a wide variety of models in the pre-trained model family, we have only chosen models from the ResNet family [20].

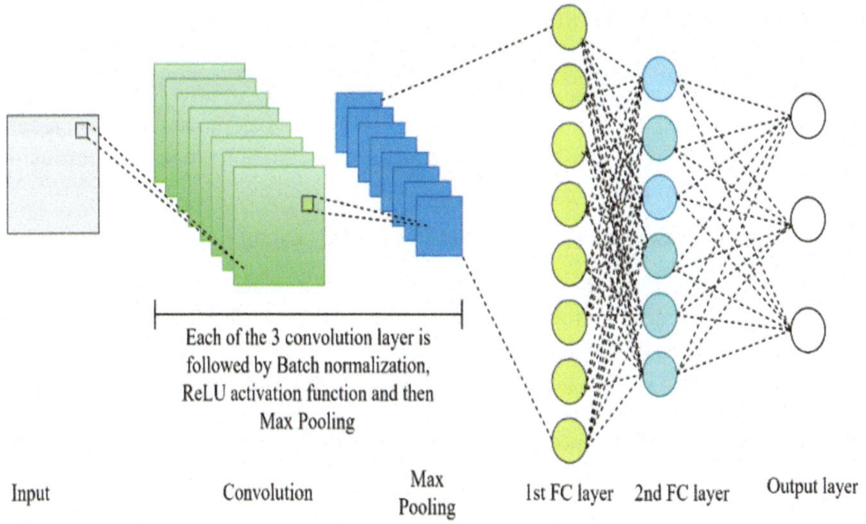

Fig. 3 Three-layer base CNN model simplified diagram

3.4 Model Evaluation

A tabular overview of the number of accurate and inaccurate predictions made by a classifier is called a confusion matrix. It is used to gauge the effectiveness of a model. It can be used to evaluate the performance of a classification model by examining performance metrics such as accuracy, precision, recall, and F1-score that are provided in the classification report. The execution times of each model are taken into account in order to identify the most efficient model.

$$Accuracy = TP + TN/(TP + TN + FP + FN) \tag{1}$$

$$Precision = TP/(TP + FP) \tag{2}$$

$$Recall = TP/(TP + FN) \tag{3}$$

$$F1 - score = (2 * Precision * Recall)/(Recall + Precision) \tag{4}$$

4 Results and Analysis

This section presents and analyses each model's output using a variety of performance criteria. Every model ran on Kaggle with a GPU T100 accelerator. Precision, recall, and F1-score tables were constructed based on the classification reports and confusion matrices for each model.

We arrive at the following conclusions after carefully examining all the results:

- Figure 4 makes clear that, in contrast to ResNet models, the base CNN model exhibits the greatest volatility and only reaches above 80% accuracy (90.66%) in the final segment of epochs between 45 and 50. We see abrupt spikes in the figure because the base CNN model has to learn all of the visual attributes from scratch, which causes the fluctuations.
- The accuracy of all the models, with the exception of the base CNN, was limited to 70%.
- Table 4 displays the confusion matrix of the best model, or base CNN, showing the classifications the model made that were accurate and inaccurate.
- Table 5 displays all of the models' accuracy and execution times. For every dataset class, the base CNN model produced higher f1-score, recall, and accuracy values (Tables 6, 7 and 8).

Thus, taking into account the model execution times and the aforementioned analysis, we can conclude that the base CNN performs optimally when taking into account all performance criteria.

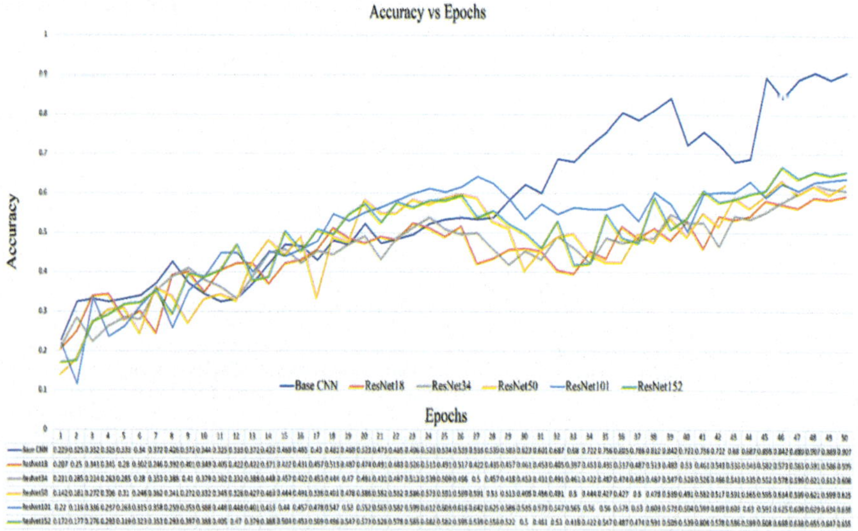

Fig. 4 Three-layer base CNN model simplified diagram

Table 4 Confusion matrix table for base CNN model

	Predicted classes												
	Classes	G1_ Q1	G1_ Q2	G1_ Q3	G1_ Q4	G2_ Q1	G2_ Q2	G2_ Q3	G2_ Q4	G3_ Q1	G3_ Q2	G3_ Q3	G3_ Q4
Actual classes	G1_Q1	17	1	1	0	0	0	0	0	0	0	0	0
	G1_Q2	0	18	0	0	0	0	0	0	0	0	0	0
	G1_Q3	0	0	17	1	0	1	0	0	0	0	0	0
	G1_Q4	0	0	1	17	1	0	0	0	0	0	0	0
	G2_Q1	0	0	0	0	17	1	0	0	0	0	0	0
	G2_Q2	0	0	0	0	0	17	2	0	0	0	0	0
	G2_Q3	0	0	0	0	0	2	17	0	0	0	0	0
	G2_Q4	0	0	0	0	0	0	0	16	1	0	1	1
	G3_Q1	0	0	0	0	0	0	1	0	18	0	0	0
	G3_Q2	0	0	0	0	0	0	1	0	0	16	2	0
	G3_Q3	0	0	0	1	0	0	0	0	0	1	16	0
	G3_Q4	0	0	0	0	0	1	0	0	0	0	0	18

Table 5 Execution times of different compared models

Models	Base CNN	ResNet18	ResNet34	ResNet50	ResNet101	ResNet152
Accuracy (%)	90.66	59.48	60.78	62.5	63.79	65.52
Execution time	3 min 18 s	4 min 50 s	4 min 56 s	6 min	9 min 30 s	12 min 48 s

Table 6 Precision values of different compared models

Models	Base CNN	ResNet18	ResNet34	ResNet50	ResNet101	ResNet152
G1_Q1	1	0.18	0.4	0.3	0.5	0.36
G1_Q2	0.9473	0.64	0.75	0.5	0.65	0.93
G1_Q3	0.89474	0.3	0.56	0.47	0.53	0.58
G1_Q4	0.89474	0.46	0.5	0.42	0.45	0.42
G2_Q1	0.9444	0.58	0.64	0.65	0.72	0.8
G2_Q2	0.77273	0.37	0.4	0.58	0.42	0.58
G2_Q3	0.80952	0.59	0.58	0.82	0.75	0.78
G2_Q4	1	0.54	0.59	0.52	0.58	0.78
G3_Q1	0.94737	0.71	0.58	0.76	0.83	0.81
G3_Q2	0.94118	0.76	0.74	0.68	0.79	0.64
G3_Q3	0.84211	0.61	0.82	0.45	0.67	0.6
G3_Q4	0.94737	0.56	0.5	0.53	0.58	0.53

Table 7 Recall values of different compared models

Models	Base CNN	ResNet18	ResNet34	ResNet50	ResNet101	ResNet152
G1_Q1	0.89474	0.16	0.42	0.16	0.42	0.26
G1_Q2	1	0.78	0.83	0.39	0.83	0.72
G1_Q3	0.89474	0.16	0.47	0.37	0.47	0.58
G1_Q4	0.89474	0.58	0.53	0.53	0.68	0.58
G2_Q1	0.9444	0.79	0.74	0.79	0.68	0.84
G2_Q2	0.89474	0.53	0.53	0.58	0.58	0.58
G2_Q3	0.89474	0.53	0.37	0.74	0.63	0.74
G2_Q4	0.84211	0.68	0.68	0.58	0.74	0.74
G3_Q1	0.94737	0.53	0.74	0.84	0.79	0.81
G3_Q2	0.84211	0.72	0.78	0.83	0.61	0.78
G3_Q3	0.8889	0.58	0.47	0.47	0.42	0.47
G3_Q4	0.94737	0.28	0.39	0.56	0.39	0.56

Table 8 F1-score values of different compared models

Models	Base CNN	ResNet18	ResNet34	ResNet50	ResNet101	ResNet152
G1_Q1	0.9444	0.17	0.41	0.21	0.46	0.3
G1_Q2	0.9729	0.7	0.79	0.44	0.73	0.81
G1_Q3	0.89474	0.21	0.51	0.41	0.5	0.58
G1_Q4	0.89474	0.51	0.51	0.47	0.54	0.49
G2_Q1	0.9444	0.67	0.68	0.71	0.7	0.82
G2_Q2	0.8292	0.43	0.45	0.58	0.49	0.58
G2_Q3	0.8499	0.56	0.45	0.78	0.69	0.76
G2_Q4	0.9142	0.6	0.63	0.55	0.65	0.76
G3_Q1	0.9473	0.61	0.65	0.8	0.81	0.85
G3_Q2	0.8888	0.74	0.76	0.75	0.69	0.7
G3_Q3	0.8648	0.59	0.6	0.46	0.52	0.53
G3_Q4	0.9473	0.37	0.44	0.54	0.47	0.54

5 Conclusions and Future Work

Creating a categorization and grading model that is both extremely accurate and efficient was the primary goal of this research project. To do this, a baseline CNN model is created, and the outcomes are contrasted with pre-trained ResNet models that are currently in use. To address issues with data imbalance and inadequacy, an appropriate dataset was selected. There were 1080 photos in the collection overall, with equal representations of each class. To improve accuracy, the dataset was subjected to data augmentation techniques such as rotation, random vertical flip, random horizontal flip, and centre crop. For the classification and grading image processing tasks, a thorough comparison of the models was conducted. Both the classification report and the confusion matrix were created. The models were compared using the following performance evaluation criteria: F1-score, accuracy, precision, and recall.

On the test dataset, the accuracy of the base CNN model was 90.7%, whereas pre-trained ResNet models had accuracy below 70%. When it comes to all performance metrics taken into account, including accuracy, precision, recall, F1-score, and execution time, base CNN performs the best out of all the models. The study's image processing models can be applied in real-time image processing systems to classify and grade fruits in order to make decisions about supply chain management. Based on the results of the created image processing model, which categorizes products according to their quality, dynamic pricing can be implemented. This will ultimately result in a decrease in fruit waste and the monetary losses that go along with it, opening the door for industrial automation in fruit quality control and inspection.

The models outlined in this research can be adapted and compared with other pre-trained models, such as AlexNet, VGG, SqueezeNet, DenseNet, and others, for future studies.

References

1. Kumar A, Agrawal S (2023) Challenges and opportunities for agri-fresh food supply chain management in India. Comput Electron Agric 212. https://doi.org/10.1016/j.compag.2023.108161
2. FAO (2009) How to feed the world 2050. https://doi.org/10.5822/978-1-61091-885-5
3. Salinas-Hernández RM, González-Aguilar GA, Tiznado-Hernández ME (2015) Utilization of physicochemical variables developed from changes in sensory attributes and consumer acceptability to predict the shelf life of fresh-cut mango fruit. J Food Sci Technol 52(1):63–77. https://doi.org/10.1007/s13197-013-0992-0
4. Palumbo M, Cefola M, Pace B, Attolico G, Colelli G (2023) Computer vision system based on conventional imaging for non-destructively evaluating quality attributes in fresh and packaged fruit and vegetables. Postharvest Biol Technol 200(June 2022):112332. https://doi.org/10.1016/j.postharvbio.2023.112332
5. He Y, Xiao Q, Bai X, Zhou L, Liu F, Zhang C (2022) Recent progress of nondestructive techniques for fruits damage inspection: a review. Crit Rev Food Sci Nutr 62(20):5476–5494. https://doi.org/10.1080/10408398.2021.1885342
6. Wu S, Zhang M, Yu Q, Mujumdar AS, Yang C (2023) Fresh food quality deterioration detection and labeling: a review of recent research and application in supply chain. Food Bioprocess Technol 0123456789. https://doi.org/10.1007/s11947-023-03197-9
7. Kakani V, Nguyen VH, Kumar BP, Kim H, Pasupuleti VR (2020) A critical review on computer vision and artificial intelligence in food industry. J Agric Food Res 2(February):100033. https://doi.org/10.1016/j.jafr.2020.100033
8. Chakraborty SK et al (2023) Development of an optimally designed real-time automatic citrus fruit grading–sorting machine leveraging computer vision-based adaptive deep learning model. Eng Appl Artif Intell 120(November 2022):105826. https://doi.org/10.1016/j.engappai.2023.105826

9. Kumar A, Agrawal S (2024) A quality-based sustainable supply chain architecture for perishable products using image processing in the era of industry 4.0. J Clean Prod 450(March):141910. https://doi.org/10.1016/j.jclepro.2024.141910
10. Katarzyna R, Paweł M (2019) A vision-based method utilizing deep convolutional neural networks for fruit variety classification in uncertainty conditions of retail sales. Appl Sci 9(19). https://doi.org/10.3390/app9193971
11. Zhang Y, Lian J, Fan M, Zheng Y (2018) Deep indicator for fine-grained classification of banana's ripening stages. EURASIP J Image Video Process
12. Sciortino R, Micale R, Enea M, La Scalia G (2016) A webGIS-based system for real time shelf life prediction. Comput Electron Agric 127:451–459. https://doi.org/10.1016/j.compag.2016.07.004
13. Yu Y, Zhang K, Yang L, Zhang D (2019) Fruit detection for strawberry harvesting robot in non-structural environment based on Mask-RCNN. Comput Electron Agric 163(April):104846. https://doi.org/10.1016/j.compag.2019.06.001
14. Sakib S, Ashrafi Z, Siddique MAB (2019) Implementation of fruits recognition classifier using convolutional neural network algorithm for observation of accuracies for various hidden layers, pp 8–11. http://arxiv.org/abs/1904.00783
15. Lu Y (2016) Food image recognition by using convolutional neural networks (CNNs), pp 1–6. http://arxiv.org/abs/1612.00983
16. Kumar A, Agrawal S (2024) Enhancing quality-based classification of perishable products: a convolutional neural network approach with statistical hyperparameter optimization. Multimed Tools Appl 0123456789. https://doi.org/10.1007/s11042-024-19700-z
17. Kumar A, Tiwari S, Agrawal S (2024) Convolutional neural network based image processing model for supply chain management. Lecture Notes Mech Eng (March):113–123. https://doi.org/10.1007/978-981-99-7445-0_11
18. Zeng G (2017) Fruit and vegetables classification system using image saliency and convolutional neural network. In: Proceedings of 2017 IEEE 3rd information technology mechatronics engineering conference ITOEC 2017, vol 2017-January, pp 613–617. https://doi.org/10.1109/ITOEC.2017.8122370
19. Kumar RA, Rajpurohit VS, Jirage BJ (2018) Pomegranate fruit quality assessment using machine intelligence and wavelet features. J Hortic Res 26(1):53–60. https://doi.org/10.2478/johr-2018-0006
20. Kaiming He JS, Zhang X, Ren S. Deep residual learning for image recognition

Securing Tele-Dermatology: Leveraging Deep Learning for Remote Melanoma Diagnosis

Gadde Ashok, N. Ruthvik, N. Sri Charan, M. Dushyanth, and T. Gireesh Kumar

Abstract Tele-dermatology allows for the remote diagnosis of melanoma by utilizing cutting-edge technology to facilitate meetings and exchange images. Utilized 43,141 dermoscopic images from the two massive datasets, HAM10000 and ISIC2020. Lesion_id, image_name, age, sex, and area were significant attributes that were consistent across the two datasets. A 'target' trademark was included throughout the model construction process to provide trustworthy melanoma differentiating evidence. During the preprocessing phase, the Differentiation Restricted Versatile Histogram Evening out (CLAHE) technique was applied to obtain the best possible difference expansion and sound decrease. Morphological opening was introduced to the CLAHE-handled images to emphasize qualities even more. After a review of many image segmentation techniques, region-based division was shown to be the most effective due to its ease of use and high diagnostic yield for melanoma. Careful alignment was completed in order to work on model layout and make sure the goal variable circulation was achieved with greater efficiency. Work done on transfer learning combined with deep learning models, such as DenseNet, ResNet, Inception, Xception and EfficientNet. After moving forward, the suggested half-and-half engineering achieved impressive accuracy, demonstrating its genuine potential to advance melanoma diagnosis in tele-dermatology.

G. Ashok · N. Ruthvik · N. Sri Charan · M. Dushyanth · T. G. Kumar (✉)
Department of Computer Science and Engineering, Amrita School of Computing, Amrita Vishwa Vidyapeetham, Coimbatore, India
e-mail: t_gireeshkumar@cb.amrita.edu

G. Ashok
e-mail: cb.en.u4cse20020@cb.students.amrita.edu

N. Ruthvik
e-mail: cb.en.u4cse20442@cb.students.amrita.edu

N. Sri Charan
e-mail: cb.en.u4cse20640@cb.students.amrita.edu

M. Dushyanth
e-mail: cb.en.u4cse20037@cb.students.amrita.edu

A. C. Frery et al. (eds.), *Computational Intelligence and Data Analytics*, Lecture Notes on Data Engineering and Communications Technologies 236,
https://doi.org/10.1007/978-981-96-0451-7_26

Keywords Tele-dermatology · Melanoma diagnosis · Deep learning · Image segmentation · Transfer learning · CLAHE · Morphological opening · Threshold-based segmentation · HAM10000 · ISIC2020

1 Introduction

One of the most dangerous types of skin cancerous growths is melanoma, which is also a serious medical disease in general. Early detection and precise discovery are necessary for a workable treatment plan and improved comprehension of the outcomes. Tele-dermatology, a kind of telemedicine, has become widely used as a possible method of remote skin condition diagnosis. By permitting patients and dermatologists to counsel and trade dermoscopic pictures, it lessens the requirement for in-person meetings [1]. This strategy would be particularly useful for the people who live in devastated or country regions where admittance to a few dermatological medicines is restricted.

Late advancements in mechanized imaging and profound learning have enormously expanded the capability of the discipline of tele-dermatology. High-level man-made reasoning calculations can possibly break down high-goal dermoscopic pictures and exactly analyze and arrange skin wounds, matching the degree of accuracy of dermatologists, if they are arranged [2]. Convolutional neural networks (CNNs) are one class of profound learning models that have shown remarkable execution in picture acknowledgment tests. CNNs are a reasonable fit for grasping hazy clinical pictures. These models are valuable devices in the recognizable proof cycle since they can recognize characteristics and circumstances connected to melanoma. Still, there are a few obstructions in the approach to joining profound learning with tele-dermatology. The sum and nature of dermoscopic pictures, data security concerns, and the prerequisite for justifiable and sensible models are the primary hardships that should be settled. Consequently, the formation of solid and generally material models relies upon the utilization of huge, profoundly shifted, and completely portrayed datasets [3]. In this technique, standardized datasets, for example, HAM10000 and ISIC2020, which incorporate a critical number of made sense dermoscopic pictures, ought to be utilized to develop and check profound learning models. Tele-dermatology may incredibly upgrade the early distinctive proof and investigation of melanoma by utilizing these assets and lessening the obstructions, which will ultimately chip away at super durable results.

Deep learning in tele-dermatology can increment end precision and change treatment procedures. Profound learning calculations might perceive little highlights and qualities that the natural eye could miss overwhelmingly of visual information. This degree of accuracy takes into consideration a more exact order of skin sores, which is basic for recognizing harmless and perilous circumstances. Besides, reliably keeping up with and understanding huge datasets may most likely diminish indicative time, permitting treatment to start prior [4]. These progressions are projected to impact dermatological treatment as tele-dermatology spreads, especially where admittance to specific medications is restricted.

Regardless of its various purposes, tele-dermatology's broad use isn't without troubles. Safeguarding the privacy and security of touchy patient information is vital, particularly while managing complex pictures that are immediately communicated and put away. Moreover, the gave material's variety and kind decide how profound learning models are introduced. Inclinations in datasets could give deceiving discoveries by losing models course [5]. To diminish these dangers, groundbreaking work projects focus on improving data gathering philosophies, performing precise computations, and building tight data security assessments. Tele-dermatology, with its broad learning base, is ready to turn into a foundation of present-day dermatological treatment, it is perceived and treated to reform the manner in which skin issues.

2 Related Works

Tele-dermatology has made huge advances as of late, with deep learning getting the hang of arising as a crucial procedure for working on the identification of skin problems, outstandingly melanoma. The utilization of profound learning calculations on dermoscopic pictures has yielded empowering results in dependably distinguishing melanoma, which is basic for early mediation and treatment. Esteva et al. [2] demonstrated the way that a profound brain organization could order skin malignant growth with dermatologist-level precision, featuring the capability of arising innovations to further develop clinical independent direction. Likewise, Tschandl et al. [3] featured the commitment of convolutional brain organizations (CNNs) in recognizing skin sores, underscoring the meaning of preparing these models on huge, clarified datasets.

One of the main issues in tele-dermatology is picture quality fluctuation, which requires the utilization of vigorous picture preprocessing approaches. As indicated by Reza et al. [6], Contrast Limited Adaptive Histogram Equalization (CLAHE) is widely used to work on the difference of dermoscopic pictures without expanding commotion. This preprocessing step is basic for expanding the perceivability of sore qualities, which can affect the exhibition of ensuing picture handling procedures. Moreover, Gonzalez et al. [7] portray the utilization of morphological cycles like opening and closing to improve picture division. These methodologies help in isolating the sore from the encompassing skin, which is a significant stage in right finding. Dermoscopic picture examination depends vigorously on picture division. Different methodologies have been investigated, with limit-based division techniques ending up especially powerful for double assignments like melanoma recognition. Otsu's strategy, a generally utilized thresholding method, has been demonstrated to be viable in isolating the sore from the foundation in light of pixel force values, as exhibited by Otsu [8].

Its straightforwardness and viability settle on it a typical decision for the first division in tele-dermatology applications. The presentation of profound learning has changed the area of clinical picture examination, with CNNs being the most frequently utilized model for picture arrangement. CNNs have been utilized effectively in tele-dermatology to order skin sores. Models like DenseNet, ResNet, Initiation, and Xception have been produced for melanoma analysis by utilizing move figuring out how to calibrate pre-prepared networks utilizing dermatological datasets. Move gaining empowers these models to profit from data gained from huge, general picture datasets, bringing about better execution on specific errands like melanoma arrangement. For example, Liu et al. [9] proved the usefulness of transfer learning using Inception-v3 in identifying skin lesions, with high accuracy rates. Ashok et al. [10] present a classification strategy based on machine learning models to improve the security of IoT networks by properly recognizing various forms of assaults, which is critical for maintaining the integrity and dependability of these linked devices. Chanchal et al. [11] have out a detailed study of neural network-based visual data concealing techniques. Their study presents an in-depth review of several strategies for concealing information inside photographs, highlighting both advances and problems in the field of secure data exchange and steganography. Poorna et al. [12] explored the use of PC vision and AI techniques to dissect dermoscopic pictures for early melanoma discovery. The review underscores the extraction of ABCD qualities (asymmetry, border, color, and diameter) as basic features for melanoma location.

Shobha Rani et al. [13] propose a bunch of preprocessing strategies intended to work on the nature of harmed record pictures, which is basic for applications like Optical Character Recognition (OCR). The methodologies laid out effectively manage many types of clamors, for example, nonuniform lighting and pen markings, utilizing commotion debasement models and local standardization to further develop pictures. Even with these developments, there are still a number of obstacles to overcome before deep learning can be fully applied in tele-dermatology. Security and privacy of data are important issues, particularly when handling sensitive patient data. Deep learning-enhanced tele-dermatology has great promise for improving melanoma and other skin disease detection. Improvements in segmentation, classification, and image preprocessing methods have made diagnosis more precise and effective. But for these technologies to be widely used in clinical practice, it will be imperative to solve the issues of data privacy, model interpretability, and dataset biases.

The three main objectives of this study are:

1. Firstly, to merge the HAM10000 and ISIC2020 datasets into a unified dataset with consistent attributes such as lesion_id, image_name, age, sex, and site. A standardized 'target' attribute for melanoma identification is created to facilitate comprehensive analysis across datasets.
2. The second objective is to identify the most effective preprocessing and segmentation methods for melanoma detection. The study examines the use of CLAHE for contrast enhancement and threshold-based segmentation for accurate lesion delineation.
3. The third objective is to implement and evaluate the use of transfer learning with deep learning models to improve melanoma classification. The study explores the impact of pre-trained weights on both original and segmented images.

3 Methodology

The proposed architecture on melanoma detection was designed after the performance of two prominent datasets: HAM10000 and ISIC 2020. The HAM10000 dataset provides over 10,000 dermatoscopic images of skin lesions, including melanoma and various benign nevi. The ISIC 2020 dataset offers an additional 33,126 dermoscopic images from over 2,000 patients, featuring both benign and malignant lesions. Each image in the ISIC 2020 dataset is associated with a unique patient identifier, with malignancies confirmed through histopathology and benign diagnoses established by expert interpretation or histopathology. To standardize current approach, common attributes such as lesion_id, image_name, age, sex, and site across both datasets were integrated.

In the HAM10000 dataset, a 'target' attribute was introduced for binary classification, assigning a value of 1 to melanoma cases and 0 to other skin conditions. The ISIC 2020 dataset already included a similar 'target' attribute, facilitating our analysis and integration efforts. Proposed methodology as shown in Fig. 1 involved a multi-stage image processing workflow. Initially, Contrast Limited Adaptive Histogram Equalization (CLAHE) was applied to enhance the local contrast of the original images, reducing lighting inconsistencies. The enhanced images underwent morphological opening transformation to refine lesion boundaries and remove unwanted artifacts. Then threshold-based segmentation was performed to obtain a set of segmented images.

Deep learning models, namely, DenseNet, ResNet, Inception, Xception, and EfficientNet were used for model training. Performance measurements were generated

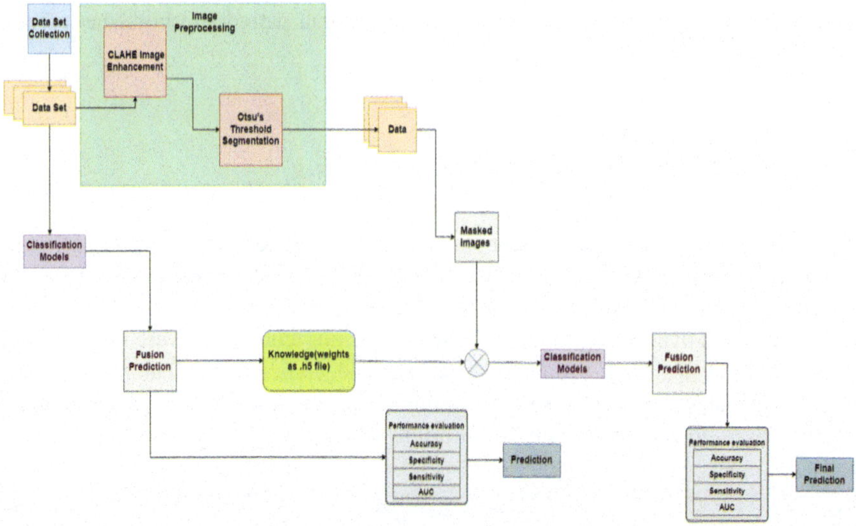

Fig. 1 Proposed Melanoma detection architecture

when the models were first trained using the original photos. We used pre-trained weights from the first training phase in order to apply transfer learning and increase accuracy even more. These weights were used to the segmented pictures in order to optimize the deep learning models. As this strategy, as shown in Fig. 1, sought to improve precision by utilizing the fine-tuned data from the first training to create an efficient architecture for melanoma diagnosis.

3.1 Dataset Integration and Attribute Standardization

Coordinating and normalizing information is a basic move toward melanoma identification, as it establishes the groundwork for creating models equipped for recognizing harmful examples in the dataset. The incorporation cycle included consolidating two huge datasets: ISIC2020, which contains 33,126 pictures, and HAM10000, with 10,015 pictures. Challenges in adjusting the datasets were tended to, bringing about a consistent combination and a complete data set for examination.

3.1.1 Dataset Merging

The combining methodology was facilitated by distinguishing and matching comparative highlights, bringing about a uniform dataset that incorporates a wide assortment of dermatoscopic pictures. To keep up with consistency, qualities, for example, 'lesion_id,' 'image_name,' 'age,' 'sex,' and 'site' were normalized across both datasets. This normalization was basic in guaranteeing quality consistency, for example, the count of non-melanoma and melanoma patients in both the test and train datasets, as found in Fig. 2.

3.2 Target Variable Distribution Balancing

To create a balanced and reliable model, the combined dataset was split into training and test sets, containing 9,890 and 4,240 images, respectively, from the original dataset. This split was carefully executed to maintain a ratio that helps in avoiding model overfitting. In the training set, various techniques were implemented to equalize the distribution of the target variable, indicating melanoma and non-melanoma images. A new folder for training images was established, which included 2,376 images, to accomplish this equilibrium efficiently, as depicted in Fig. 3.

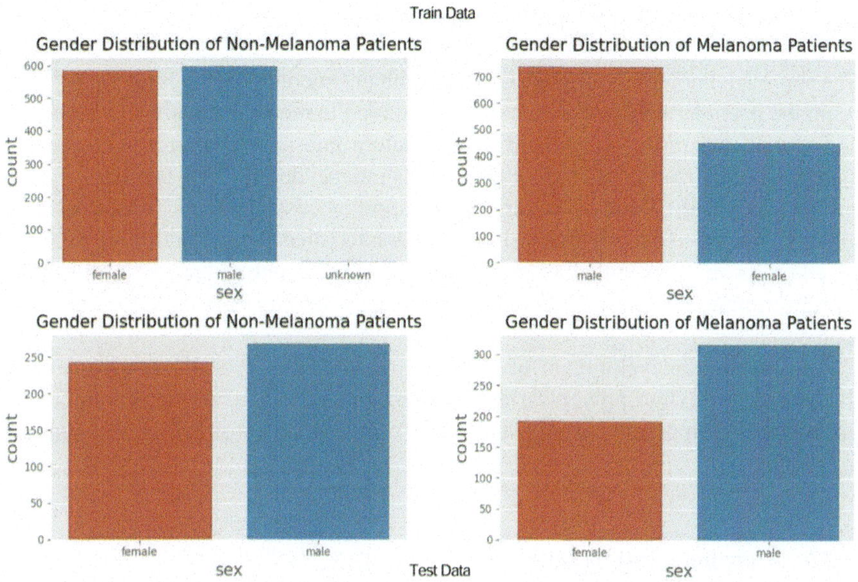

Fig. 2 Normalization of count in train and test dataset

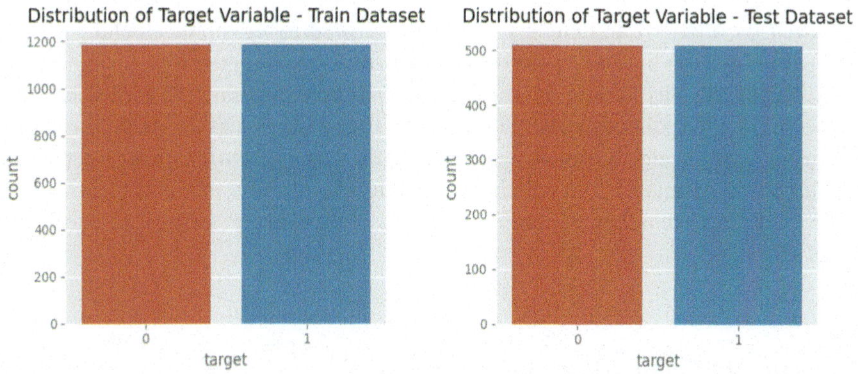

Fig. 3 Equal distribution of target variable

3.3 Image Processing

The principal phase of cycle involves working on the nature of dermoscopic pictures to extricate basic attributes expected for dependable melanoma distinguishing proof.

3.3.1 Adaptive Contrast Enhancement

The approach known as Adaptive Contrast Enhancement (ACE) is essential for rectifying the irregular brightness commonly observed in medical imaging, especially in dermoscopic pictures. After dividing the picture into smaller parts, ACE executes histogram equalization to each of those parts independently. This localized method solves issues with global histogram equalization by ensuring that contrast enhancement is tailored to each region's distinct characteristics. The selection of ACE was based on its ability to enhance local contrast, which is particularly helpful in highlighting minute details in dermatological images that are essential for identifying melanoma.

To reduce the development of clamor in moderately level districts depicted in the image, a flexible power variation approach is used in which the amount of differentiation upgrading is adjusted based on the severity of brilliance change in a location.

3.3.2 Morphological Smoothing

Morphological smoothing is an image processing method that minimizes noise while enhancing the structure of objects within an image, particularly when working with binary or grayscale pictures. It performs two operations as follows: opening (erosion followed by dilation) and closing (dilation followed by erosion). Opening is very useful in removing little things or disturbance from the background. Closing is ideal for reducing the small cracks or gaps in patterns and smoothing the forms.

Morphological smoothing occurs when the following processes are performed on a request. First, extend the image, which makes the items within it appear larger. Then, cycle back to disintegration, which reduces the size of these items to their original shape but with smoother edges. Here erosion followed by dilation is performed meticulously.

3.4 Image Segmentation

Image Segmentation is a critical necessary step in this process, which involves the separation process of images that are digital/computerized into particular regions or locations to consolidate only the required items.

3.4.1 Threshold-Based Segmentation

For the process of segmentation Otsu's thresholding [8], an approach for automated image thresholding technique in the analysis of medical images is used. This method separates pixels into two classes (foreground and background) based on their intensity

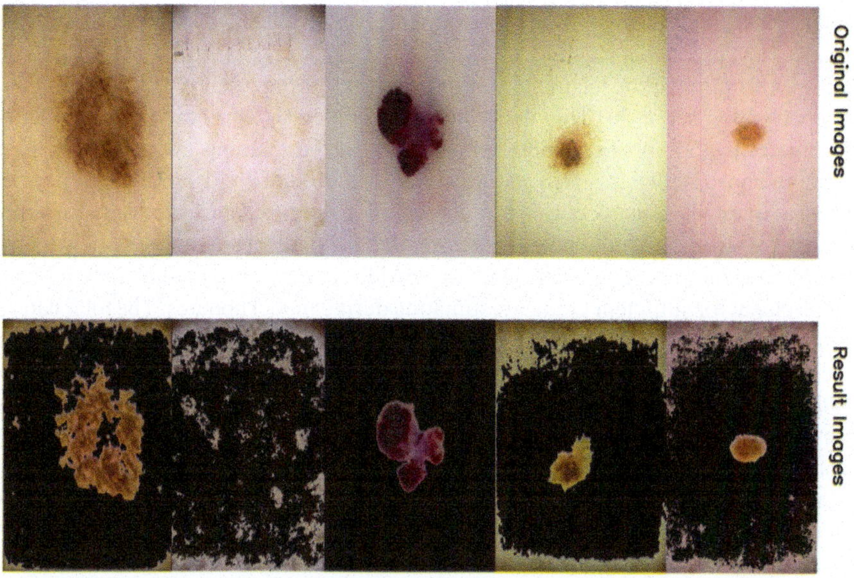

Fig. 4 Result images for input to model training

values, aiming to find an optimal threshold that minimizes intra-class variance and maximizes inter-class variance.

After applying ACE and morphological smoothing as preprocessing steps, Otsu's thresholding is performed to generate a binary image that accentuates regions of interest, as depicted in Fig. 4. Then extending the functionality by saving the segmented images as JPG files in an output folder after analyzing and extracting valuable information from the medical images as shown in Fig. 4.

4 Results and Discussion

4.1 Transfer Learning and Experimental Setup

Firstly, deep learning models were trained on the original images, leveraging the inherent characteristics of each dataset. Concurrently, initiated transfer learning, utilizing the weights obtained from our initial training to adapt the models to the nuances of melanoma detection. A global max pooling layer followed the last convolutional layer, leading to dense layers that incorporated rectified linear unit (ReLU) activation, dropout, batch normalization, and leaky ReLU, enhancing feature representation and regularization. For model training, the Adam optimizer was employed with a learning rate of 1e-4 and a binary cross-entropy loss function, suitable for binary classification tasks. Model performance was evaluated based on metrics such as accuracy, precision, ROC AUC, and true negatives.

Several optimization techniques were employed during training where the number of epochs was set at 20. This decision was informed by observations that running the model for 50 epochs resulted in a plateau in the learning curve starting from the 20th epoch, indicating that additional training did not yield significant improvements. An epoch is one complete cycle where the entire dataset is passed forward and backward through the neural network. A batch size of 128 was chosen to optimize the learning process, balancing the need for computational efficiency with the requirement for sufficient gradient estimation during model updates. The total dataset comprised 3,394 images, which were divided into a training set of 2,374 images and a testing set of 1,020 images. DenseNet's accuracy increased from 85.88% to 89.93% with transfer learning as shown in Table 1, while Xception maintained high accuracy, slightly dipping from 87.86% to 86.98%. InceptionV3's accuracy rose notably from 88.75% to 93.25%, and ResNet's performance remained robust, decreasing marginally from 93.72% to 93.59% after transfer learning. EfficientNetV2B0 showed progress from an initial accuracy of 80.32%–82.09%, illustrating the overall efficacy of transfer learning in enhancing melanoma detection.

Table 1 Model performance with and without pre-trained weights

Model	Metric	Before transfer learning	After transfer learning
DenseNet	Accuracy	0.8588	0.8993
	Precision	0.8535	0.8914
	Recall	0.8517	0.8989
Xception	Accuracy	0.8786	0.8698
	Precision	0.8786	0.8698
	Recall	0.8786	0.8698
Inception V3	Accuracy	0.8875	0.9305
	Precision	0.8868	0.9307
	Recall	0.8816	0.9275
ResNetV2	Accuracy	0.9372	0.9359
	Precision	0.9337	0.9388
	Recall	0.9376	0.9376
EfficientNetV2B0	Accuracy	0.8032	0.8209
	Precision	0.8003	0.8146
	Recall	0.7990	0.7998

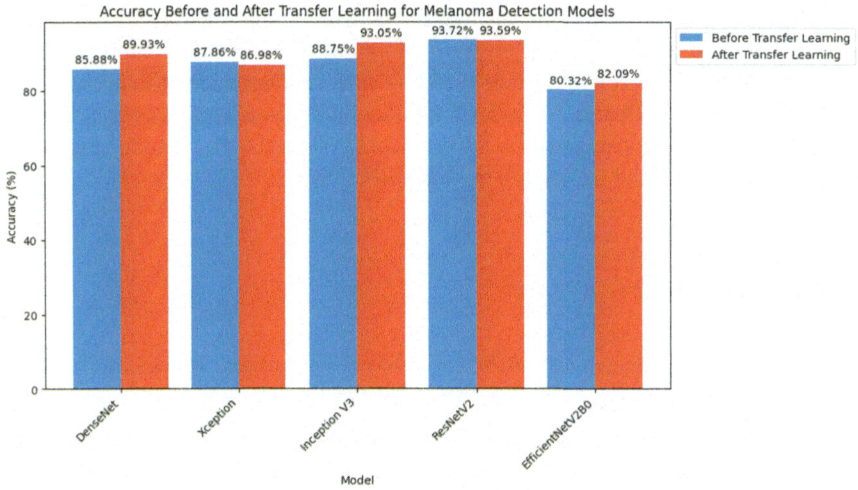

Fig. 5 Performance analysis before and after transfer learning

4.2 *Evaluation Metrics and Analysis*

Upon clear observation reveals that transfer learning enhanced model accuracy, precision, and recall, validating the efficacy of pretrained weights. Model architecture was optimized to reduce overfitting, with a pooling layer to condense spatial information and dense layers activated by ReLU for better feature detection, complemented by dropout and batch normalization to ensure model robustness. The use of the Adam optimizer and binary cross-entropy, alongside metrics like accuracy and recall, allowed for effective model evaluation. Transfer learning utilized pretrained ImageNet weights to fine-tune the models, while class weights corrected for data imbalances. Overfitting treatments included dropout layers, regularization, and information expansion, which, along with learning rate approaches, improved the model's reliable presentation. Below graph in Fig. 5 provided further insights into each model's learning path, highlighting rise in efficacy before and after Transfer Learning. Table 1 and Fig. 5 may be used to facilitate the understanding about model execution, characteristics, and benefits of weights file (.h5) in training the model with weights gained from initial training of models on images before preprocessing.

5 Conclusion and Future Work

The comprehensive evaluation of deep learning models for melanoma detection, highlightened the effectiveness of CNNs for binary classification using models like DenseNet and Xception were explored. Future work may include methods such as ensemble predictions and architectural modifications to further enhance model performance. Additionally, exploring hyperparameter tuning and integrating explainable AI module called Class Activation Mapping (CAM) could improve model interpretability and precision, especially for tasks requiring exact localization by capturing multi-scale features besides the Federative Framework for image privacy and protection. These avenues present significant potential for advancing the field of image classification and enhancing model prediction.

References

1. Trettel A, Eissing L, Augustin M (2018) Telemedicine in dermatology: findings and experiences worldwide—a systematic literature review. J Eur Acad Dermatol Venereol 32(2):215–224. https://doi.org/10.1111/jdv.14341
2. Esteva A, Kuprel B, Novoa R et al (2017) Dermatologist-level classification of skin cancer with deep neural networks. Nature 542:115–118. https://doi.org/10.1038/nature21056
3. Tschandl P, Rosendahl C, Kittler H (2018) The HAM10000 dataset, a large collection of multi-source dermatoscopic images of common pigmented skin lesions. Sci Data 5:180161. https://doi.org/10.1038/sdata.2018.161
4. Han SS, Kim MS, Lim W, Park GH, Park I, Chang SE (2018) Classification of the clinical images for benign and malignant cutaneous tumors using a deep learning algorithm. J Investig Dermatol 138(7):1529–1538. https://doi.org/10.1016/j.jid.2018.01.028
5. Adamson AS, Smith A (2018) Machine learning and health care disparities in dermatology. JAMA Dermatol 154(11):1247–1248. https://doi.org/10.1001/jamadermatol.2018.2348
6. Reza AM (2004) Realization of the contrast limited adaptive histogram equalization (CLAHE) for real-time image enhancement. J VLSI Signal Process-Syst Signal, Image, Video Technol 38:35–44. https://doi.org/10.1023/B:VLSI.0000028532.53893.82
7. Gonzalez RC, Woods RE (2002) Digital image processing, 2nd ed. Prentice Hall
8. Otsu N (1979) A threshold selection method from gray-level histograms. IEEE Trans Syst Man Cybern 9:62–66. https://doi.org/10.1109/TSMC.1979.4310076
9. Liu Y, Jain A, Eng C, Way DH, Lee K, Bui P, Kanada K, de Oliveira Marinho G, Gallegos J, Gabriele S, Gupta RK, Singh P, Natarajan P, Hofmann-Wellenhof R, Corrado GS, Peng LH (2020) A deep learning system for differential diagnosis of skin diseases. Nat Med 26(6):900–908
10. Ashok G, Serath K, Gireesh Kumar T (2024) A multi-class classification for detection of IoT network attacks using machine learning models. In: Devismes S, Mandal PS, Saradhi VV, Prasad B, Molla AR, Sharma G (eds) Distributed computing and intelligent technology. ICDCIT 2024. Lecture Notes in Computer Science, vol 14501. Springer, Cham. https://doi.org/10.1007/978-3-031-50583-6_11
11. Chanchal M, Malathi P, Kumar TG (2020) A comprehensive survey on neural network-based image data hiding scheme. In: Proceedings of the fourth international conference on I-SMAC (IoT in Social, Mobile, Analytics and Cloud), Palladam, India, pp 1245–1249. https://doi.org/10.1109/I-SMAC49090.2020.9243579
12. Poorna SS, Reddy M, Kiran R, Akhil N, Kamath S, Mohan L, Anuraj K, Pradeep HS (2020) Computer vision aided study for Melanoma detection: a deep learning versus conventional supervised learning approach. In: Advanced computing and intelligent engineering, advances in intelligent systems and computing, vol 1082. Springer, pp 75–83
13. Shobha Rani N, Sajan Jain A, Kiran HR (2019) A unified preprocessing technique for enhancement of degraded document images. Lecture Notes Comput Vision Biomech 30:221–233

IOT, Robotics, Security and Soft Computing

A Modified Word Saliency-Based Adversarial Attack on Text Classification Models

Hetvi Waghela, Sneha Rakshit, and Jaydip Sen

Abstract This paper introduces a novel adversarial attack method targeting text classification models, termed the Modified Word Saliency-based Adversarial Attack (MWSAA). The technique builds upon the concept of word saliency to strategically perturb input texts, aiming to mislead classification models while preserving semantic coherence. By refining the traditional adversarial attack approach, MWSAA significantly enhances its efficacy in evading detection by classification systems. The methodology involves first identifying salient words in the input text through a saliency estimation process, which prioritizes words most influential to the model's decision-making process. Subsequently, these salient words are subjected to carefully crafted modifications, guided by semantic similarity metrics to ensure that the altered text remains coherent and retains its original meaning. Empirical evaluations conducted on diverse text classification datasets demonstrate the effectiveness of the proposed method in generating adversarial examples capable of successfully deceiving state-of-the-art classification models. Comparative analyses with existing adversarial attack techniques further indicate the superiority of the proposed approach in terms of both attack success rate and preservation of text coherence.

Keywords Adversarial text attack · Word saliency · Semantic similarity · Perturbation · Lexical correctness · Large language model · ALBERT · BERT · ROBERTa

H. Waghela · S. Rakshit · J. Sen (✉)
Praxis Business School, Kolkata 700104, India
e-mail: jaydip.sen@acm.org

H. Waghela
e-mail: hetvi.mahendra.waghela_ds23fall@praxistech.school

S. Rakshit
e-mail: sneha.rakshit_ds23fall@praxistech.school

A. C. Frery et al. (eds.), *Computational Intelligence and Data Analytics*, Lecture Notes on Data Engineering and Communications Technologies 236,
https://doi.org/10.1007/978-981-96-0451-7_27

1 Introduction

In the realm of machine learning, the threat of adversarial attacks has emerged as a formidable challenge, highlighting vulnerabilities across various domains. Initially observed in image classification tasks, adversarial examples with subtle modifications designed to mislead models have since permeated into Natural Language Processing (NLP) systems. The evolution of adversarial attacks began with seminal works in image classification by Szegedy et al. [1] and Goodfellow et al. [2], illuminating the susceptibility of deep neural networks (DNNs) to imperceptible alterations. These discoveries prompted extensive research efforts, yielding many attack methodologies and corresponding defense mechanisms.

While the landscape of adversarial attacks in computer vision witnessed an array of advancements, the exploration of adversarial examples in NLP remained relatively nascent. The adaptation of attack strategies from the image domain posed inherent challenges in the discrete nature of textual data. Initial forays into NLP adversarial attacks focused on character or word-level modifications, with endeavors by Liang et al. [3], Samanta and Mehta [4], and Alzantot et al. [5] illuminating the potential vulnerabilities in text-based classifiers. Among the emerging methods, Probability Weighted Word Saliency (PWWS) stood out as a promising approach. PWWS, proposed by Li et al., integrated word saliency and classification probability to generate adversarial examples while preserving lexical correctness and semantic compatibility [6]. This method showcased substantial efficacy in perturbing text classifiers, prompting further exploration and refinement.

However, the landscape of NLP adversarial attacks remained ripe for innovation. Despite the progress made by existing methods, challenges persisted in achieving robustness and maintaining human imperceptibility in adversarial perturbations. It is within this context that novel contributions emerge, seeking to address the limitations of current approaches and fortify NLP systems against adversarial threats.

In this work, we present a robust and effective word substitution-based adversarial text attack known as Modified Word Saliency-Based Adversarial Attack (MWSAA). Building upon the foundational principles of the PWWS approach to word substitution, the proposed scheme identifies the substituting words based on the contextual information and semantic consistency between the original and the adversarial texts. The use of contextual information makes the proposed substitution strategy produce a higher level of syntactic consistency, while the imposed semantic consistency constraint ensures semantic similarity between the input and the generated texts. Through extensive experimental results, the efficacy and effectiveness of the proposed adversarial text attack are demonstrated. A comparative analysis of the performance of the proposed scheme MWSAA with the original PWWS attack on some well-known datasets is also presented to illustrate the superiority of the former.

The paper is organized as follows. In Sect. 2, some existing works on adversarial text attacks in the literature are discussed. In Sect. 3, the details of the proposed scheme are presented. Experimental results on the performance of the proposed scheme and a comparative analysis of the performance of the proposed scheme with the PWWS approach are presented in Sect. 4. Section 5 concludes the paper while highlighting some future works.

2 Related Work

Boucher et al. delve into the realm of imperceptible attacks within NLP [7]. The study investigates the ability to subtly alter text in a manner that deceives NLP models while remaining undetectable to human observers. By introducing a novel framework, the authors demonstrate the creation of these imperceptible adversarial examples and showcase their efficacy across a spectrum of NLP tasks. By exploiting linguistic features and employing strategic perturbations, these attacks capitalize on vulnerabilities inherent in NLP models, resulting in misclassification or manipulation of output. Extensive experimentation conducted by the authors highlights the effectiveness of these attacks across various NLP benchmarks, underscoring the significant challenges they pose to model robustness and security.

Malfa and Kwiatkowska explore the crucial aspect of robustness within the realm of NLP [8]. The authors argue that existing measures for assessing robustness in NLP often lack precise definitions and fail to provide comprehensive evaluations. They propose a fresh framework for evaluating robustness in NLP tasks, aiming to overcome the limitations of current methods. Through the introduction of precise definitions and metrics, the authors offer a more refined understanding of robustness, facilitating improved comparisons and advancements in NLP models.

Xu et al. carry out a comprehensive survey on adversarial attacks and defenses across different domains, namely images, graphs, and text [9]. It discusses the methods employed to craft adversarial examples and their potential risks to machine learning models. Furthermore, various defense mechanisms are scrutinized, highlighting their effectiveness and shortcomings in countering adversarial attacks. By conducting a systematic analysis, the authors aim to provide insights into the evolving landscape of adversarial techniques and defense strategies. Overall, this paper serves as a valuable reference for researchers and practitioners seeking to comprehend and tackle the challenges posed by adversarial manipulation across various data formats.

Zhou et al. present a strategy aimed at fortifying text classification models against adversarial attacks by teaching them to differentiate between perturbations [10]. The work introduces an innovative framework that incorporates adversarial training to bolster the resilience of these classifiers. Through the inclusion of a perturbation discriminator during training, the model gains the ability to distinguish authentic inputs from those that have been tampered with in an adversarial manner, thus reducing the impact of such attacks. Experimental findings underscore the efficacy of this approach in thwarting various adversarial strategies, thereby enhancing the robustness and trustworthiness of text classification systems.

Zhang et al. present a thorough examination of adversarial attacks aimed at deep-learning models for NLP [11]. The authors meticulously survey a range of techniques utilized to manipulate NLP models, elucidating both the methodologies employed and the vulnerabilities exposed. These attacks are classified based on their goals, encompassing evasion attacks, poisoning attacks, and backdoor attacks, thereby providing insight into the multifaceted strategies adversaries employ. Furthermore, the paper delves into the potential ramifications of these attacks on practical NLP applications and offers recommendations for bolstering model resilience against such threats.

Jin et al. present a robust benchmark for assessing BERT's susceptibility to adversarial attacks [12]. Through rigorous experimentation, the authors demonstrate BERT's vulnerability to meticulously crafted adversarial examples, which significantly diminish its effectiveness. These findings underscore the importance of evaluating the robustness of NLP models like BERT and highlight the pressing need for fortified defenses against adversarial intrusions in text-based applications. In summary, this work provides valuable insights into BERT's limitations and suggests potential avenues for bolstering its resilience in practical settings.

Liu et al. present a novel strategy for efficiently crafting word-level adversarial attacks [13]. The approach involves framing the task as a combinatorial optimization problem, to maximize attack success while minimizing perturbations to the original text. By integrating optimization and NLP techniques, they demonstrate the effectiveness and efficiency of their method across various benchmark datasets.

Li et al. present a novel approach to crafting adversarial examples in NLP tasks [14]. The method involves introducing subtle changes to the input text using contextualized perturbations, which take advantage of contextual information from pretrained language models. This strategy aims to maintain semantic coherence while evading detection by classifiers. Extensive experiments across different tasks such as sentiment analysis and text classification are carried out, and the results demonstrate the efficacy of the method.

Gao et al. propose a strategy for generating adversarial text sequences designed to outsmart deep learning classifiers without requiring internal classifier knowledge [15]. The method, functioning as a black-box approach, iteratively tweaks input text to create misclassified sequences while retaining semantic similarity to the original text. Through experimentation across different text classification tasks, the authors show the efficacy of the technique.

Li et al. present an innovative strategy for attacking BERT, by employing BERT itself [16]. Utilizing the contextual embeddings of BERT, the authors demonstrate the generation of adversarial examples capable of misleading BERT's predictions. Their method utilizes BERT's gradients to optimize perturbations on input tokens, resulting in crafted examples that can evade detection by BERT. Through extensive experimentation, the authors demonstrate the effectiveness of their proposition in generating adversarial samples with high success rates.

Zhang et al. introduce a novel approach to attacking text at the word level, portraying it as a combinatorial optimization challenge [17]. The technique aims to craft adversarial examples that maintain semantic coherence while outsmarting NLP models. Through extensive experiments, the authors demonstrate the efficacy of their proposed method across diverse NLP tasks. The work significantly contributes to advancing the state-of-the-art of adversarial attacks in NLP, offering valuable insights into potential weaknesses in contemporary NLP systems.

Waghela et al. propose an enhancement to the BERT-Attack [16] by incorporating Projected Gradient Descent (PGD) to improve its effectiveness and robustness [18]. The original BERT-Attack, which generates adversarial examples for BERT-based models, has limitations such as a fixed perturbation budget and insufficient attention to semantic similarity. The scheme proposed by the authors overcomes these limitations by using PGD to iteratively produce adversarial examples that are both imperceptible and semantically similar to the original input.

3 The Modified Word Saliency-Based Attack

Ren et al. proposed an innovative text attack known as the probability-weighted word saliency (PWWS) that creates adversarial examples in NLP tasks. The method modifies input text to influence the output of a target language model while ensuring that the lexical correctness and semantic coherence of the modified text are maintained. The PWWS attack focuses on altering words in the input text based on their significance as determined by a saliency measure. PWWS is based on a word replacement strategy that involves the selection of synonyms and deciding on the order of replacement of the target words. In a nutshell, the scheme involves the following steps:

(1) *Computation of word saliency*: Before initiating replacements, the saliency of each word in the input text is computed. The saliency metric reflects the contribution of each word to the model's prediction or output.
(2) *Selection of replacement words*: Words slated for replacement are chosen based on their saliency scores. Words with higher saliency scores are given precedence for replacement because changing them is more likely to impact the model's output.
(3) *Generation of replacement options*: For each selected word, a range of replacement options is generated. These options encompass synonyms, antonyms, words with similar meanings, or words with different connotations.
(4) *Weighting by probability*: Each replacement option is assigned a probability weight considering its saliency score and similarity to the original word. Replacement options resembling the original word more closely and possessing higher saliency scores are assigned higher probabilities for replacement.
(5) *Randomized selection*: Replacement options are randomly chosen based on their probability weights. Options with higher weights are more likely to be chosen for replacement. This randomized selection ensures variability in the altered text while still prioritizing more significant words for replacement.

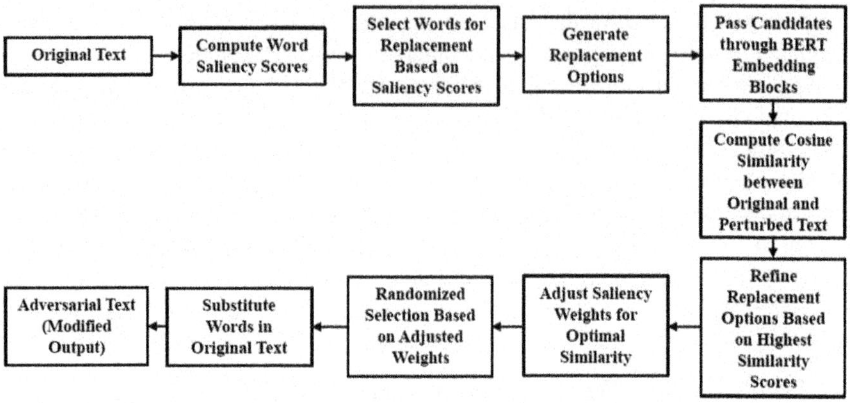

Fig. 1 The schematic representation of the proposed MWSAA scheme for adversarial text attack

(6) *Word substitution*: The selected words in the input text are substituted based on the selected substitute. This process is repeated for each word designated for replacement, resulting in a modified version of the input text containing alternative words based on their significance and similarity to the original words.

By employing this word replacement strategy, the PWWS attack aims to generate adversarial examples that deceive the target model while retaining the natural flow and coherence of the text. This strategy ensures that the replacements are both impactful in terms of the model's prediction and plausible within the context of natural language.

The effectiveness of the PWWS attack relies on multiple factors, including its implementation, the characteristics of the target model, the accuracy of saliency estimation, and the constraint applied during the text perturbation process. While it is found to be effective in generating adversarial examples capable of deceiving large language models, its efficacy can vary depending on the specific context and task at hand.

We present a modification of the PWWS scheme so that its effectiveness is improved in generating adversarial text attacks. The schematic representation of the proposed scheme is depicted in Fig. 1. Our proposition includes the following two features in the word substitution strategy: (i) integration of contextual embeddings and (ii) imposing semantic consistency constraints. In the following, we parent the details of these modifications.

3.1 Integration of Contextual Embeddings

Instead of relying solely on word-level saliency measures, we propose to incorporate contextual embedding using BERT embeddings, to capture richer semantic and contextual information. We argue that these embeddings can provide PWWS with a more nuanced understanding of word importance within the context of the entire sentence or paragraph. BERT is built on the transformer architecture, which consists of multiple layers of self-attention mechanisms and feed-forward neural networks. This architecture enables BERT to comprehend long-range dependencies and contextual associations among words within sentences. In its pre-training phase, BERT employs a masked language modeling objective. In this phase, a portion of the input token is randomly masked, and the model learns to predict these masked tokens based on their surrounding context. By predicting masked tokens from both left and right contexts, BERT learns to grasp word meanings with the context. BERT also can predict the next sentence by analyzing the relationships between sentences and the broader contextual usage of words.

Further, BERT generates contextual embeddings for each token in the input text. These embeddings are created by passing the input tokens through multiple layers of transformer blocks. Each token attends to all other tokens in the input sequence, enabling BERT to dynamically adjust the representation of each token based on its context. Since BERT embeddings capture contextual information by pre-training on extensive text datasets using masked language modeling and next-sentence prediction objectives, this enables BERT to generate comprehensive representations for words that consider their surrounding contexts. These features of BERT embeddings have been exploited in the proposed scheme for choosing the best candidate substitute word from the list produced by the PWWS scheme.

3.2 *Imposing Semantic Consistency Constraints*

To guarantee a high semantic similarity between the original and the perturbed texts, we introduce a further constraint. This is done by incorporating semantic similarity metrics to measure the coherence between words in the input text and their replacements. By enforcing semantic consistency, our proposed scheme ensures that the modified text remains coherent and natural, enhancing the stealthiness of the adversarial examples. While several semantic similarity measures can be used for this purpose including cosine similarity, word movers distance, Jaccard similarity, and so on, in the current work, we use the cosine similarity measure to compute the semantic similarity between the original text and its adversarial counterpart.

In the proposed scheme, instead of relying solely on saliency measures to select the replacement words, first, the candidate words from the PWWS scheme are passed through the BERT embedding block to identify the words that best fit into the context of the text. This list is further refined based on the cosine similarity between the original text and the perturbed when these words are substituted. The substitution that yields the highest similarity score is chosen as the final substitute. This ensures that the replacements are semantically similar to the original words, maintaining the coherence and meaning of the text. Finally, the saliency weighting or replacement strategy is adjusted to optimize for higher cosine similarity between the original and adversarial text. This feedback loop to the PWWS block makes the proposed scheme more effective while retaining a very high level of lexical correctness in the perturbed text and maintaining semantic similarity with the original text.

4 Performance Results

The performance of the proposed MWSAA scheme is evaluated by launching adversarial text attacks on three large language models and three diverse datasets. The language models are (i) BERT, (ii) ALBERT, and (iii) ROBERTa. These pre-trained language models are applied to three well-known datasets: (i) IMDB, (ii) AG News, and (iii) SST. The following seven victim language models are used to test the attack efficacy of the proposed MWSAA scheme: (i) ALBERT_AG, (ii) ALBERT_IMDB, (iii) ALBERT_SST, (iv) BERT_SST, (v) ROBERTa_AG, (vi) ROBERTa_IMDB, and (vii) ROBERTa_SST.

BERT is a groundbreaking advancement in NLP which has introduced bidirectional context comprehension, enabling it to grasp intricate linguistic nuances and relationships within text in both directions [19]. Equipped with the powerful transformer architecture, BERT has large-scale training and contextual word embeddings. The ALBERT model represents a variant of the BERT that seeks to overcome certain limitations in BERT, particularly its large size and computational demands while maintaining or sometimes even surpassing performance across a spectrum of NLP tasks [20]. ROBERTa is a language model that builds upon BERT architecture but incorporates several modifications and enhancements to improve its performance and robustness [21]. Some of the salient features of ROBERTa are the use of dynamic masking during pre-training, a more extensive and diverse pre-training approach, training data augmentation, and optimized hyperparameters.

The IMDB dataset comprises movie reviews alongside associated sentiment labels, denoting whether each review expresses a positive or negative sentiment. The dataset contains 25,000 training and 25,000 test samples, labeled as positive or negative. The AG News dataset encompasses news articles sourced from the AG's web-based corpus covering a spectrum of four broad classes of topics: world affairs, sports, business, and science & technology. Each class contains 30,000 training samples and 1900 test samples. The SST dataset stands as a prevalent benchmark dataset within NLP tasks related to sentiment analysis. It comprises sentences extracted from movie reviews, alongside human-annotated sentiment labels.

Table 1 presents the success rate of the proposed MWSAA adversarial attack on seven pre-trained text classifiers used as the victim language model. For each victim language model, 200 test samples are used to evaluate the attack performance. It is observed that while the ALBERT model on the IMDB dataset exhibited the highest robustness against MWSAA with an attack success rate of 42.50%, the ROBERTa model on the AG News dataset is found to be the most vulnerable to the attack, yielding an attack success rate of 95.50%.

Table 2 presents MWSAA attack efficiency on the victim language models. The attack efficiency is evaluated by two metrics: (i) average running time and (ii) average

Table 1 The attack success rate of the proposed MWSAA scheme

Victim language model	Total attacked instances	Successful attacked instances	Attack success rate (%)
ALBERT_AG	200	130	65.00
ALBERT_IMDB	200	85	42.50
ALBERT_SST	200	117	58.50
BERT_SST	200	160	80.00
ROBERTa_AG	200	145	72.50
ROBERTa_IMDB	200	191	95.50
ROBERTa_SST	200	144	72.00

number of queries made on the victim model. For a better attack, the values of both these measures should be low. MWSAA took the minimum average time on the AG News dataset for the ROBERTa model. However, the lowest average number of queries was needed for the IMDB dataset for the ROBERTa model.

Tables 3 and 4 present a comparative analysis of the performance of the two schemes PWWS and MWSAA. It may be noted that since the Yahoo! Answers dataset is no longer active, it could not be considered in the analysis. Following the work of PWWS, the pre-trained language models considered for the comparative study are (i) Word-CNN [22], (ii) Bi-directional LSTM, and (iii) Char-CNN [23]. The word-based CNN model comprises an embedding layer followed by a 1D convolutional layer with 250 filters of size 3. Subsequently, a 1D max-pooling layer followed by two fully connected layers is used. The Bi-directional LSTM model comprises a 128-dimensional embedding layer, followed by a bi-directional LSTM layer consisting of 64 LSTM units. The output layer is fully connected. The char-based CNN consists of two CNNs, each of which is constructed with a depth of 9 layers, 6 of which are convolutional layers and 3 full-connected layers.

Table 3 shows that both PWWS and MWSAA have adverse effects on the original classification accuracies of the model. For both IMDB and AG News datasets, the effect of MWSAA is much more severe compared to the PWWS. Table 4 depicts

Table 2 The running time and number of queries made by the proposed MWSAA scheme

Victim language model	Average running time (in second)	Total number of queries exceeded	Average victim model queries
ALBERT_AG	0.0069733	0	124.66
ALBERT_IMDB	0.0074853	0	126.12
ALBERT_SST	0.0069764	0	124.94
BERT_SST	0.0064753	0	122.31
ROBERTa_AG	0.0040825	0	123.36
ROBERTa_IMDB	0.0040906	0	120.10
ROBERTa_SST	0.0041479	0	122.93

Table 3 The comparison of classification accuracy of PWWS and MWSAA

Dataset	Model	Original (%)	PWWS (%)	MWSAA (%)
IMDB	Word-CNN	86.55	5.50	4.80
	Bi-directional LSTM	84.86	2.20	1.90
AG News	Char-CNN	89.70	56.20	48.97
	Word-CNN	90.56	56.72	50.45

Table 4 The comparison of word replacement rate of PWWS and MWSAA

Dataset	Model	PWWS (%)	MWSAA (%)
IMDB	Word-CNN	3.81	3.27
	Bi-directional LSTM	3.38	3.21
AG News	Char-CNN	18.93	15.74
	Word-CNN	16.76	16.24

that the percentage of word replacement involved in MWSAA is lower than that of PWWS, signifying that MWSAA requires, on average, fewer words to be substituted in comparison to PWWS to launch a successful attack.

Tables 5, 6, and 7 exhibit sample adversarial examples on ROBERTa on IMDB, ALBERT on AG News, and BERT on the SST dataset.

Table 5 Adversarial attack on the ROBERTa classifier on the IMDB dataset

Original prediction	Adversarial prediction	Perturbed texts
Positive Conf = 99.67%	Negative Conf = 84.69%	Just the toil (labor) postulate (involved) in creating the superimposed (layered) fertility (richness) of the mental (imagery) in this chiaroscuro of foolishness (madness) and visible (light) is astonishing

Table 6 Adversarial attack on the ALBERT classifier on the AG News dataset

Original prediction	Adversarial prediction	Perturbed texts
Sci/Tech Conf = 79.43%	Sports Conf = 54.84%	The careen (Rock) is destine (destined) to be the twenty-first one (21st Century) 's raw (new) Conan and that he's exit (going) to spend (make) a splashing (splash) even peachy (greater) than Arnold Schwarzenegger, jean-claud vanguard (Van) damme or Steven george (Segal)

Table 7 Adversarial attack on the BERT classifier on the SST News dataset

Original prediction	Adversarial prediction	Perturbed texts
Negative Conf = 84.86%	Positive Conf = 70.89%	You 'd think by now America would have had enough of feisty (plucky) British eccentrics with hearts of gold

5 Conclusion

The paper presented a modified word substitution strategy for launching a very robust and effective adversarial text attack. Motivated by the probability-weighted word saliency approach to word substitution, the proposed scheme integrated a word embedding-based contextual word substitution coupled with a semantic similarity check in identifying the substituting words. Extensive evaluation has been done to demonstrate the efficacy and effectiveness of the proposed scheme. The results exhibited a significantly superior performance of the proposed scheme in comparison to the PWWS scheme. The future work includes designing a reinforcement learning-based scheme for optimizing the design of the word substitution scheme.

References

1. Szegedy C, Zaremba W, Sutskever I, Bruma J, Erhan D, Goodfellow IJ, Fergus R (2014) Intriguing properties of neural networks. In: Proceedings of ICLR, Poster Track
2. Goodfellow IJ, Shlens J, Szegedy C (2015) Explaining and harnessing adversarial examples. In: Proceedings of ICLR, Poster Track
3. Liang B, Li H, Su M, Bian P, Li X, Wenchang S (2018) Deep text classification can be fooled. In: Proceedings of international joint conference on artificial intelligence, pp 4208–4215
4. Samanta S, Mehta S (2017) Towards crafting text adversarial samples. arXiv:1707.02812 [cs.LG]
5. Alzantot M, Sharma Y, Elgohary A, Ho B-J, Srivastava M, Chang K-W (2018) Generating natural language adversarial examples. In: Proceedings of EMNLP, pp 2890–2896
6. Ren S, Deng Y, He K, Che W (2019) Generating natural language adversarial examples through probability weighted word saliency. In: Proceedings of the 57th annual meeting of the association for computational linguistics, pp 1085–1097
7. Boucher N, Shumailov I, Anderson R, Papernot N (2022) Bad characters: imperceptible NLP attacks. In: Proceedings of the IEEE SP'22. IEEE, pp 1987–2004
8. Malfa EL, Kwiatkowska M (2022) The king is Naked: on the notion of robustness for natural language processing. In: Proceedings of the 36th association for the advancement of artificial intelligence, pp 11047–11057
9. Xu H, Ma Y, Liu H-C, Deb D, Liu H, Tang J-L, Jain AK (2020) Adversarial attacks and defenses in images, graphs and text: a review. Int J Autom Comput 17:151–178
10. Zhou Y, Jiang J-Y, Chang K-W, Wang W (2019) Learning to discriminate perturbations for blocking adversarial attacks in text classification. In: Proceedings of the EMNLP-IJCNLP, pp 4904–4913
11. Zhang WE, Sheng QZ, Alhazmi A, Li C (2020) Adversarial attacks on deep-learning models in natural language processing: a survey. ACM Trans Intell Syst Technol 11(3):1–41

12. Jin D, Jin Z, Zhou JT, Szolovits P (2020) Is BERT really robust? A strong baseline for natural language attack on text classification and entailment. In: Proceedings of the 34th AAAI conference on artificial intelligence (AAAI'20), pp 8018–8025
13. Liu S, Lu N, Chen C, Tang K (2021) Efficient combinatorial optimization for word-level adversarial textual attack. IEEE/ACM Trans ASLP 30:98–111
14. Li D, Zhang Y, Peng H, Chen L, Brockett C, Sun M-T, Dolan B (2021) Contexualized perturbation for textual adversarial attack. In: Proceedings of the conference of the North American chapter of the association for computational linguistics, pp 5053–5069
15. Gao J, Lanchantin J, Soffa ML, Qi Y (2018) Black-box generation of adversarial text sequences to evade deep learning classifiers. In: Proceedings of the IEEE security and privacy workshops, San Francisco, CA, USA, pp 50–56
16. Li L, Ma R, Guo Q, Xue X, Qiu X (2020) BERT-Attack: adversarial attack against BERT using BERT. In: Proceedings of EMNLP, pp 6193–6202
17. Zhang Y, Qi F, Yang C, Liu Z, Zhang M, Liu Q, Sun M (2020) Word-level textual adversarial attacking as combinatorial optimization. In: Proceedings of the 58h annual meeting of the association for computational linguistics, pp 6066–6080
18. Waghela H, Sen J, Rakshit S (2024) Enhancing adversarial text attacks on BERT models with projected gradient descent. In: Proceedings of IEEE 4th international conference for intelligent technologies (CONIT'24), Hubbali, India, June 21–23
19. Devlin J, Chang M-W, Lee K, Toutanova K (2019) BERT: pre-training of deep bidirectional transformers for language understanding. In: Proceedings of the 2022 conference of the North American chapter of the association for computational linguistics: human language technologies, pp 4171–4186
20. Lan Z, Chen M, Goodman S, Gimpel K, Sharma P, Soricut R (2020) ALBERT: a lite BERT for self-supervised learning of language representations. In: ICLR Poster Track
21. Liu Y, Ott M, Goyal N, Du J, Joshi M, Chen D, Levy O, Lewis M, Zettlemoyer L, Stoyanov V (2019) RoBERTa: a robustly optimized BERT pretraining approach. arXiv:1907.11692 [cs.CL]
22. Kim Y (2014) Convolutional neural networks for sentence classification. In: Proceedings of conference on empirical methods in natural language processing (EMNLP), pp 1746–1751
23. Zhang X, Zhao J, Y (2015) Character-level convolutional networks for text classification. In: Proceedings of advances in neural info processing systems

EEG-Based Emotion Classification Through Multi-Objective Hyperparameter Search

Vaishnavi Vats, Ivan Fenyom, Oladayo S. Ajani, Daison Darlan, and Rammohan Mallipeddi

Abstract This study leverages electroencephalogram (EEG) data for the classification of emotions. To enhance the accuracy of the classification process, a multi-objective hyperparameter search is employed, specifically tailored to optimize the performance of the Multi-Layer Perceptron (MLP) classifier. A pivotal aspect of this methodology lies in effectively managing the trade-off between accuracy and computational complexity. This challenge is effectively addressed through the implementation of the Non-dominated Sorting Genetic Algorithm (NSGA-II) as the optimization framework. By adopting this approach, the study offers a robust and efficient solution for emotion classification, showcasing its relevance and effectiveness within the domain of affective computing. Moreover, the findings contribute significantly to the broader understanding of how to navigate the intricate balance between accuracy and computational resource demands in the realm of machine learning applications.

Keywords Multi-objective optimization · Hyperparameter search · Emotion classification · EEG Signal Processing · Multi-Layer Perceptron (MLP) · NSGA-II · Affective computing

1 Introduction

Emotion classification, a key section in the domain of affective computing and human-computer interaction, underscores the significance of incorporating emotions into technological advancements [1]. Emotions are essential for human development, efficient communication, and making decisions, hence both humans and machines must naturally and intuitively interact with each other. To understand and interpret these emotions, two primary models are predominantly used: the Discrete Model [2, 3] and the Dimensional model [2, 3]. The Discrete Model states that a fixed

V. Vats (✉) · I. Fenyom · O. S. Ajani · D. Darlan · R. Mallipeddi
Department of Artificial Intelligence, School of Electronics Engineering, Kyungpook National University, Daegu, South Korea
e-mail: vaishnavivats1999@gmail.com

A. C. Frery et al. (eds.), *Computational Intelligence and Data Analytics*, Lecture Notes on Data Engineering and Communications Technologies 236,
https://doi.org/10.1007/978-981-96-0451-7_28

number of fundamental emotions are common to all humans. This model was proposed by Paul Ekman in the 1960s, identifying six primary emotions: fear, anger, disgust, happiness, sadness, and surprise [4]. Emotions are classified based on their intensity level in the Dimensional Model. According to Russell, [5], all affective states originate from a two-dimensional, valence-arousal space. Arousal refers to the degree of emotional intensity, ranging from low to high, whereas valence covers the full spectrum of emotions, from negativity to positivity. The combination of these two dimensions is used to depict the various states of emotions in the Dimensional Model [6].

As the focus shifts towards interpreting emotional states in Human-Computer Interaction (HCI), EEG-based emotion classification has emerged as a key technology, given its ability to capture neural signals associated with emotions [7]. Artificial intelligence provides the basis to recognize human emotions and enhances the human-computer relationship by allowing computers to understand and respond accordingly. This technology has profound implications in various fields. In the medical domain, it aids in identifying mental health issues, while in the corporate world, it enhances customer service by understanding and responding to customer emotions. Several methodologies have been proposed for classifying emotions, encompassing modalities such as speech, skin conductance [8], and facial expressions. However, physiological signals, particularly those derived from EEG, may offer a more valuable contribution to machine intelligence than visual or auditory signals [9]. This is due to their direct correlation with the individual's emotional state, providing a more accurate and reliable measure for emotion recognition.

Several models have been proposed to address the fundamental challenges within EEG-based emotion recognition. The Hierarchical Attention-based Temporal Convolutional Network (HATCN) [10] strategically leverages temporal convolutional networks (TCNs) and spectrogram representations, mitigating the time-consuming nature of LSTM-RNNs. A hierarchical attention mechanism enhanced emotional information aggregation, resulting in superior performance on the DEAP dataset [11], accuracy across the four emotional dimensions is 0.716. In the study, Tripathi et al. [12] explored the effectiveness of Deep Neural Network (DNN) and Convolutional Neural Network (CNN) models for emotion classification. Although the DNN model achieved notable accuracy, the CNN model surpassed it, emphasizing the potential of CNNs in enhancing emotion recognition precision. Additionally, an algorithm featuring dimensionality reduction and discrete wavelet transform-based feature extraction, unified with a Long Short-Term Memory (LSTM) model, demonstrates promising results [13]. Another study introduced a multi-layer perceptron (MLP) framework for EEG-based emotion classification, emphasizing the potential of combining neural networks, spectral features, and optimization techniques [7]. These models are effective in improving the interpretability and ease of implementation by focusing primarily on a single objective, such as accuracy. In machine learning, Single-objective models concentrate on optimizing a specific metric, such as accuracy, precision, or recall, which simplifies the model design process and improves interpretability. However, these models may have limitations in real-life applications. Their emphasis on a single metric may restrict their adaptability when multiple metrics are important. They may overfit the training data to maximize the chosen metric, resulting in poor generalization on unseen data. Single-objective models may encounter challenges in situations requiring a balance between different metrics. Additionally, their performance is heavily reliant on the chosen metric, and if the metric does not align well with the real-world objective, the model's performance may be suboptimal.

Single-objective models, while offering simplicity and interpretability, have limited real-world applicability due to their failure to account for potential trade-offs and externalities, resulting in suboptimal solutions. Real-world challenges often involve multiple objectives and conflicting priorities, necessitating a more comprehensive approach. Multi-objective Evolutionary Algorithms (MOEAs) [14], such as genetic programming and evolutionary algorithms, have become essential for addressing optimization challenges characterized by multiple conflicting objectives. MOEAs are crucial in situations where identifying a single optimal solution is impractical due to inherent trade-offs between objectives. They are excellent at identifying solutions that are Pareto-optimal, effectively outlining the optimal trade-offs between conflicting objectives, and offering a diverse array of viable options to decision-makers. While MOEAs are pivotal for managing the heightened complexity of multi-objective optimization problems and delivering diverse high-quality solutions, a notable trade-off exists between accuracy and computational complexity. The efficacy of MOEAs lies in their ability to optimize multiple conflicting objectives simultaneously, overcome the limitations of single-objective methods, and expand the scope of optimization. The field of Multi-objective Evolutionary Algorithms (MOEAs) has evolved significantly with the development of various algorithms. It began with the Nondominated Sorting Genetic Algorithm (NSGA) [15], then progressed with its enhanced version, the NSGA-II [16]. The MOEA/D [17] algorithm, based on decomposition, offers a unique approach to multi-objective optimization. The latest advancement is the ISDE+ [18, 19] algorithm, which promotes convergence and diversity. Each algorithm, from NSGA to ISDE+, has expanded the scope of multi-objective optimization, providing diverse solutions to complex problems. NSGA-II's refined features, addressing challenges such as computational complexity and the absence of an elitism strategy, demonstrate the adaptability and versatility of MOEAs, contributing to the broader understanding of evolutionary algorithms in optimization and paving the way for future advancements in this field, making it a good fit for our research.

This paper aims to explore the application of a multi-objective approach to hyperparameter search in the context of emotion classification using an MLP model. The optimization process is guided by the NSGA-II, a popular method for multi-objective optimization. NSGA-II generates a 'Pareto front' of optimal solutions, each representing a different trade-off between multiple objectives. In the case of an MLP, these objectives could include various performance metrics such as accuracy, and computational complexity. The parameters of the MLP, like the count of hidden layers and the count of neurons within each layer, are modified to determine the combination of parameters that result in the optimal performance based on the Pareto front. This approach considers the often complex and conflicting demands of emotion classification tasks, allowing for a nuanced optimization process. Utilizing NSGA-II and the

Pareto front, this study navigates performance metric trade-offs to identify the optimal MLP model for emotion classification. It has been investigated to determine if emotional states can be categorized based on the range of valence from low to high (sad to happy) and the range of Arousal from low to high (boring to excited) [20]. Additionally, emotions have been classified into quad-class: Low Valence Low Arousal (LVLA), Low Valence High Arousal (LVHA), High Valence Low Arousal (HVLA), and High Valence High Arousal (HVHA) [20].

The other sections of this paper are outlined as follows: Sect. 2 gives the overview of the methodology of the emotion classification based on EEG and details regarding the multi-objective hyperparameter search through NSGA II taking a tradeoff between accuracy and computational complexity. Specifically, an overview hyperparameter search is presented as well as the problem objectives. Experimental setup and results based on the DEAP dataset are displayed and evaluated in Sect. 3. Conclusions with future work are presented in Sect. 4.

2 Emotion Classification Based on EEG

This section introduces a comprehensive framework for EEG-based emotion classification, incorporating multi-objective hyperparameter search. The framework includes signal pre-processing, feature extraction, and emotion classification. The classification task is achieved through the use of advanced techniques and optimization strategies, including a multi-objective hyperparameter search. This approach, evaluated using the DEAP dataset [11], aims to enhance the accuracy of classifying emotions, we also consider computational efficiency, ensuring that our model is not only accurate but also resource-efficient.

2.1 Pre-Processing of EEG Signal

The analysis of the EEG data was begun by removing the pre-trial baseline of 3 s. Following this, the data underwent low-pass filtering at 45 Hz. Independent Component Analysis was then used to remove any artifacts present in the data. The final stage of data processing involved normalizing the EEG data through the subtraction of the common average reference.

2.2 Feature Extraction

Several studies have been dedicated to identify the EEG channels that are most efficient in recognizing human emotional states. It's been noted that the left as well as right frontal regions in the brain are significantly comprised in the expression of emotions [21, 22]. In this experiment, four EEG electrodes, named Fp1, Fp2, F3, and F4, were employed to capture emotional extraction. Functional neuroimaging studies have revealed distinct activation patterns in the FP1, FP2, F3, and F4 channels corresponding to various emotional states, providing empirical support for their utility in EEG-based emotion classification research. Figure 1 illustrates the electrode placement system of 10-20 electrodes. This approach utilized EEG signals for the more dependable and accurate extraction of emotions. The selected four channels were evaluated using Welch's average periodogram method [23]. This method involves dividing the time-domain signal information into next sections. A periodogram is then formed for each block or section. This approach effectively reduces the impact of noise within the signal, enhancing the reliability of the analysis.

The five primary frequency bands in brain waves are as follows: delta (0.5–4 Hz), theta (4–7.5 Hz), alpha (8–13 Hz), beta (13–30 Hz), and gamma (30–45 Hz). During this study, the delta waves, which are mainly associated with deep sleep [24], were excluded. The focus of this analysis was thus on the theta, alpha, beta, and gamma frequency bands. This decision was based on the relevance of these frequency bands to the awake and alert states typically associated with emotion recognition. In this study, the average power for each frequency range associated with every channel was calculated as per Eq. 1.

$$Mean(\mu) = \frac{1}{D}\Sigma_{i=1}^{D} Y_i \tag{1}$$

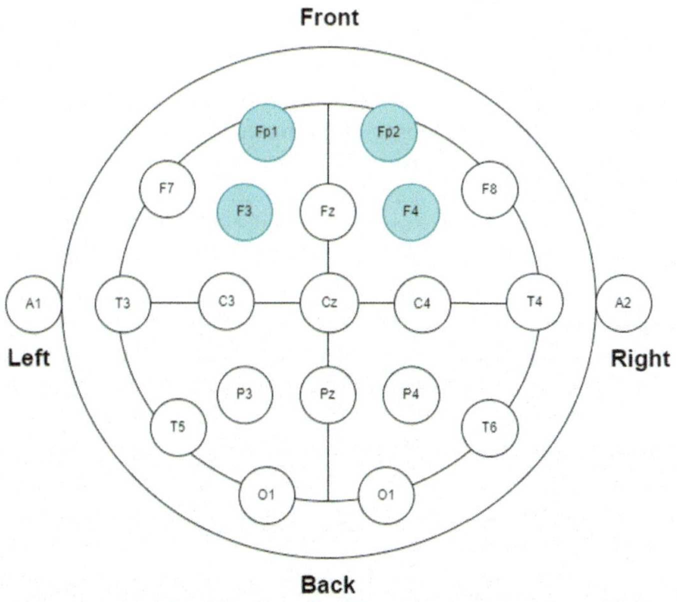

Fig. 1 Electrode placement system of 10–20 electrodes

where Y_i denotes the spectral data derived from the Power Spectral Density corresponding to a specific frequency band and channel. For a particular frequency band and channel, D represents the number of spectral data points. μ is employed to represent the average power of the frequency range. These computations serve as statistical features in our analysis which allows us to capture the essential characteristics of the data in a more structured and quantifiable manner.

In this EEG study, four distinct frequency bands are identified for each electrode. The mean power of each band is extracted for every channel, leading to the derivation of the feature vector for each channel. The feature vector is represented in a specific manner for each trial.

$$\begin{aligned}\text{Feature Vector} = (&Fp1_{mean\_theta}, Fp2_{mean\_theta}, F3_{mean\_theta}, F4_{mean\_theta},\\ &Fp1_{mean\_alpha}, Fp2_{mean\_alpha}, F3_{mean\_alpha}, F4_{mean\_alpha},\\ &Fp1_{mean\_beta}, Fp2_{mean\_beta}, F3_{mean\_beta}, F4_{mean\_beta},\\ &Fp1_{mean\_gamma}, Fp2_{mean\_gamma}, F3_{mean\_gamma}, F4_{mean\_gamma}).\end{aligned}$$

The extraction of features for each subject is based on a matrix of (40 trials * 16 features). This methodical approach allows for a systematic analysis and interpretation of the EEG data.

2.3 *Multi-Objective HyperParameter Search for Classification*

To classify the emotions based on the aforementioned features, an optimized MLP model is employed. MLP was chosen as the model of interest due to its relatively low computational complexity, particularly compared to other recent computational methods. This makes MLP a practical and efficient choice for our research. In terms of optimizing the hyperparameters of the MLP model, NSGA-II was employed. NSGA-II is a multi-objective optimization algorithm with the ability to identify multiple Pareto-optimal solutions within each generation. Specifically, the approach focuses on balancing model accuracy and hardware complexity for both binary (Low/High Valence Arousal) and quad-class (HVLA/HVHA/LVLA/LVHA) emotion classification tasks. To this aim separate optimization processes were conducted for valence and arousal classification, resulting in two sets of optimized binary classifiers. To achieve quad-class classification (HVLA, HVHA, LVLA, LVHA), a third set of classifiers was constructed by forming the Cartesian product of the binary classifier sets. Each element within this third set effectively combines two binary classifiers, one for valence and one for arousal. The quad-class accuracy of each element was assessed by pairing the outputs of its constituent classifiers and comparing them to the true four-class labels. The complexity of each element was calculated as the sum of the complexities of its two binary classifiers. Non-dominated sorting was applied to the elements within this third set to identify the final set of Pareto-optimal solutions. A summary of this process can be seen as depicted in Fig. 2.

The NSGA-II algorithm starts by initializing a set of random solutions. It then iteratively improves the performance of these solutions on a set of fitness functions until some stopping criteria. These solutions should be an accurate representation of

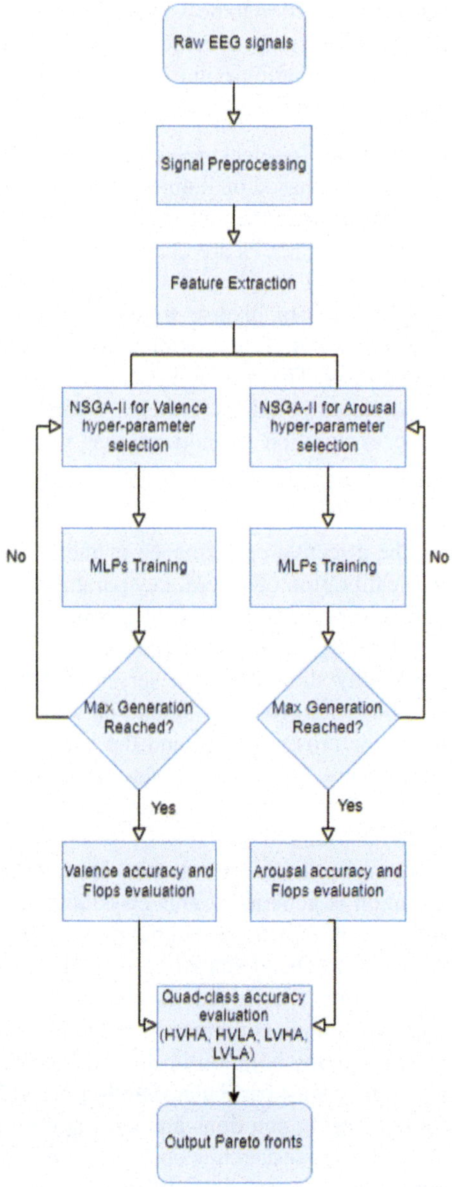

Fig. 2 Proposed multi-objective-based hyperparameter search for emotion classification

the parameters to be optimized. In this study, the aim was to optimize a two-layer MLP where the MLP parameters that were optimized were the number of neurons per layer, the optimizer, and the activation function. Based on this, the solution was represented by a 4-element vector $S = [n_1, n_2, a, o]$ where $n_1 = \{n \in \mathbb{Z} : 2 \leq n \leq 100\}, n_2 = \{n \in \mathbb{Z} : 2 \leq n \leq 50\}, a = \{ReLU$ [25], $Sigmoid, Tanh, Linear\}$, and $o = \{SGD, ADAM$ [26], $RMSprop$ [27]$\}$. Here, n_1 and n_2 represent the number of neurons in the first and second hidden layers of the MLP, respectively, a represents the activation function to be used, and o is the weight optimization technique. As for the fitness function, the accuracy of the models as well as the computational complexity in terms of floating-point operations (FLOPS) were used. The optimization itself consists of three main parts:

- ***Selection***: To select individuals in NSGA-II, different weights are assigned to each individual based on their dominance status in the population. The MLP classifier is trained with the given parameters to determine the fitness value:
- ***Crossover***: MLPs are combined using the crossover operator, which focuses on exchanging weights between pairs of parent solutions. This process enables the exploration of diverse weight configurations, thus contributing substantially to optimization. Single-point crossover is the method used for recombination.
- ***Mutation***: The mutation operator is utilized to introduce minor perturbations into individual solutions. It helps to further diversify the population and facilitates the exploration of more regions within the parameter space. In this study, the polynomial mutation operator was used due to its ability to improve the robustness of the search process.

The effective performance of the NSGA-II algorithm hinges upon the meticulous calibration of several key parameters. These include the crossover probability, mutation probability, and the population size and the number of generations. As Dejong [28] suggested, to prevent the search from becoming random, the mutation probability should be kept very low for the proposed population size. The rate of parameter for crossover should be within the range of 0 to 1. In the context of this work, the crossover rate and the mutation probability were set at 0.95 and 0.001, respectively. These values were selected to balance exploration and exploitation in the search process. The algorithm was run for 20 generations with a population size of 20 individuals.

Using optimization solutions for valence and arousal, a quad-class emotion classification is generated. This classification includes the High Valence Low Arousal (HVLA), High Valence High Arousal (HVHA), Low Valence Low Arousal (LVLA), and Low Valence High Arousal (LVHA) classes as shown in Fig. 2. To achieve the combination, the valence-arousal space is divided into four quadrants each representing a unique combination of high/low valence and arousal. These quadrants are mapped to any solution in the multi-objective optimization so that every solution is assigned to a particular emotion class. It provides a comprehensive framework for classifying emotions and can capture a large range of emotional states. A key aspect of this research is optimizing the underlying models for valence and arousal prediction to improve the effectiveness of this approach.

3 Experimental Setup and Results

The experimental setup for the classification of emotional states using a Multilayer Perceptron (MLP) optimized with Non-dominated Sorting Genetic Algorithm II (NSGA-II) is described as follows:

3.1 Dataset

The research utilizes Electroencephalogram (EEG) signals from the DEAP dataset [11]. The signals were captured using 32 active electrodes following the 10-20 international system. The dataset contains data from 32 subjects who were exposed to one-minute music videos, and the emotional states were rated post-session using the Self Assessment Manikin (SAM) [29] scale. The SAM scale measures valence, arousal, dominance, and liking.

The EEG signals were initially sampled at a frequency of 512 Hz and then downsampled to 128 Hz for efficient data management. Liking and dominance ratings are excluded primarily from the analysis, which seeks to assess valence and arousal. The subjective scale for valence and arousal ranges from 1 to 9, and the ratings are categorized into two levels according to valence and arousal states, using a threshold of 5. The methodology introduced by Koelstra [11] in the DEAP dataset is followed to obtain this binary mapping, which enhances data interpretability and offers a nuanced understanding of valence and arousal states based on subjective evaluations.

3.2 Setup

In this research, a systematic approach is adopted for the classification of emotional states using an optimized Multilayer Perceptron (MLP) model. The dataset was initially divided using an 80–20 split, allocating 80% of the data for training purposes and 20% for testing purposes. This ensured a robust evaluation of the model. The MLP model was implemented using PyTorch [30], and the Non-dominated Sorting Genetic Algorithm II (NSGA-II) was utilized for multi-objective hyperparameter search using pymoo[31]. These computations were performed on a high-performance computer equipped with an NVIDIA GeForce RTX 4090 GPU, 32GB of RAM, and an Intel Core i9-13900K CPU. This combination of dataset separation, advanced algorithm implementation, and powerful computing resources facilitated the effective training and optimization of our model for emotion classification based on EEG signals. The use of an 80–20 split, coupled with the implementation of the algorithms in PyTorch and Pymoo, allowed for efficient computation and optimization. The high-performance computer specifications enabled the handling of large datasets and complex models. This setup is expected to yield reliable and insightful results in the domain of affective computing, thereby contributing significantly to the body of knowledge in this area.

3.3 Results and Discussions

The results of the study are presented in Figs. 3 and 4, which show the Pareto fronts for all 32 subjects defining the quad classification of emotions (High Valence High Arousal, High Valence Low Arousal, Low Valence High Arousal, Low Valence Low Arousal). Pareto fronts refer to solutions that cannot be improved in one area without compromising another. In this case, the two objectives are accuracy and computational complexity, which are shown through X and Y axes, respectively. The figure shows that there exists a clear trade-off between accuracy and hardware complexity. The solutions on the left side of the Pareto front have higher accuracy but are also more computationally expensive. Conversely, the solutions on the right side of the Pareto front have lower accuracy but are less computationally expensive. The NSGA-II algorithm can find a diverse set of solutions that are well-distributed along the Pareto front. In this way, the user can choose the best solution according to the relative importance of accuracy and hardware complexity. For example, in case accuracy is the most important factor, the user might choose the left-side solution, even though it is more computationally expensive. Conversely, if computational complexity is the most important factor, then the user might choose the right-side solution of the Pareto front, even though it has lower accuracy.

4 Conclusion

In conclusion, our research has shown the effectiveness of utilizing a multi-objective hyperparameter search approach for the classification of emotional states based on EEG signals using a base MLP classifier. Our findings illustrate the benefits of this approach in improving the classification accuracy and performance of the model for emotional state categorization. The multi-objective approach adopted in this work optimizes the model not solely for accuracy but also for additional objectives, such as minimizing hardware complexity. The results indicate that the optimized MLP model can successfully classify emotional states into Low/High Valence and Low/High Arousal categories. The findings of this study contribute to the growing understanding in the field of affective computing and hold promise for diverse applications, spanning from mental health to entertainment and human-computer interaction. In the future, our intention is to extend the application of the multi-objective hyperparameter search to other classifiers for the same task, incorporating additional datasets and detailed benchmarking against alternative techniques for emotion classification.

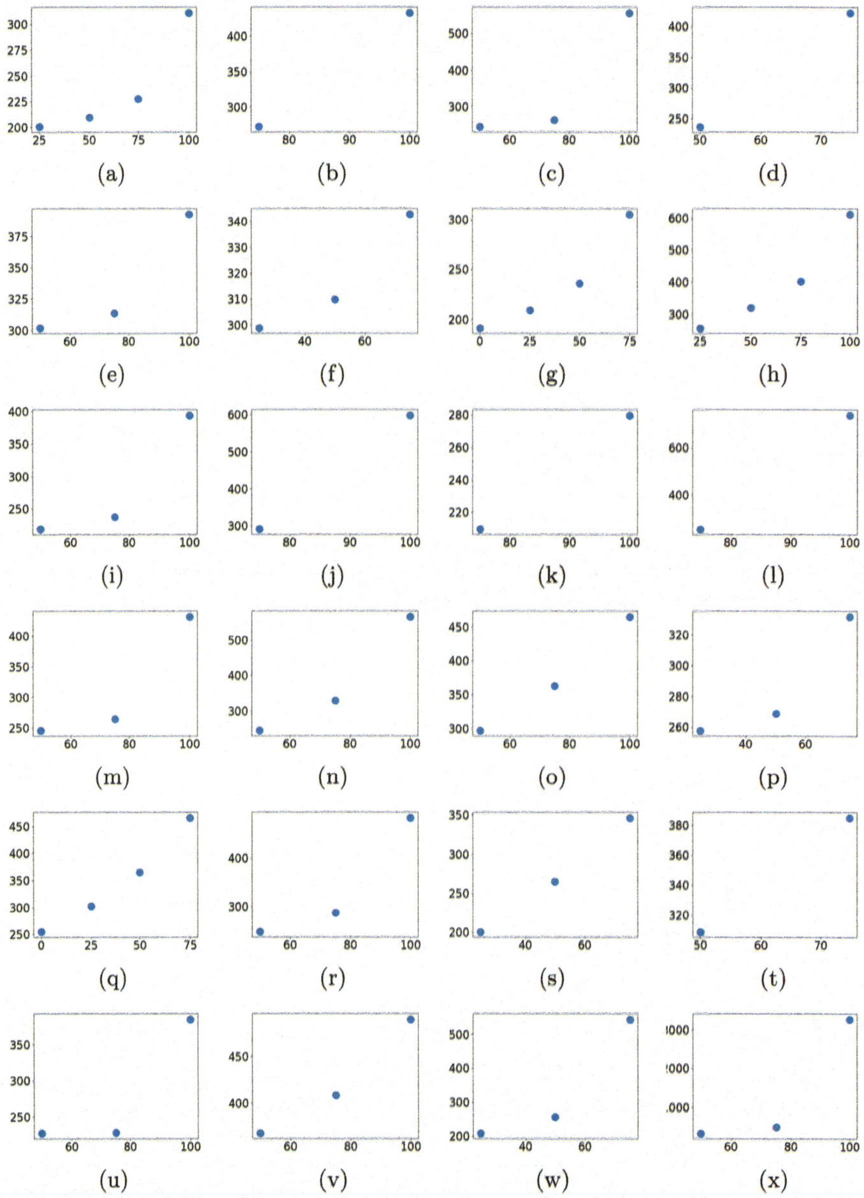

Fig. 3 Pareto fronts for subjects 1–24: X-axis shows the accuracy and Y axis shows the computational complexity in FLOPS

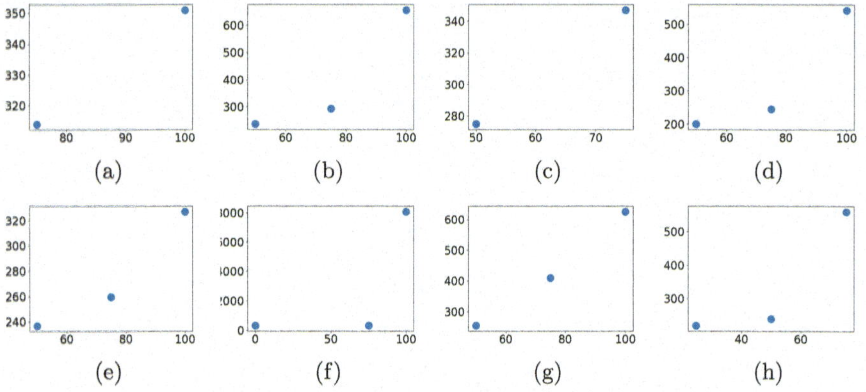

Fig. 4 Pareto fronts for subjects 25 to 32: X-axis shows the accuracy and the Y-axis shows the computational complexity in FLOPS

Acknowledgements This research was supported by the Core Research Institute Basic Science Research Program through the National Research Foundation of Korea (NRF) funded by the Ministry of Education (2021R1A6A1A03043144).

References

1. Mühl C, Allison BZ, Nijholt A, Chanel G (2014) A survey of affective brain computer interfaces: principles, state-of-the-art, and challenges
2. Shu L, Xie J, Yang M, Li Z, Li Z, Liao D, Xu X, Yang X (2018) A review of emotion recognition using physiological signals. Sensors (Basel, Switzerland) 18
3. Alarcão SM, Fonseca MJ (2019) Emotions recognition using EEG signals: a survey. IEEE Trans Affect Comput 10(3):374–393
4. Ekman P (1992) An argument for basic emotions. Cogn Emotion 6(3–4):169–200
5. Russell J (1980) A circumplex model of affect. J Pers Soc Psychol 39:1161–1178
6. Raheel A, Anwar SM, Majid M (2019) Emotion recognition in response to traditional and tactile enhanced multimedia using electroencephalography. Multimedia Tools Appl 78(10):13971–13985
7. Marjit S, Talukdar U, Hazarika SM (2021) EEG-based emotion recognition using genetic algorithm optimized multi-layer perceptron. In: 2021 international symposium of Asian control association on intelligent robotics and industrial automation (IRIA), pp 304–309
8. Darlan D, Ajani OS, Parque V, Mallipeddi R (2023) Recognizing social touch gestures using optimized class-weighted CNN-LSTM networks. In: 2023 32nd IEEE international conference on robot and human interactive communication (RO-MAN), pp 2024–2029
9. Picard RW, Vyzas E, Healey J (2001) Toward machine emotional intelligence: analysis of affective physiological state. IEEE Trans Pattern Anal Mach Intell 23(10):1175–1191
10. Li C, Chen B, Zhao Z, Cummins N, Schuller BW (2021) Hierarchical attention-based temporal convolutional networks for EEG-based emotion recognition. In: ICASSP 2021–2021 IEEE international conference on acoustics, speech and signal processing (ICASSP), pp 1240–1244
11. Koelstra S, Muhl C, Soleymani M, Lee J-S, Yazdani A, Ebrahimi T, Pun T, Nijholt A, Patras I (2012) Deap: a database for emotion analysis; using physiological signals. IEEE Trans Affect Comput 3(1):18–31

12. Tripathi S, Acharya SG, Sharma RD, Mittal S, Bhattacharya S (2017) Using deep and convolutional neural networks for accurate emotion classification on deap dataset. In: AAAI conference on artificial intelligence
13. Garg A, Kapoor A, Bedi AK, Sunkaria RK (2019) Merged LSTM model for emotion classification using EEG signals. In: 2019 international conference on data science and engineering (ICDSE), pp 139–143
14. Deb K (2001) Multiobjective optimization using evolutionary algorithms. Wiley, New York, Jan 2001
15. Srinivas N, Deb K (2000) Multiobjective function optimization using nondominated sorting genetic algorithms, 2 June 2000
16. Deb K, Pratap A, Agarwal S, Meyarivan T (2002) A fast and elitist multiobjective genetic algorithm: Nsga-II. IEEE Trans Evol Comput 6(2):182–197
17. Zhang Q, Li H (2007) Moea/d: a multiobjective evolutionary algorithm based on decomposition. IEEE Trans Evol Comput 11(6):712–731
18. Pamulapati T, Mallipeddi R, Suganthan PN (2019) i_{sde} +-an indicator for multi and many-objective optimization. IEEE Trans Evol Comput 23(2):346–352
19. Ajani OS, Mallipeddi R, Sri Srinivasa Raju M (2023) Icsde+-an indicator for constrained multi-objective optimization
20. Petrantonakis PC, Hadjileontiadis LJ (2009) Emotion recognition from EEG using higher order crossings. IEEE Trans Inf Technol Biomed 14(2):186–197
21. Petrantonakis PC, Hadjileontiadis LJ (2010) Emotion recognition from EEG using higher order crossings. IEEE Trans Inf Technol Biomed 14(2):186–197
22. Menezes MLR, Samara A, Galway L, Sant'Anna A, Verikas A, Alonso-Fernandez F, Wang H, Bond R (Aug 2017) Towards emotion recognition for virtual environments: an evaluation of EEG features on benchmark dataset. Personal Ubiquitous Comput 21(6):1003–1013
23. Welch P (1967) The use of fast fourier transform for the estimation of power spectra: a method based on time averaging over short, modified periodograms. IEEE Trans Audio Electroacoust 15(2):70–73
24. Abhang P, Gawali B, Mehrotra S (2016) Technological basics of EEG recording and operation of apparatus, pp 19–50, Dec 2016
25. Agarap AF (2018) Deep learning using rectified linear units (relu). arXiv:1803.08375
26. Kingma DP, Ba J (2014) Adam: a method for stochastic optimization. arXiv:1412.6980
27. Tieleman T, Hinton G et al (2012) Lecture 6.5-rmsprop: divide the gradient by a running average of its recent magnitude. COURSERA: Neural Netw Mach Learn 4(2):26–31
28. De Jong KA (1975) An analysis of the behavior of a class of genetic adaptive systems. PhD thesis, USA. AAI7609381
29. McEvoy LK, Smith ME, Gevins A (2000) Test-retest reliability of cognitive EEG. Clin Neurophysiol 111(3):457–463
30. Paszke A, Gross S, Massa F, Lerer A, Bradbury J, Chanan G, Killeen T, Lin Z, Gimelshein N, Antiga L et al (2019) Pytorch: an imperative style, high-performance deep learning library. Adv Neural Inf Process Syst 32
31. Blank J, Deb K (2020) PYMOO: multi-objective optimization in python. IEEE Access 8:89497–89509

Optimal Design of Car Side Impact Using Improved GA

Deepika Kaushik and Mohammad Nadeem

Abstract The complex design problems constitute numerous design variables and several constraints. Car Side Impact design (CSID) Problem is an optimization problem that aims to ensure a vehicle's crash-worthiness during the side impact by ensuring international safety standards. In the proposed work, a novel version of the Genetic Algorithm (GA) named Batch Genetic Algorithm (BGA) minimizes a car's weight by ensuring crash-worthiness. This problem constitutes 11 impact factors and 10 constraints. The results obtained are compared with 12 metaheuristic algorithms ensuring the superiority of the proposed approach.

Keywords Car side impact design problem · Batch metaheuristic · Batch GA · GA · Optimization problems

List of Abbreviations

GA	Genetic Algorithm
BGA	Batch Genetic Algorithm
CSID	Car Side Impact design
BM	Batch Metaheuristic

1 Introduction

Optimization problems are pervasive, every sphere of life contains challenging problems accompanying numerous constraints and design variables. These problems are NP-hard in nature and require special methods to solve them. Metaheuristics [1] is

D. Kaushik (✉) · M. Nadeem
Department of Computer Science, Aligarh Muslim University, Aligarh, India
e-mail: deepikakaushikaligarh@gmail.com

A. C. Frery et al. (eds.), *Computational Intelligence and Data Analytics*, Lecture Notes on Data Engineering and Communications Technologies 236,
https://doi.org/10.1007/978-981-96-0451-7_29

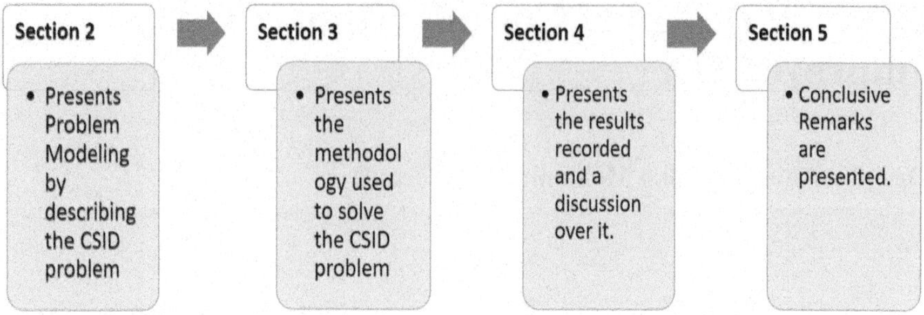

Fig. 1 Arrangement of paper

a problem-solving methodology that has proved immensely successful in solving optimization problems in diverse fields like engineering, economics, business, etc. Therefore, researchers are trying to solve optimization problems using metaheuristics and their variants like parallel, hybrid, etc. CSID problem is one such optimization problem belonging to the design industry. This problem aims to ensure a vehicle's crash-worthiness during the side impact by fulfilling the international safety standards. In the proposed work, a novel version of the GA named BGA is used to solve CSID, which contains 11 design variables and 10 constraints. The arrangement of the paper is depicted in Fig. 1.

2 Problem Modeling

CSID problem is a design problem that has been solved extensively in the literature. The work [2] presents a comprehensive and multi-faceted study over CSID. This problem ensures the optimum design of a vehicle that could ensure protection against the side impact and incur the least loss (see Fig. 2). To ensure safety, two international standards are considered, namely, European Enhanced Vehicle-Safety Committee (EEVC) side impact procedures for European vehicles and National Highway Traffic Safety Administration (NHTSA) side impact procedures for Federal Motor Vehicle Safety Standard (FMVSS) and Canadian Motor Vehicle Safety Standard (CMVSS).

This problem is solved by considering 11 design variables, namely

1. y_1: thickness of inner B-pillar,
2. y_2: B-pillar built-up,
3. y_3: interior of floor side,
4. y_4: cross members,
5. y_5: beam of the door,
6. y_6: built-up of beltline of door,
7. y_7: roof rail,
8. y_8: B-pillar inner material,
9. y_9: inner floor side,

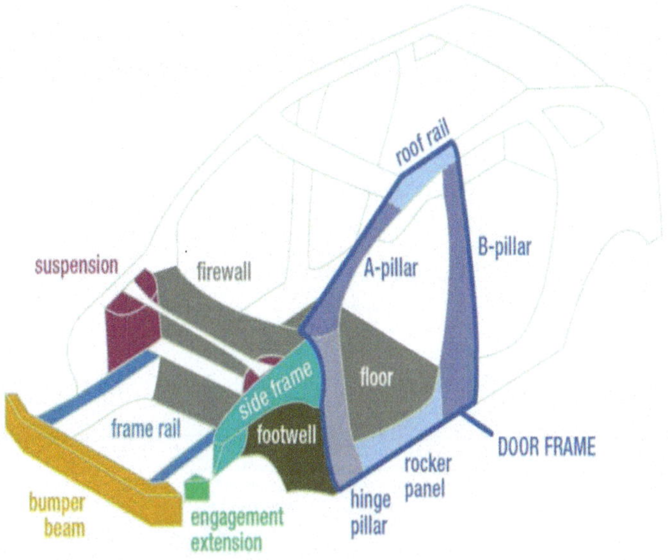

Fig. 2 Schematic diagram of CSID [3]

10. y_{10}: height of barrier,
11. y_{11}: hitting position.

The contemporary literature solves CSID for minimizing the weight of vehicles by considering design variables and satisfying associated constraints. A similar approach is adopted in the proposed work, in the CSID car's weight is minimized by considering design variables and associated constraints. The constraints are handled using a simple scalar penalty function to penalize. Mathematically, it could be modeled as follows

Minimize,

$$weight, f(y) = 1.98 + 4.9y_1 + 6.67y_2 + 6.98y_3 + 4.01y_4 + 1.78y_5 + 2.73y_7 \quad (1)$$

Subjected to

$$\text{Abdominal Force}, c_1 = 1.16 - 0.3717y_2y_4 - 0.00931y_2y_{10} + 0.484y_3y_9 + 0.01343y_6y_{10} \leq 1\text{kNewton} \quad (2)$$

$$\text{Dummy's Upper Chest}, \; c_2 = 0.261 + 0.261 - 0.0159y_1y_2 - 0.188y_1y_8 - 0.019y_2y_7 + 0.0144y_3y_5 + 0.0008757y_5y_{10} + 0.08045y_6y_9 + 0.00139y_8y_{11} + 0.00001575y_{10}y_{11} \leq 0.32\,\text{m/s} \quad (3)$$

$$\begin{aligned}\text{Dummy's Mid Chest, } c_3 =& 0.214 + 0.00817y_5 - 0.131y_1y_8 - 0.0704y_1y_9 \\ &+ 0.03099y_2y_6 - 0.018y_2y_7 + 0.0208y_3y_8 \\ &+ 0.121y_3y_9 - 0.00364y_5y_6 + 0.0007715y_5y_{10} \\ &- 0.0005354y_6y_{10} + 0.00121y_8y_{11} \\ &+ 0.00184y_9y_{10} - 0.02(y_2)^2 \le 0.32\,\text{m/s}\end{aligned} \tag{4}$$

$$\begin{aligned}\text{Dummy's Lower Chest, } c_4 =& 0.74 - 0.61y_2 - 0.163y_3y_8 + 0.001232y_3y_{10} \\ &- 0.166y_7y_9 + 0.227y_2^2 \le 0.32\text{m/s}\end{aligned} \tag{5}$$

$$\begin{aligned}\text{Upper Rib Deflection, } c_5 =& 28.98 + 3.818y_3 - 4.2y_1y_2 + 0.00207y_5y_{10} \\ &+ 6.63y_6y_9 - 7.7y_7y_8 + 0.32y_9y_{10} \le 32\,\text{mm}\end{aligned} \tag{6}$$

$$\begin{aligned}\text{Mid Rib Deflection, } c_6 =& 33.86 + 2.95y_3 + 0.1792y_{10} - 5.05y_1y_2 \\ &- 11y_2y_8 - 0.0215y_5y_{10} - 9.98y_7y_8 + 22y_8y_9 \le 32\,\text{mm}\end{aligned} \tag{7}$$

$$\begin{aligned}\text{Lower Rib Deflection, } c_7 =& 46.36 - 9.9y_2 - 12.9y_1y_8 \\ &+ 0.1107y_3y_{10} \le 32\,\text{mm}\end{aligned} \tag{8}$$

$$\begin{aligned}\text{Pubic Force, } c_8 =& 4.72 - 0.5y_4 - 0.19y_2y_3 - 0.0122y_4y_{10} \\ &+ 0.009325y_6y_{10} + 0.000191y_{11}^2 \le 4\text{kN}\end{aligned} \tag{9}$$

$$\begin{aligned}\text{V-Pillar velocity at mid-point, } c_9 =& 10.58 - 0.674y_1y_2 - 1.95y_2y_8 \\ &+ 0.02054y_3y_{10} - 0.0198y_4y_{10} \\ &+ 0.028y_6y_{10} \le 9.9\,\text{mm/ms}\end{aligned} \tag{10}$$

$$\begin{aligned}\text{V-Pillar front door velocity, } c_{10} =& 16.45 - 0.489y_3y_7 - 0.843y_5y_6 \\ &+ 0.0432y_9y_{10} - 0.0556y_9y_{11} \\ &- 0.000786y_{11}^2 \le 15.7\,\text{mm/ms}\end{aligned} \tag{11}$$

where, $y_i \in [0.5, 1.5], i = 1, 2, \ldots, 7$; $y_i \in [0.192, 0.345], i = 8, 9$ and $y_i \in [-30, 30], i = 10, 11$.

3 Methodology Used

The proposed work solves CSID problem by using the concept of Batch Metaheuristic (BM) [4]. It is a general framework, utilizing the island approach, i.e. dividing the population into sub-populations. Still, unlike the island approach, it does not use the concept of migration among sub-populations. BM is designed for population-based metaheuristic algorithms. In the proposed work, GA is embedded with BM framework; initially randomly generated population is split into equal-sized sub-populations (batches) each of which then utilizes different crossover and mutation rate values, then these batches are merged to generate improved population. Also, the best solution out of all batches say e_i is preserved (elitism is followed). In the next iteration, the population is again divided into multiple batches where e_i replicated into each batch, these batches are again exposed to different GA parameters then elitism is being done and again the batches are merged. This process of splitting, evolution, and merging is repeated until *maxIteration* is achieved. The flowchart of the process is depicted in Fig. 3.

3.1 Experimental Setup

The study was executed using MATLAB in an Octave environment on a system having an Intel i7-3770 processor with 3.40GHz frequency and 8.00 GB RAM. The population size considered is 30, and *maxIteration* considered is 1000. Moreover, the study was conducted by dividing the population into two and three batches. The values of parameters considered for the study are specified in Table 1.

4 Results and Discussion

This section presents the results obtained by using BGA to solve CSID, further comparison is also presented with state-of-the-art algorithms, and the insights drawn from the results are also presented.

4.1 Observations Recorded

The results recorded are presented in Table 2, which presents the values of design variables and the minimum fitness value (weight) recorded by using GA, BGA(2), and BGA(3).

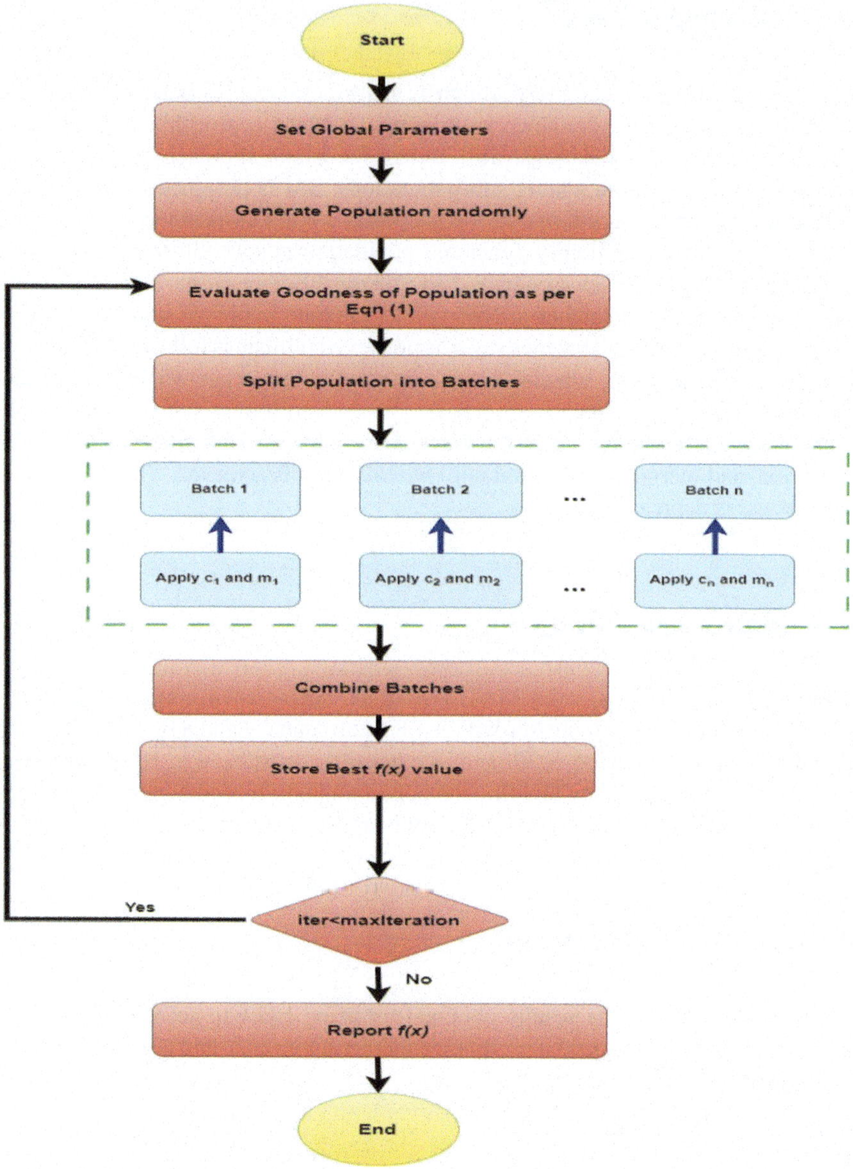

Fig. 3 Flowchart of BGA

Table 1 Parameter values for BGA

Number of batches	Batch size	Batch id	Crossover probability	Mutation rate
1	60 (GA)	*B*(1) 1	0.7	0.3
		B(1) 2	0.8	0.2
		B(1) 3	0.9	0.1
2	30 BGA(2)	*B*(2) 1	0.7	0.3
			0.8	0.2
		B(2) 2	0.8	0.2
			0.9	0.1
		B(2) 3	0.7	0.3
			0.9	0.1
3	20 BGA(3)	*B*(3) 1	0.7	0.3
			0.8	0.2
			0.9	0.1

Table 2 Results obtained by using BGA to solve CSID

Variable	GA	BGA(2)	BGA(3)
y_1	0.5000	0.5001	0.5602
y_2	0.9561	1.0351	1.0277
y_3	0.5157	0.5000	0.5147
y_4	0.5074	0.5000	0.5001
y_5	0.5059	0.5013	0.5093
y_6	1.4252	1.5000	0.6573
y_7	0.5100	0.5052	0.5012
y_8	0.3450	0.3449	0.3434
y_9	0.2893	0.2772	0.3449
y_{10}	−29.9674	−29.9995	−29.9034
y_{11}	−4.8259	7.2959	−4.0135
Best value, $f(y)$	20.0446	**19.6761**	20.0069

From Table 2, it could be inferred that the minimum $f(y)$ value recorded is **19.6761**; it is obtained with BGA(2), i.e. the population is split up into two batches. It ascertains that the BM framework is superior to the standard version of GA.

The convergence analysis of the conducted study is presented in Fig. 4.

The convergence analysis of Fig. 4 proves the righteousness of BGA over the standard version of GA; it could be inferred that GA performs least effectively,

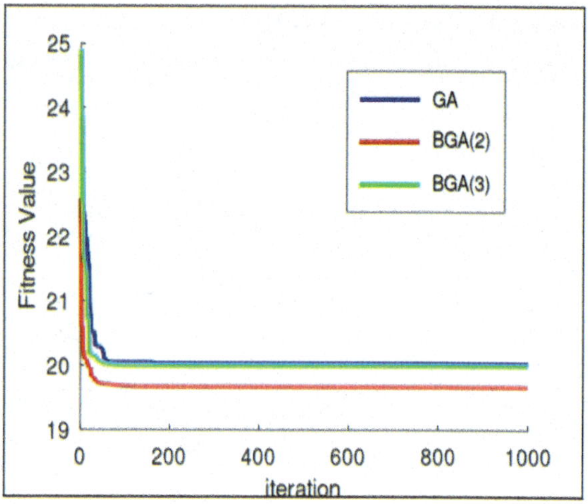

Fig. 4 Convergence behavior of BGA

while both versions of BGA show good results out of which BGA(2) gives the best values out of all.

4.2 Comparison with State-of-Art

The results recorded are compared with 11 metaheuristic algorithms whose results are presented in Table 3; containing values of decision variables and the best fitness value.

From Table 3, it can be seen that out of all algorithms based on vivid concepts BM enabled GA, i.e. BGA performs significantly better out of which BGA(2) outperforms all algorithms with a value of **19.6761** by complying with all constraints.

The results are also analyzed statistically as presented in Table 4. From this, it could be ascertained that BGA(2) possesses the **best fitness value**, **best average fitness value**, and the **least bad fitness value**. Thus, affirming the superiority of BGA over other algorithms. Further, it could also be observed that the worst fitness values for BGA(2) and BGA(3), i.e. **20.6003** and **20.8457** are better than the best fitness value for most of the algorithms. This justifies the use of BM framework for solving CSID problem.

Table 3 Comparison with state-of-art

Variable	PSO [5]	DE [6]	FA [7]	CS [8]	GWO [9]	ALO [10]	MFO [11]
y_1	0.5000	0.5000	0.5000	0.5000	0.5000	0.5001	0.5602
y_2	1.17925	1.1502	1.1169	1.1673	0.9561	1.0351	1.0277
y_3	0.5	0.5000	0.5000	0.5000	0.5157	0.5000	0.5147
y_4	1.2357	1.2567	1.3014	1.2427	0.5074	0.5000	0.5001
y_5	0.5	0.5	0.5	0.5000	0.5059	0.5013	0.5093
y_6	1.4923	1.5	1.5	1.5000	1.4252	1.5000	0.6573
y_7	0.5	0.5	0.5	0.5000	0.5100	0.5052	0.5012
y_8	0.6297	0.9121	0.9919	0.3450	0.7548	0.9461	0.9998
y_9	0.3888	0.6723	0.128	0.1920	0.4494	0.0492	0.00001
y_{10}	−8.32	−13.6159	−19.4716	−10.4492	−29.9674	−29.9995	−29.9034
y_{11}	1.0770	0.00014	−0.0512	−0.0010	0.5202	1.0513	−0.000007
Best value, f (y)	22.8567	22.85517	22.843184	22.944377	22.84647	22.84622	22.84297
	WOA [12]	RAO-3 [13]	MGO [14]	GTO [15]	GA[16]	BGA(2)	BGA(3)
y_1	0.5000	0.5003	0.5000	0.5000	0.5002	0.5000	0.5000
y_2	1.0592	1.1583	1.1186	1.2785	1.1182	1.1170	1.1165
y_3	0.5	0.5003	0.5000	0.5000	0.5001	0.5000	0.5000
y_4	1.3058	0.5003	1.2985	1.2113	1.2996	1.3020	1.3019
y_5	0.5	0.5003	0.5	0.5000	0.5003	0.5000	0.5000
y_6	1.4917	1.5008	1.5	1.5000	1.4990	1.5	1.5
y_7	0.5	0.5003	0.5	0.5000	0.4999	0.5	0.5
y_8	0.5698	0.4005	0.9535	0.3360	0.3450	0.3449	0.3434
y_9	0	0.32	0.3228	0.1953	0.2893	0.2772	0.3449

(continued)

Table 3 (continued)

	WOA [12]	RAO-3 [13]	MGO [14]	GTO [15]	GA[16]	BGA(2)	BGA(3)
y_{10}	−30	−29.983	−19.161	−8.3945	−19.1954	−19.4517	−19.5329
y_{11}	−0.4102	−1.6447	−0.0000007	−0.51169	−4.8259	7.2959	−4.0135
Best value, $f\ (y)$	23.19722	23.59435	22.84324	20.63769	20.0446	**19.6761**	20.0069

Table 4 Statistical analysis with state-of-art

Algorithm	Best	Average	Worst	Standard Deviation	Algorithm	Best	Average	Worst	Standard Deviation
PSO	22.85673	23.6124	25.73265	1.379658E-01	MFO	22.84297	23.12039	23.48124	5.064973E-02
DE	22.85517	23.12172	23.39516	4.866515E-02	WOA	23.19722	24.33408	26.34446	2.075613E-01
FA	22.84318	23.75213	26.195	1.659508E-01	RAO-3	23.23209	24.37533	28.4253	2.087261E-01
CS	22.94438	24.32517	26.36105	2.520966E-01	MGO	22.84324	23.14917	23.3004	5.585492E-02
GWO	22.84647	23.10978	23.49137	4.807361E-02	GTO	20.63769	20.8988	21.1748	4.767195E-02
ALO	22.84622	23.39635	23.89138	1.004395E-01	GA	20.0446	20.5815	21.5106	5.28288E-01
BGA(2)	**19.6761**	**20.1277**	**20.6003**	3.49099E-01	BGA(3)	**20.0069**	**20.52238**	**20.8457**	3.07879E-01

5 Conclusion

This paper solves CSID problem with an improved version of GA named BGA. CSID is a problem associated with designing a car that can withstand the side impact by complying with the international safety standards in the form of constraints. For the study, CSID problem aims at minimizing the weight of a vehicle comprising 11 design variables and 10 constraints; by using population metaheuristic-based '*Batch Metaheuristic*' framework. This framework is general and could be embedded with any population-based metaheuristic algorithm. In this study, GA is embedded with BM and named as BGA. Here, BGA is implemented with two and three batches. The results yielded are good and encouraged to conduct a comparative study with state-of-the-art algorithms. Upon comparing, BGA(2) provided the best results out of all 12 state-of-the-art algorithms. Moreover, the statistical comparative analysis affirms the superiority of BGA. Thus, it could be concluded that BM is a beneficial concept that has the potential to solve optimization problems.

References

1. Talbi E-G (2009) Metaheuristics: from design to implementation. John Wiley & Sons.
2. Youn BD, Choi KK (2004) A new response surface methodology for reliability-based design optimization. Comput Struct 82(2–3):241–256
3. Dhiman G, Kaur A (2019) Hkn-rvea: a novel many-objective evolutionary algorithm for car side impact bar crashworthiness problem. Int J Veh Des 80(2–4):257–284
4. Kaushik D, Nadeem M, Adil Mohsin S (2023) Batch metaheuristic: a migration-free framework for metaheuristic algorithms. Evol Intell 1–33
5. Kennedy J, Eberhart R (1995) Particle swarm optimization. In: Proceedings of ICNN'95-international conference on neural networks, vol 4, pp 1942–1948. IEEE
6. Storn R, Price K (1997) Differential evolution—a simple and efficient heuristic for global optimization over continuous spaces. J Global Optim 11:341–359
7. Yang X-S (2009) Firefly algorithms for multimodal optimization. In: International symposium on stochastic algorithms. Springer, pp 169–178
8. Yang X-S, Deb S (2009) Cuckoo search via l´evy flights. In: 2009 world congress on nature & biologically inspired computing (NaBIC). IEEE, pp 210–214
9. Mirjalili S, Mirjalili SM, Lewis A (2014) Grey wolf optimizer. Adv Eng Softw 69:46–61
10. Mirjalili S (2015) The ant lion optimizer. Adv Eng Softw 83:80–98
11. Mirjalili S (2015) Moth-flame optimization algorithm: a novel nature-inspired heuristic paradigm. Knowl-Based Syst 89:228–249
12. Mirjalili S, Lewis A (2016) The whale optimization algorithm. Adv Eng Softw 95:51–67
13. RVJIJoIEC Rao (2020) Rao algorithms: three metaphor-less simple algorithms for solving optimization problems. Int J Ind Eng Comput 11(1):107–130
14. Abdollahzadeh B, Gharehchopogh FS, Khodadadi N, Mirjalili S (2002) Mountain gazelle optimizer: a new nature-inspired metaheuristic algorithm for global optimization problems. Adv Eng Softw 174:103282
15. Sadeeq HT, Abdulazeez AM (2022) Giant trevally optimizer (GTO): a novel metaheuristic algorithm for global optimization and challenging engineering problems. IEEE Access 10:121615–121640
16. Holland JH (1992) Genetic algorithms. Sci Am 267(1):66–73

Parameter Sensitivity Analysis for Branch-And-Bound Algorithm-Based CFAsT-Match Approach for Shoe Detection in Collaborative Robotics

Sayantan Chakraborty, Pritam Paral, and Amitava Chatterjee

Abstract In collaborative robotics, human tracking problems are considered as one of the major problems. In some of our recent works, we demonstrated the ways to address this problem by the use of CFAsT-Match algorithm in our real-world shoe detection problem. This paper focuses on a study of performance of CFAsT-Match algorithm with respect to the change of parameters values used in CFAsT-Match. Instead of using a particular scaling factor for dimensions of affine transformations, we set four different parameters taking different values, and tried to find how sensitive are those parameters for CFAsT-Match algorithm, while dealing with our challenging and characteristically diverse datasets. Study of performance of the algorithm for different values of parameters and for different characteristics of images shows the varying sensitivity of those parameters under different challenging scenarios.

Keywords CFAsT-Match · CSAD error · DBSCAN · Center location error · Overlap ratio

1 Introduction

As robots become more integrated into people's lives, research into human–robot interaction [1–9] has become increasingly important. Human detection and tracking is a critical aspect of collaborative robotics, and the *leader–follower* mechanism is a crucial component of such human detection and tracking problems. In our recent works [1–4, 7], we have presented those ongoing researches. We have shown how a general-purpose template matching approach can be successfully employed to solve visual object (in those studies, human shoes) detection problem, and also demon-

S. Chakraborty (✉) · A. Chatterjee
Department of Electrical Engineering, Jadavpur University, Kolkata 700032, India
e-mail: sayantanchakrabortyju@gmail.com

P. Paral
Department of Electrical Engineering, Indian Institute of Engineering Science and Technology, Shibpur, Howrah 711103, India

A. C. Frery et al. (eds.), *Computational Intelligence and Data Analytics*, Lecture Notes on Data Engineering and Communications Technologies 236,
https://doi.org/10.1007/978-981-96-0451-7_30

strated successful implementation of a contemporary template matching algorithm, called CFAsT-Match, in this regard [1]. CFAsT-Match [9] is a modified version of the FAsT-Match algorithm [8, 10], which is used to match templates fast and effectively. However, when matching color images, the images are converted to grayscale, resulting in the loss of color information. This can lead to errors in areas that have different colors but similar grayscale values. To address this issue, an improved FAsT-Match algorithm, called CFAsT-match is proposed [9]. CFAsT-Match uses all three RGB channels to construct *color sum-of-absolute differences* (CSAD) [9] and improve the sum-of-absolute differences (SAD) distance used in FAsT-Match. This method first categorizes each pixel in the image by clustering them using *density-based spatial clustering of applications with noise* (DBSCAN) [11] algorithm over the RGB vector. Then, the number of pixels in each category and the cumulative RGB values for each RGB channel are calculated to identify the centroid of each category. The RGB vector centroid is used as the *CSAD decision criteria*, and the inverse of the number of pixels in each category is used as the differentiating coefficient to construct a new similarity measure. Here we have considered the real-world human shoe detection problem and used the CFAsT-Match algorithm to solve the problem, similar to our previous works [1]. Constructing the six-dimensional (x and y translations, x and y scales, and two rotations) grid of affine transformations is an important step in CFAsT-Match [1]. In our earlier paper [1], we have segmented each dimensions using an accuracy parameter scaled by a constant $\sqrt{2}$, which is discussed in Sect. 2.5 of this paper. But this value may not be the best value for all real-world problems. So, we replace this value and set four different parameters. In this paper, our focus is to perform an analysis that how these parameters can affect the performance of the CFAsT-Match algorithm for our real-world visual shoe data, acquired in different laboratory environments.

In Sect. 2, we showcase an illustrious overview of the CFAST-Match system for detecting shoes, based on our previous research [1]. In Sect. 3, we proposed the way to parameterize the grid of transformations for sensitivity analysis, it's a slightly modification of Sect. 2.5. Section 4 presents real-life experimental results and performance evaluations. Section 5 concludes the paper.

2 Overview of CFAsT-Match Algorithm for Shoe Detection

2.1 Automatic Pre-processing Of Images

During the tracking process, the robot captures an initial shoe frame. Subsequently, a rectangular shoe template is extracted from the initial (reference) frame and pre-processed similarly to the methodology explicated in our paper [1].

2.2 Template Image Clustering

After pre-processing the template image, the *density-based spatial clustering of applications with noise* (DBSCAN), a density-based clustering algorithm, is used to form clusters of the pixels of template image [1]. These clusters contain pixels with similar color intensity values. This technique is capable of grouping densely populated point regions while also isolating any outliers. In this study, we cluster the pixels with respect to their normalized R, G, B intensity values. In one of our papers, we investigated a comparative study of the performance of the CFAsT-Match algorithm by using different distance metrics [7].

2.3 Calculation of Score Coefficients, RGB Vector Centroids, and CSAD Errors

Using clusters, formed in Sect. 2.2, we compute the centroids of the RGB vectors for every cluster, along with their respective score coefficients. This data is a prerequisite for computing the CSAD distance. Target images are also pre-processed similar to template images [1]. Afterward, we perform the matching process to identify the best affine transformations [1]. The CSAD distance between the target shoe images and the template shoe image is then evaluated using the score coefficients. The detailed algorithm to calculate the score coefficients and RGB vector centroids is discussed in [1]. CFAsT-Match proposes a similarity distance by utilizing the difference of intensity between two images in every color channel (i.e., R,G,B). Let I_1 with dimension $(w_1 \times h_1)$ and I_2 with dimension $(w_2 \times h_2)$ are template and target images, respectively, then Color SAD distance between target and template images is calculated as follows [9]:

$$\begin{cases} \delta_\tau(I_1, I_2) = \frac{1}{w_1 \times h_1} \sum\limits_{p \epsilon I_1} \begin{pmatrix} |I_1^R(p) - I_1^R(\tau(p))| + \\ |I_1^R(p) - I_1^R(\tau(p))| + \\ |I_1^R(p) - I_1^R(\tau(p))| \end{pmatrix} * \Delta_s(p) \forall p \in I_1 \text{ and } \tau(p) \in I_2; & \text{if } dist(C(I_1(p)), I_2(p)) \leq \zeta \\ = 1; & \text{if } dist(C(I_1(p)), I_2(p)) > \zeta \end{cases} \tag{1}$$

where $\tau(p)$ is the transformations such that $\tau(p) : I_1 \rightarrow I_2$ and $\Delta_s(p)$ is the score coefficient which is defined as a differentiating coefficient between p and $\tau(p)$ and the function [9].

$$dist\big(C(I_1(p)), I_2(\tau(p))\big) = \begin{pmatrix} |C(I_1(p))^R - I_2(\tau(p))^R| + \\ |C(I_1(p))^G - I_2(\tau(p))^G| + \\ |C(I_1(p))^B - I_2(\tau(p))^B| \end{pmatrix} \tag{2}$$

It is important to check whether or not the function in 2 is crossing the similarity threshold ζ, where $C(I_1(p))$ is the RGB value of centroid of cluster containing p in I_1. For our purpose, the similarity threshold is set as 2 [1]. After performing this step we will have to move toward the formation of grid of affine transformations [1].

2.4 Initialize the Grid of Transformation

In the field of image processing, an affine transformation is a prevalent mathematical operation that involves a combination of two rotations, each with one degree of freedom, denoted as r_1 and r_2; *two* degrees of freedom translation in the x- and y-directions, represented as t_x and t_y; and *two* degrees of freedom scaling in the x- and y-directions, given by s_x and s_y. The search range for the aforementioned operations is a crucial aspect of the transformation process. Specifically, the range for the first and second rotations is commonly set to $[-\pi, \pi]$, while the range for the x and y translation is determined by the dimensions of the target and template images. On the other hand, scaling is usually restricted to a range of $[d, 1/d]$, where d is a positive constant [1]. For our shoe detection purpose d is set as 2 [1].

2.5 Parameterization of The Grid of Transformations

Here a grid $\mathcal{G}_\varepsilon$, made of small affine transformations, is constructed maintaining an accuracy parameter ε $(0 < \varepsilon < 1)$ such that for two arbitrary transformation T and T', the $l_\infty(T, T') < \Theta(\varepsilon)$ [8]. While constructing the grid of transformations $\mathcal{G}_\varepsilon$ we have to discretize the space of affine transformation properly. This space is a six-dimensional space with $(t_x, t_y, r_1, r_2, s_x, s_y)$. Now in discretization phase, each of six dimensions is split into $\Theta(\varepsilon)$ number of equal segments such that two consecutive transformations on any of the dimensions differ by at most $\Theta\big(\varepsilon.max(w_1, h_1)\big)$, where w_1 and h_1 are width and height of the shoe template, respectively. This equal segments corresponding to a grid $\mathcal{G}_{\varepsilon_i}$, $(i = 1, 2, \ldots)$, denoted by stp_i, are represented as follows [1]:

$$\begin{cases} stp_i|_{t_x} = \frac{\varepsilon_i w_1}{\sqrt{2}} \\ stp_i|_{t_y} = \frac{\varepsilon_i h_1}{\sqrt{2}} \\ stp_i|_{r_1, r_2} = \sqrt{2}\varepsilon_i \\ stp_i|_{s_x, s_y} = \frac{\varepsilon_i}{\sqrt{2}} \end{cases} \tag{3}$$

and the accuracy parameter ε_i associated with grid $\mathcal{G}_{\varepsilon_i}$, $(i = 1, 2, \ldots)$ is improved as: $\varepsilon_i = \alpha\varepsilon_{i-1}$, where α $(0 < \alpha < 1)$ is a constant factor. For our problem, we have considered the initial accuracy parameter $\varepsilon_0 = 0.25$, similar to [1].

2.6 Generate the Initial Grid of Transformations

Based on the initial stepsizes stp_0 and search range information for different dimensions, we set up an initial six-dimensional grid of affine transformations. Each dimension represents one of the six parameters involved in an affine transformation: t_x, t_y, s_x, s_y, r_1, and r_2. Let the initial grid of transformations $\mathcal{G}_{\varepsilon_0}$ (for initial precision ε_0) be represented as $[t_{x_0}, t_{y_0}, s_{x_0}, s_{y_0}, r_{1_0}, r_{2_0}]$. We assume that the complete set of affine transformations in the grid $\mathcal{G}_{\varepsilon_0}$ is denoted as $\mathcal{S}_0$ with its cardinality $\mathcal{N}_0$.

2.7 Branch-And-Bound Scheme

Now, as the α is less than 1, when the value of ε is decreased, the size of the grid increases rapidly. We formulate the *branch-and-bound* scheme [1] to make sure that complexity will not increase with the rise in the size of the grid. The idea is to eliminate all transformations from the script set S_i that produce errors greater than a specified *threshold*. This approach drastically reduces the number of affine transformations to be evaluated, while preserving the precision, as demonstrated in our research paper [1].

3 Proposed Parameterization: The Grid of Transformation For Sensitivity Analysis

In Sect. 2.5, we have discussed how to parameterize the grid of transformations. In paper [1], a factor $\sqrt{2}$ was used in Equation (3). Now in this paper, we are going to change the value 2 by replacing them with four different variables a, b, c, d in Equation (3). For our interest to study the sensitivity of a parameter, we keep three parameters fixed at their original value 2 [1] and one of the four parameters is varied in range [1.5 : 0.5 : 3]. For each value of a parameter, we obtain a *center location error* (CLE) and an *overlap ratio* (OR)[1]. CLE is the Euclidean distance between the center of the ground-truth bounding box and the center of the bounding box that we obtained as an output of the CFAsT-Match algorithm. OR is the ratio of intersection and union of the ground-truth bounding box and the bounding box obtained with CFAsT-Match. We have modified Equation (3) as follows:

$$\begin{cases} stp_i|_{t_x} = \frac{\varepsilon_i w_1}{\sqrt{a}} \\ stp_i|_{t_y} = \frac{\varepsilon_i h_1}{\sqrt{b}} \\ stp_i|_{r_1, r_2} = \sqrt{c}\varepsilon_i \\ stp_i|_{s_x, s_y} = \frac{\varepsilon_i}{\sqrt{d}} \end{cases} \tag{4}$$

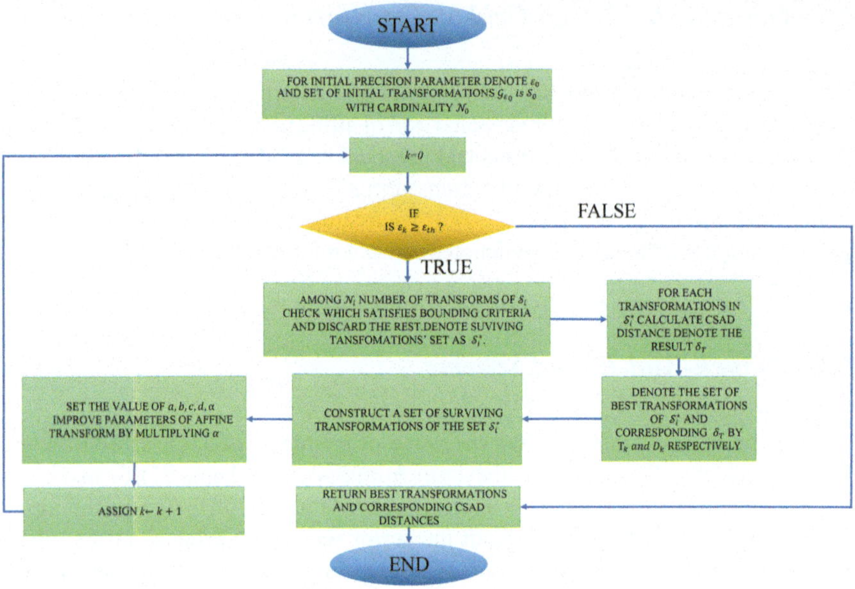

Fig. 1 The flowchart of branch-and-bound-based CFAsT-Match algorithm

In Fig. 1, we have presented the flowchart of our proposed approach to construct the grids of transformations within the framework of CFAsT-Match using a new set of parameters $\{a, b, c, d\}$, and study the impact of variation of those parameters on the real-life shoe detection results.

4 Experimental Result

4.1 Experimental Setup

The experimental setup is similar to what we had adopted in our earlier works [1, 4]. We employ a differentially driven wheeled mobile robot, which is designed and developed at a minimal expense in the measurement and instrumentation laboratory at Jadavpur University. The Robot is equipped with a Raspberry Pi (RPi) 3 model B+ as its onboard computer, and also houses a RPi camera model v2 for image acquisition purposes. A visual representation of the robot is showcased in Fig. 2.

For the purposes of experimental evaluation, we use images of two real-world shoe-image datasets, namely, SET-I and SET-II, acquired, respectively, in two different laboratory environments, during human following [3]. The shoe templates corresponding to two datasets are shown in Fig. 3.

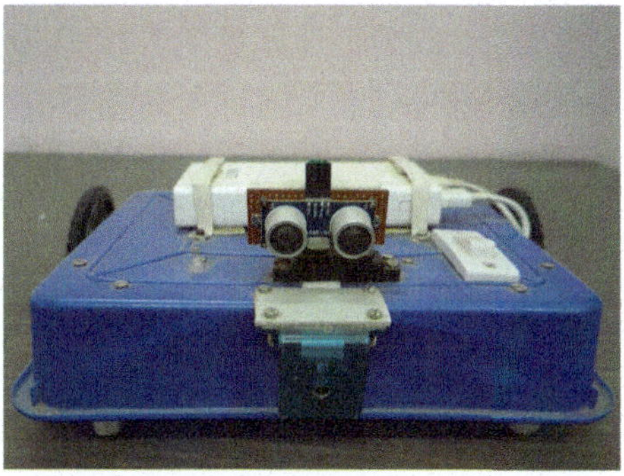

Fig. 2 Mobile robot used in the experimental work

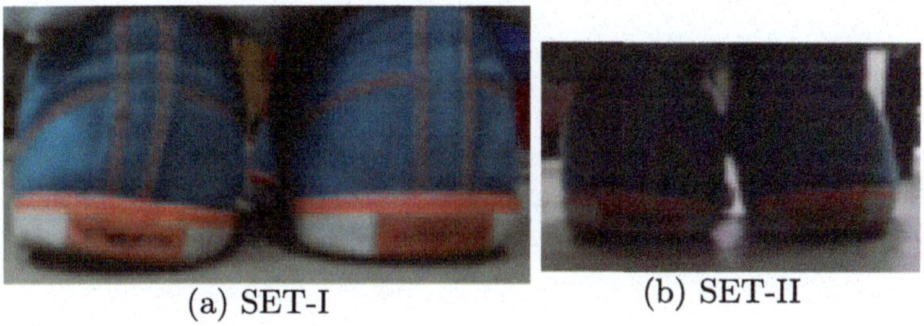

Fig. 3 Template shoe images [1]

4.2 Sensitivity Analysis

In this study, we consider characteristically diverse *four* images from the real-world datasets SET-I and SET-II [3]. The performance of the algorithm is measured using two quantitative metrics, namely, *center location error* (CLE) and *overlap ratio* (OR). The results are described in Tables 1, 2, 3, and 4. Second row of each table reports the CLE and OR values corresponding to original parameter settings in [1]. As can be seen from the tables, the change of parameter values makes significant impact on the algorithm's performance.

CLE and OR results, which showed overall best responses, are shown in bold letter in the tables. From Table 1, it is clear that $a = 3$ emerges as overall superior to other values for different images. Table 2 reveals that $b = 1.5$ and $b = 2.5$, with respect to the OR and CLE results, respectively, outperforms other values for different images. In case of c, as can be seen from Table 3, the value $c = 1.5$ has been found to be a good performer, surpassing other values and delivering satisfying results across all the images under consideration. For d, we should consider the value $d = 3.0$ in Table 4 as better than others. Based on the analysis of the tabulated results, it can be observed that the parameters a and b exhibit comparatively lower sensitivity for the IMG-I of SET-I, whereas the parameters c and d demonstrate greater sensitivity

Table 1 Performance of CFAsT-Match algorithm for different values of *a*

Values of *a*	SET-I				SET-II				Average Values	
	IMG-I		IMG-II		IMG-I		IMG-II			
	CLE	OR	CLE	OR	CLE	OR	CLE	OR	CLE	OR
1.5	2.828	76.814	25.019	46.091	14.317	57.833	41.725	6.521	20.972	46.796
2.0	2.236	77.424	29.428	41.725	30.000	25.337	164.879	0.000	56.635	36.121
2.5	3.162	71.524	33.000	35.137	17.204	55.349	25.298	29.059	19.666	46.760
3.0	**2.000**	**81.693**	**5.099**	**76.837**	**10.440**	**55.854**	**20.808**	**22.411**	**9.520**	**59.198**

Table 2 Performance of CFAsT-Match algorithm for different values of *b*

Values of *b*	SET-I				SET-II				Average Values	
	IMG-I		IMG-II		IMG-I		IMG-II			
	CLE	OR	CLE	OR	CLE	OR	CLE	OR	CLE	OR
1.5	**5.385**	**72.327**	**5.385**	**77.751**	**13.601**	**53.018**	**30.594**	**39.461**	**13.741**	**60.639**
2.0	2.236	77.424	29.428	41.725	30.000	25.337	164.879	0.000	56.635	36.121
2.5	5.385	62.549	3.000	67.510	16.155	45.243	20.808	22.411	11.337	49.428
3.0	2.236	84.355	3.605	74.812	32.388	30.896	133.06	0.000	42.822	47.515

Table 3 Performance of CFAsT-Match algorithm for different values of *c*

Values of *c*	SET-I				SET-II				Average Values	
	IMG-I		IMG-II		IMG-I		IMG-II			
	CLE	OR	CLE	OR	CLE	OR	CLE	OR	CLE	OR
1.5	**5.385**	**67.976**	**5.000**	**77.279**	**54.083**	**51.815**	**26.833**	**24.23**	**22.825**	**55.324**
2.0	2.236	77.424	29.428	41.725	30.000	25.337	164.879	0.000	56.635	36.121
2.5	**27.074**	**27.685**	**3.605**	**68.072**	**30.083**	**17.541**	**31.241**	**21.818**	**23.000**	**33.779**
3.0	2.236	83.61	6 3.162	71.442	18.248	56.686	155.000	0.000	59.667	52.934

for the same image, i.e., the parameters r_1, r_2, s_x, and s_y are more sensitive in this case. For the IMG-II of SET-I, each parameter shows one abrupt change. It is noteworthy that the parameter a delivers a superior outcome at $a = 3.0$, while the results for parameters b, c, and d at the values 1.5, 2.5, 3.0, respectively, are better than those corresponding to $a = b = c = d = 2.0$. Regarding the IMG-I of SET-II, the parameter setting $a = 2.0$ provides a suboptimal result compared to other values of a. The parameter d indicates similar behavior, while the parameters b and c demonstrate high sensitivity for this image. Finally, for the IMG-II of SET-II, each parameter exhibits high sensitivity, necessitating the fine-tuning of parameters to achieve optimal results (Figs. 4, 5, 6, and 7).

Table 4 Performance of CFAsT-Match algorithm for different values of d

Values of	SET-I				SET-II				Average	
d	IMG-I		IMG-II		IMG-I		IMG-II		Values	
	CLE	OR	CLE	OR	CLE	OR	CLE	OR	CLE	OR
1.5	2.236	79.955	5.099	77.914	17.204	53.647	31.622	14.389	14.040	56.467
2.0	2.236	77.424	29.428	41.725	30.000	25.337	164.879	0.000	56.635	36.121
2.5	32.062	24.554	1.000	68.895	7.211	55.634	24.515	33.197	16.197	45.570
3.0	**2.828**	**82.364**	**6.082**	**75.160**	**12.000**	**58.792**	**6.324**	**56.132**	**6.808**	**68.112**

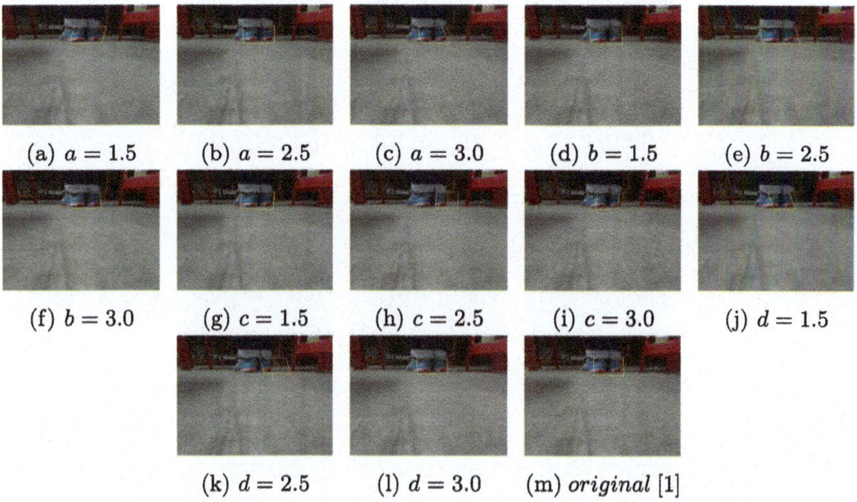

Fig. 4: Experimental results for SET-I, IMG-I

Fig. 4 Experimental results for SET-I, IMG-I

5 Conclusion

This paper presents a parameter sensitivity analysis of the parameters used to construct the grid of affine transformations. The study also investigates how CFAsT-Match can be modified to suitably address this study using the shoe detection problem. The experimental results demonstrate that changing the parameter values can have a significant impact on the results. It is noteworthy that sometimes a different parameter value leads to better results than the previous value, while in other cases, it yields unsatisfactory results. So, one should judiciously tune the parameters of the CFAsT-Match algorithm for a specific purpose.

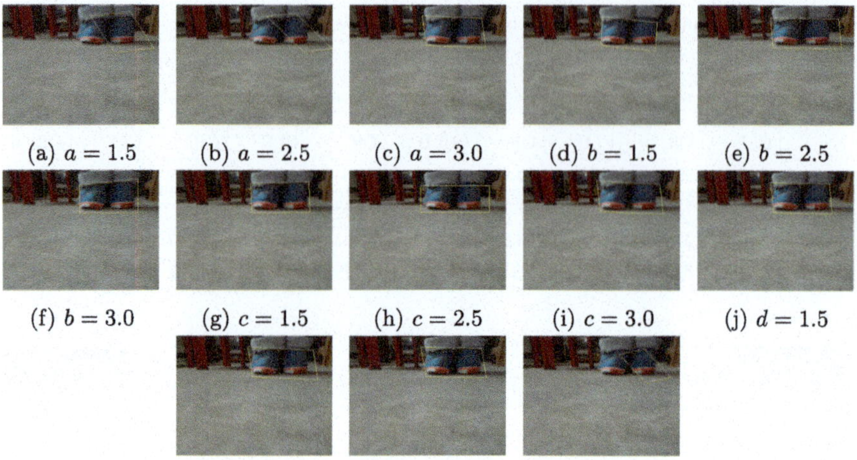

(a) $a = 1.5$ (b) $a = 2.5$ (c) $a = 3.0$ (d) $b = 1.5$ (e) $b = 2.5$

(f) $b = 3.0$ (g) $c = 1.5$ (h) $c = 2.5$ (i) $c = 3.0$ (j) $d = 1.5$

(k) $d = 2.5$ (l) $d = 3.0$ (m) $original$[1]

Fig. 5 Experimental results for SET-I, IMG-II

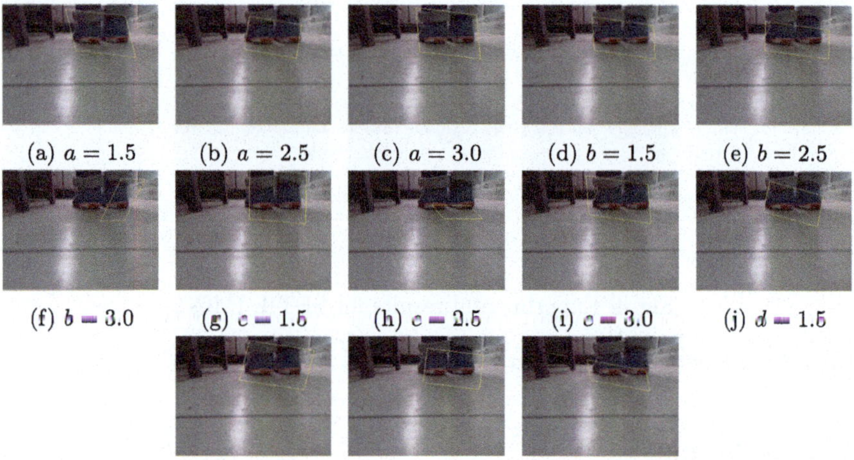

(a) $a = 1.5$ (b) $a = 2.5$ (c) $a = 3.0$ (d) $b = 1.5$ (e) $b = 2.5$

(f) $b = 3.0$ (g) $c = 1.5$ (h) $c = 2.5$ (i) $c = 3.0$ (j) $d = 1.5$

(k) $d = 2.5$ (l) $d = 3.0$ (m) $original$ [1]

Fig. 6 Experimental results for SET-II, IMG-I

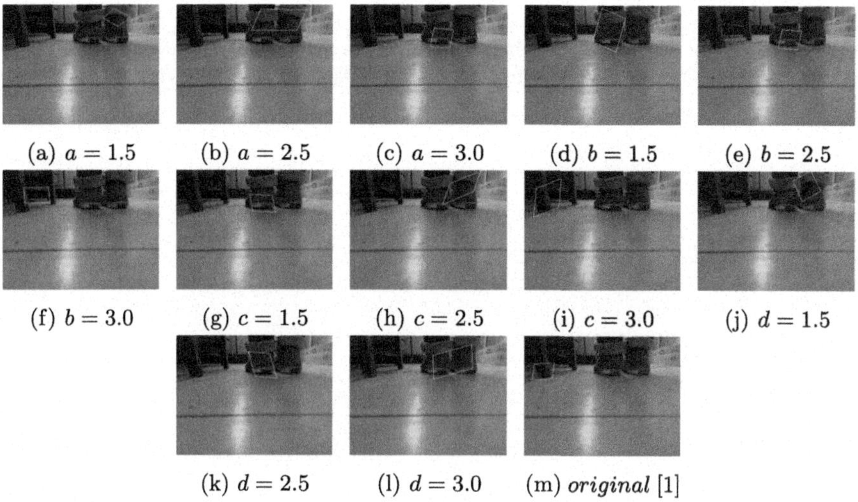

(a) $a = 1.5$ (b) $a = 2.5$ (c) $a = 3.0$ (d) $b = 1.5$ (e) $b = 2.5$ (f) $b = 3.0$ (g) $c = 1.5$ (h) $c = 2.5$ (i) $c = 3.0$ (j) $d = 1.5$ (k) $d = 2.5$ (l) $d = 3.0$ (m) *original* [1]

Fig. 7 Experimental results for SET-II, IMG-II

References

1. Paral P, Chatterjee A, Rakshit A (2019) Vision sensor-based shoe detection for human tracking in a human-robot coexisting environment: a photometric invariant approach using DBSCAN algorithm. IEEE Sens J 19(12):4549–4559
2. Paral P, Gosh S, Chatterjee A, Pal SK (2023) Automatic relevance determination kernel-embedded gaussian process regression for sonar-based human leg localization with a mobile robot. IEEE Sens Lett 7(1):1–4
3. Paral P, Chatterjee A, Rakshit A (2020) Chaos-based random sampling for photometric invariant shoe detection with vision sensor in human-robot coexisting environments. IEEE Sens J 20(10):5424–5434
4. Paral P, Chatterjee A, Rakshit A, Pal SK (2023) Extended target tracking in human-robot coexisting environments via multisensor information fusion: a heteroscedastic gaussian process regression-based approach. In: IEEE transactions on industrial informatics, vol 19, no 9, pp 9877–9886, Sept 2023
5. Yuan J, Chen H, Sun F, Huang Y (2015) Multi sensor information fusion for people tracking with a mobile robot: a particle filtering approach. IEEE Trans Instrum Meas 64(9):2427–2442
6. Bellotto N, Hu H (2009) Multisensor-based human detection and tracking for mobile service robots. IEEE Trans Syst Man Cybern B Cybern 39(1):167–181
7. D. Brahma PP, Chatterjee A (2023) Choice of distance metrics in DBSCAN based color template matching applied to real-time human shoe detection. In: 2nd international conference on computational systems and communication (ICCSC). IEEE
8. Korman S, Reichman D, Tsur G, Avidan S (2013) FAsT-match: fast affine template matching. In: Proceedings of the IEEE conference on computer vision and pattern recognition (CVPR), Portland, OR, USA, Jun 2013, pp 2331–2338
9. Jia D, Cao J, Song W-D, Tang X-L, Zhu H (2016) Colour FAST (CFAST) match: fast affine template matching for colour images. Electron Lett 52(14):1220–1221, Jul 2016
10. Korman S, Reichman D, Tsur G, Avidan S (2013) Fast-Match webpage. https://engineering.tau.ac.il/~simonk/FastMatch

11. Ester M, Kriegel H-P, Sander J, Xu X (1996) A density-based algorithm for discovering clusters in large spatial databases with noise. In: Proceedings of the 2nd international conference on knowledge discovery data mining, Portland, OR, USA, pp 226–231

Secure Conditional Ambient Calculus for the IoT Context

Lamia Hamza

Abstract Nowadays, we are confronted with a spectacular increase in IoT (Internet of Things) use in all aspects of human activities. Unfortunately, this development generated new vulnerabilities, exposing these systems to increasingly significant threats representing new security problems that must be solved. Furthermore, the ubiquitous availability of resources in the IoT context has also enabled an intruder to carry out more sophisticated attacks. However, the actual protection solutions are installed haphazardly. For this reason, developing new formal techniques is of the utmost importance. The lack of consideration of the formal aspect in the IoT context prompted us to define a new algebraic calculus allowing the modeling and securing of attacks in the IoT context. This paper presents a new ambient calculus called SCAC(Secure Conditional Ambient Calculus) to model IoT attacks and to suggest protection solutions. We show that the proposed calculus offers an efficient and simple modelling by an example.

Keywords Vulnerability · Attack · Security · Internet of things · Ambient calculus

1 Introduction

The IoT brings new challenges to the ubiquitous computing age. The accessibility of everything in all places has also profited the malevolent kinds of all shades. In an unprecedented manner, being connected is synonymous with being vulnerable. Finding the optimal balance between security measures and providing IoT capabilities is inherently challenging. There are only two alternatives without a suitable middle ground: acquiring expertise in all security elements or completely ignoring them. The initial option may appear overly idealistic to the average user. The second alternative exposes all individuals to potential threats. Ensuring security within the context of IoT is a difficult, complex, and subtle task. These systems result from

L. Hamza (✉)
Laboratory of Medical Informatics (LIMED) Faculty of Exact Sciences, University of Bejaia, 06000 Bejaia, Algeria
e-mail: lamia.hamza@univ-bejaia.dz

A. C. Frery et al. (eds.), *Computational Intelligence and Data Analytics*, Lecture Notes on Data Engineering and Communications Technologies 236,
https://doi.org/10.1007/978-981-96-0451-7_31

the coexistence of different technologies and are often characterized by a highly distributed aspect. The current security solutions are mainly the result of manual analyzes and ad-hoc approaches. Unfortunately, these solutions are no longer adequate because any security attack can have considerable consequences. Therefore, there is a need for a unified security framework for modeling and securing IoT attacks. Formal approaches are well located to report such apprehensions when used for representations constructed on qualifications of system and performances. Our article proposes a new calculus to describe the interaction with a potential intruder and simulate mobility in IoT context. The present method offers the advantage of formal modeling and proposes a formal solution to the IoT attacks.

This paper is organized as follows. Section 2 reviews related research work. Section 3 introduces the syntax and semantics of our new calculus. Section 4 presents an example of an attack through a case study and proposes an experimentation for a better, invulnerable IoT. Section 5 concludes this article and gives a few prospects for future work.

2 Related Works

The IoT has infiltrated every sphere of human lives, due to their ubiquity. In exchange, cybercrimes have been on the rise as criminals and hackers continue to take advantage of IoT's vulnerabilities. The enormity of the attacks has not only been damaging to the quality of life, but it poses a disservice and an unquantifiable risk to human safety. According to [1, 5], it is possible to classify IoT attacks along different dimensions (methods, attacks, vulnerabilities, layers, orientation, risks, active, passive, internal or external). We focus on the category of attacks that can take control and steal data, more specifically the selective forwarding attack. Selective forwarding is an attack where the malicious nodes neglect to forward the messages to precise destination nodes or simply drop the messages not to propagate anymore [10]. In a selective forwarding attack [8, 11], malicious nodes behave like black hole and may refuse to forward certain messages and simply drop them, ensuring that they are not propagated any further. However, such an attacker runs the risk that neighbouring nodes will conclude that she has failed and decide to seek another route. A more subtle form of this attack is when an adversary selectively forwards packets. A lot of research [12] deals with this attack in the context of wireless sensor networks, but a no method has been explored for the detection of this attack in IoT. The only remarkable work is presented in the article [6], where the authors presented a non-cooperative zero-sum game-theoretic method for detecting intruders in the network. For this purpose, we thought of defining an algebraic calculus allowing the modelling and securing of IoT attacks. Our work is based on ambient calculus. The concept of ambient calculus was first introduced by Luca Cardelli and Andrew Gordon in [2]. An ambient is a named domain that has an interior and an exterior and can contain processes. This calculus manages to express process mobility with a simple and powerful syntax. With the appearance of modern networks (IoT, VANET, etc.) a new ambient calculus named

Table 1 Syntax of SCAC

$P, Q ::= 0 \mid P \mid Q \mid P + Q \mid !P \mid^{k}_{n} [C \rhd a.P]$
$a ::= acq^{k}_{n} \mid mig^{k}_{n} \mid in^{k}_{n} \mid out^{k}_{n} \mid sea \mid sec^{k}_{n}$

CCA (Calculus of Context-aware Ambients) [9] was developed but this calculus and its variants [4] do not deal with the security aspect. For these reasons, we have defined our calculus, named Secure Conditional Ambient Calculus (SCAC).

3 Secure Conditional Ambient Calculus

In this section, we define our calculus called SCAC (Secure Conditional Ambient Calculus). This new calculus includes primitives for safety as well as the specification of regular and irregular behavior in an IoT context. In addition to the ability to execute processes under the conditions in a particular context.
The processes of our calculus are described in Table 1.

A process that does nothing is symbolized by (0). The expression (P | Q) denotes the parallel execution of processes P and Q. The expression (P + Q) denotes the choice execution of processes P and Q. The process's unbounded replication is indicated by the symbol !P; ${}^{k}_{n}[P]$ denotes an ambient called n holding a resource protected by the key k essential for entering or exiting the protected ambient in order to interact with the resource, where $C \rhd a$ indicates a process that performs the action a if condition C is satisfied. The condition C can express any kind of condition, those related to contexts or to the thing itself.

The process capabilitie acq^{k}_{n} refers to the acquisition of the key of an ambient. The ability of a process to migrate using the proper ambient key is referred by mig^{k}_{n}; Using the proper ambient access key k, in^{k}_{n} lets an ambient to enter a collocated ambient called n, and out^{k}_{n} allows an ambient to exit its parent ambient. The sec operator enables the enforcement of resource protection within an ambient n with the key k. The sea (for search) operator permits depiction of dynamic processes that can non-deterministically select what their next action will be (case of an intruder who has the possibility to choose any sequence helping him to attract his target).

3.1 Semantics

The concept of a structural congruence between processes, denoted by $\equiv$, and a reduction relation, denoted by $\rightarrow$, constitutes the operational semantics of our calculus. In Table 2, the structural congruence is defined and used to explain how processes are equivalent in our calculus. Reflexivity, symmetry, transitivity, parallelism, choice,

Table 2 Structural congruence of SCAC

$P \equiv P$	Reflexivity
$P \equiv Q \Rightarrow Q \equiv P$	Symmetry
$P \equiv Q \wedge Q \equiv R \Rightarrow Q \equiv R$	Transitivity
$P \equiv Q \Rightarrow Q \mid R \equiv P \mid R$	Parallelism
$P \equiv Q \Rightarrow^{k}_{n} [C \triangleright a.P] \equiv^{k}_{n} [C \triangleright a.Q]$	Conditional ambient sequence
$P \mid 0 \equiv P$	Zero parallelism
$(P \mid Q) \mid R \equiv P \mid (Q \mid R)$	Parallel associativity
$P + 0 \equiv P$	Zero choice
$(P + Q) + R \equiv P + (Q + R)$	Choice associativity
$P \mid Q \equiv Q \mid P$	Commutativity
$P \equiv Q \Rightarrow !P \equiv !Q$	Replication
$!P \equiv P \mid !P$	Parallel replication
$!0 \equiv 0$	Zero replication

Table 3 Reduction relations of SCAC

$P' \rightarrow Q'$ if $P' \equiv P, P \rightarrow Q, Q \equiv Q'$	(1)
$P \mid R \rightarrow Q \mid R$ if $P \rightarrow Q$	(2)
$C \triangleright mig^{k}_{n}.P \mid {}^{k}_{n}[Q] \rightarrow {}^{k}_{n}[P \mid Q]$	(3)
${}^{k}_{n}[C \triangleright mig^{k}_{n}.P \mid Q] \rightarrow P \mid {}^{k}_{n}[Q]$	(4)
$C \triangleright sec^{k}_{n}.P \mid {}^{\delta}_{n}[Q] \rightarrow P \mid {}^{k}_{n}[Q]$	(5)
${}^{\delta}_{n}[(C \triangleright in^{k}_{m}).P \mid Q] \mid {}^{k}_{m}[R] \rightarrow {}^{k}_{m}[{}^{\delta}_{n}[P \mid Q] \mid R]$	(6)
${}^{k}_{m}[{}^{\delta}_{n}[(C \triangleright out^{k}_{m}).P \mid Q] \mid R] \rightarrow {}^{\delta}_{n}[P \mid Q] \mid {}^{k}_{m}[R]$	(7)

commutativity, and associativity are some of the more evident properties. The ambient sequence is uncomplicated. Equivalence for protected processes is expressed by the ambient sequence congruence, while zero parallelism designates the neutral element.

The reduction relation defined in Table 3 captures the process behavior in regard to its environment's protections and capabilities. Rules (1) and (2) are trivial. Rule (3) illustrates how a process can enter a protected environment if it has the necessary capabilities. Rule (4) presents the option to leave a protected environment. Rule (5) illustrates protection enforcement by simply setting the key to the new value of k, $delta$ denoting the public key without any protection. By setting key to the value $delta$, protection can also be removed. Rule (6) depicts the fact that an ambient can enter a collocated protected ambient and rule (7) depicts the fact that an ambient can leave its parent protected ambient. In case there are no conditions to check ($C = True$) we can write directly the name of the action.

4 Case Study

To demonstrate the effectiveness of our calculus, we use a straightforward system depicted in Fig. 1.

This example is composed of two logical areas: a network of connected objects vulnerable and another computer that behave like an intruder. Two nested and protected ambients serve as the representation for each object in the network. The upper ambient contains: a process representing the resources of the object, an access key publication service to the associated sub-ambient which contains a process representing the data of the routing table, this sub-ambient is designated by the letters tr followed by the name of the appropriate equipment (the same name is assigned to its process but in upper case). The initial letter of the equipment name serves as the basis for the names of ambient. For example, the ambient representing the smartphone is named *s* and contains the process *S*, so forth.

The process below indicates the system that corresponds the IoT network of Fig. 1:

$$\mathrm{S} =_{r}^{k1} [\mathrm{R} \mid_{a}^{\delta} [\mathrm{A} \mid !(acq_{tra}^{k2}.0) \mid_{tra}^{k2} [\mathrm{TRA}]] \mid_{b}^{\delta} [\mathrm{B} \mid !(acq_{trb}^{k3}.0) \mid_{trb}^{k3} [\mathrm{TRB}]] \mid_{s}^{\delta} [\mathrm{S} \mid !(acq_{trs}^{k4}.0) \mid_{trs}^{k4} [\mathrm{TRS}]]]$$

Once within the IoT network, the intruder has the freedom to pick these targets at random. In this illustration, we'll assume that the attacker choose "computer a" and "Smartphone s" as the targets. Then, at a particular time, this malicious node can receive a ROUTE_REQUEST from the "computer a" to destination "Smartphone s", then attack the routing protocol by referring to the "computer a" a ROUTE_REPLY that contains false information specifying the optimal path to the destination.

The intruder process is represented by three concurrent processes: publishing service to routing table, the routing table process running within a protected ambient, and another process whose purpose is to send the message ROUTE_REPLY. The intruder process is illustrated by the following expression.

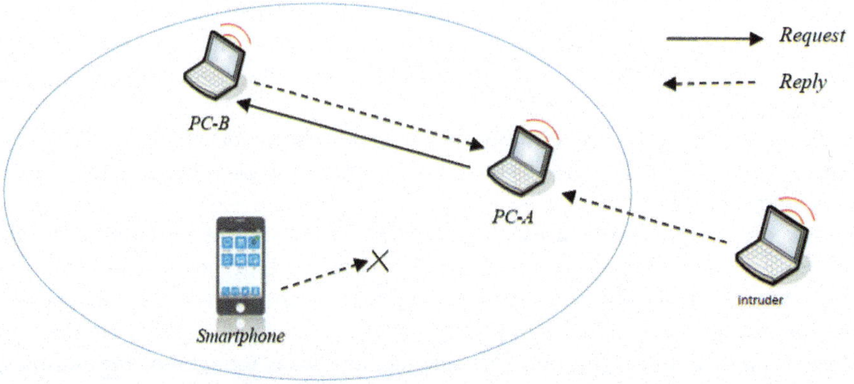

Fig. 1 Example of an IoT attack

Table 4 Stages of securing system

S'	=	$\,^{k1}_{r}[sec^{ka}_{a}.sec^{kb}_{b}.sec^{kc}_{s}.\mathrm{R}\ \vert^{\delta}_{a}[\mathrm{A}\vert!(acq^{k2}_{tra}.0)\vert^{k2}_{tra}[\mathrm{TRA}]]\vert^{\delta}_{b}[\mathrm{B}\vert!(acq^{k3}_{trb}.0)\vert^{k3}_{trb}[\mathrm{TRB}]]\vert\ ^{\delta}_{s}[\mathrm{S}\vert!(acq^{k4}_{trs}.0)\vert^{k4}_{trs}[\mathrm{TRS}]]]$
	→	$\,^{k1}_{r}[sec^{kb}_{b}.sec^{kc}_{s}.\mathrm{R}\ \vert^{ka}_{a}[\mathrm{A}\vert!(acq^{k2}_{tra}.0)\vert^{k2}_{tra}[\mathrm{TRA}]]\vert^{\delta}_{b}[\mathrm{B}\vert!(acq^{k3}_{trb}.0)\vert^{k3}_{trb}[\mathrm{TRB}]]\vert^{\delta}_{s}[\mathrm{S}\vert\ !(acq^{k4}_{trs}.0)\vert^{k4}_{trs}[\mathrm{TRS}]]]$
	→	$\,^{k1}_{r}[sec^{kc}_{s}.\mathrm{R}\ \vert^{ka}_{a}[\mathrm{A}\vert!(acq^{k2}_{tra}.0)\vert^{k2}_{tra}[\mathrm{TRA}]]\vert^{kb}_{b}[\mathrm{B}\vert!(acq^{k3}_{trb}.0)\vert^{k3}_{trb}[\mathrm{TRB}]]\vert^{\delta}_{s}[\mathrm{S}\vert\ !(acq^{k4}_{trs}.0)\vert\ ^{k4}_{trs}[\mathrm{TRS}]]]$
	→	$\,^{k1}_{r}[\mathrm{R}\vert^{ka}_{a}[\mathrm{A}\vert!(acq^{k2}_{tra}.0)\vert^{k2}_{tra}[\mathrm{TRA}]]\vert^{kb}_{b}[\mathrm{B}\vert!(acq^{k3}_{trb}.0)\vert^{k3}_{trb}[\mathrm{TRB}]]\vert^{kc}_{s}[\mathrm{S}\vert!(acq^{k4}_{trs}.0)\vert\ ^{k4}_{trs}[\mathrm{TRS}]]]$

$$\mathrm{I} = k1 \rhd in^{k1}_{r} + mig^{\delta}_{i} + mig^{\delta}_{a} + sea.\mathrm{I}'\ |!(acq^{k5}_{tri}.0)\ |^{k5}_{tri}\ [\mathrm{TRI}]$$

The intruder logs on to IoT network to explore the system. The sea operator models the manner the intruder scans the system and gains new capabilities in terms of mig, in and out actions. In this example the only condition is the knowledge of the key $k1$.

4.1 Securing the System

Modeling this attack example with our approach allowed us to see that the ROUTE_REQUEST sent by "computer a" will be received by the intruder machine. The latter emits its false ROUTE_REPLY response in order to transparently intercept selected packets exchanged between "computer a" and "spamrtphone s", and subsequently performs a denial of service (DoS). In order to defend against this attack, we have come up with a technique that uses a private key to encrypt each computer. The approach makes sure that the outdated, potentially compromised keys are not used by an attacker by updating (changing) the keys of computers a, b, and c whenever a new node enters the IoT network. The system securing procedure is achieved through the use of the action sec by each computer. The stages of system enforcement are shown in Table 4.

To demonstrate the applicability of our approach, we developed a framework called SCAC. Figure 2 illustrates the topology of a network vulnerable to selective forwarding attack.

By comparing our calculus with that of [3], we will be able to observe that the context principle has existed since the 80 s. Nevertheless, the fundamental calculus, as demonstrated by [7], does not support mobility and security. Moreover, our calculus is more straightforward than [9], which considers contextual mobility but does not provide protection. For this purpose, our calculus makes formal modeling easier to understand while providing support for mobile and sensitive contexts and solutions for security.

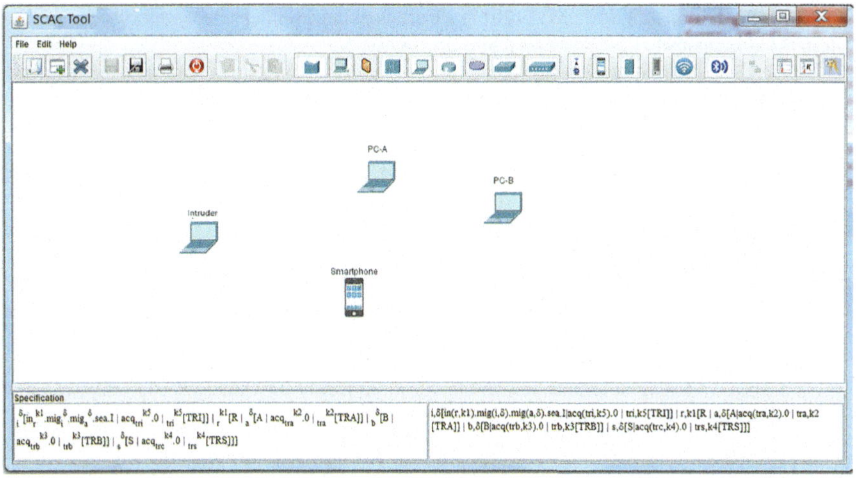

Fig. 2 Topology and attack specification

5 Conclusion

In this paper, we defined a new calculus to secure an IoT network. Our calculus allow to show that a network IoT is not secure and vulnerable to attacks. To strengthen the security of our system and prevent this attack, we proposed an alternative specification of the network represented by a protection with private keys. As perspectives, we plan to improve our calculation to include the securing of the VANET (Vehicular Ad hoc NETwork).

Acknowledgements This work has been sponsored by the General Directorate for Scientific Research and Technological Development, Ministry of Higher Education and Scientific Research (DGRSDT), Algeria.

References

1. Alferidah DK, Jhanjhi NZ (2020) A review on security and privacy issues and challenges in internet of things. Int J Comput Sci Netw Secur IJCSNS 20(4):263–286
2. Cardelli L, Gordon AD (1998) Mobile ambients. In: International Conference on foundations of software science and computation structure. Springer, Lisbon, Portugal, pp 140-155
3. Fujita H, Zualkernan I (2008) Enforcing security policies using algebraic approach. In: Proceedings of the 7th Software methodologies, tools, and techniques, vol 182, pp 84–98
4. Glushkova T, Stoyanov S, Stoyanova-Doycheva A, Ivanova V, Doukovska L (2019) AmbiNet an environment for ambient-oriented modeling. Int J Comput 18(3):331–340
5. Jing Q, Vasilakos AV, Wan J, Lu J, Qiu D (2014) Security of the Internet of Things: perspectives and challenges. Wirel Netw 20:2481–2501

6. Khanam S, Ahmedy I, Idris MYI (2017) An efficient detection of selective forwarding attacks in heterogeneous IoT Networks. FCSIT, pp 1-8
7. Milner R (1985) Lectures on a calculus for communicating systems. Seminar on concurrency, Carnegie-Mellon University, Springer, London, UK, pp 197–220
8. Shiranzaei A, Khan RZ (2018) An approach to discover the sinkhole and selective forwarding attack in IoT. J Inf Secur Res 9:107
9. Siewe F, Zedan H, Cau A (2011) The calculus of context-aware ambients. J Comput Syst Sci 77(4):597–620
10. Stephen A, Arockiam L (2021) Attacks against RPL in IoT: a survey. Ann Romanian Soc Cell Biol 9767–9786 (2021)
11. Xin-Sheng W, Yong-Zhao Z, Shu-ming X, Liang-min W (2009) Lightweight defense scheme against selective forwarding attacks in wireless sensor networks. In: International conference on cyber-enabled distributed computing and knowledge discovery. IEEE, pp 226–232
12. Zhu H, Zhang Z, Du J, Luo S, Xin Y (2018) Detection of selective forwarding attacks based on adaptive learning automata and communication quality in wireless sensor networks. Int J Distrib Sens Netw 14(11):1550147718815046

Security Analysis of an Aggregate Signcryption Scheme Unveiling Critical Vulnerabilities

S. S. Aravind Vishnu and M. Sethumadhavan

Abstract This paper analyses the security of a recently proposed signcryption scheme based on fog computing. The proposed scheme is found to be vulnerable against basic protocol level attacks and does not seem to actually satisfy their claims. We expose the vulnerabilities one-by-one and explore some of the possible attacks.

Keywords Cryptanalysis · Signcryption · Protocol vulnerability

1 Introduction

Certificateless signcryption schemes are the cryptographic constructions providing confidentiality, integrity and entity authentication in a single scheme without employing any kind of certifying authorities and thus eliminating the requirement to store and verify certificates. [2, 3, 5, 8] etc. are some such schemes introduced in the last decade. In the meanwhile, certificateless cryptography found new horizons [1, 6, 7]. Collusion resistance also has been an important security requirement in the same period. Schemes like [4, 9] shows the importance of collusion resistance. We analyse potential vulnerabilities on the Certificateless Online/Offline Aggregate Signcryption Scheme presented in [10]. In the scheme, the authors combined features from certificateless aggregate signcryption and online/offline encryption to enhance efficiency and reduce resource consumption. The authors address limitations in ad-hoc networks, particularly VANETS, and integrates fog computing and anti-collusion measures. Also the scheme avows to demonstrate unforgeability, secrecy, and resilience

S. S. Aravind Vishnu (✉)
Department of Mathematics, Amrita School of Physical Sciences, Amrita Vishwa Vidyapeetham, Coimbatore, India
e-mail: cb.sc.d.mat17003@cb.students.amrita.edu

M. Sethumadhavan
TIFAC-CORE in Cyber Security, Amrita School of Engineering, Amrita Vishwa Vidyapeetham, Coimbatore, India
e-mail: m_sethu@cb.amrita.edu

A. C. Frery et al. (eds.), *Computational Intelligence and Data Analytics*, Lecture Notes on Data Engineering and Communications Technologies 236,
https://doi.org/10.1007/978-981-96-0451-7_32

against collusion attacks in the random oracle model. The system ensures data confidentiality, real-time transmission, and robust defense against collusion without escalating computational costs, as claimed in [10]. We have found some vulnerabilities in the protocol proposed in the scheme, and we are exposing those vulnerabilities in this article.

An overview of the scheme is presented in Sect. 2. We analyse the scheme and expose some vulnerabilities in Sect. 3. Section 4 concludes the manuscript.

2 An Overview of the Scheme

The scheme mainly finds its applications in vehicular ad-hoc networks (VANET). So it involves Trusted Authority (TA), Key Generating Centre (KGC), Road Side Unit (RSU), and vehicles. The construction uses the Elliptic Curve Discrete Logarithm problem (ECDLP) defined on a q-order cyclic group G with generator P.

2.1 The Protocol

The following steps describe the protocol in the proposed scheme chronologically.

1. Set up
   - TA: $n \in_R$[1] $\mathbb{Z}_q^*$ and computes $T_{pub} = n \cdot P$.
   - KGC: $m_{sk} \in_S$[2] $\mathbb{Z}_q^*$ and computes $P_{pub} = m_{sk} \cdot P$.
   - Six hash functions are chosen.
     - $h_0 : \{0, 1\}^* \rightarrow \mathbb{Z}_q^*$
     - $h_1 : \{0, 1\}^* \rightarrow \{0, 1\}^n$, n: Bit-length of plaintexts.
     - $h_2 : \{0, 1\}^* \rightarrow \mathbb{Z}_q^*$
     - $h_3 : \{0, 1\}^* \rightarrow \mathbb{Z}_q^*$
     - $h_4 : G \rightarrow \mathbb{Z}_q^*$
     - $h_5 : G \times \{0, 1\}^* \rightarrow \mathbb{Z}_q^*$
   - KGC and TA : Publish $Params : \{q, P, P_{pub}, T_{pub}, h_0, h_1, h_2, h_3, h_4, h_5\}$
2. Vehicle pseudonym generation from the real identity RID
   - Vehicle V_i: $t_i \in_R \mathbb{Z}_q^*$, computes $PID_{i,1} = t_i \cdot P$, $O_i = t_i \cdot T_{pub} \oplus RID_i$.
   - RSU V_r: $t_r \in_R \mathbb{Z}_q^*$, computes $PID_{r,1} = t_r \cdot P$, $O_r = t_r \cdot T_{pub} \oplus RID_r$.

[1] $\in_R$ denotes random selection.

[2] $\in_S$ denotes stochastic selection.

– TA: Verifies the identities. For $j \in \{i, r\}$

$$RID_j \stackrel{?}{=} O_j \oplus n \cdot PID_{j,1} \tag{1}$$

– TA:[3] Calculates the other part of the pseudonym. For $j \in \{i, r\}$

$$PID_j = RID_j \oplus h_5(n \cdot PID_{j,1}, T_j)$$

where T_j is the effective time of the pseudonym.

– TA: Sends $PID_j = \{PID_{j,1}, PID_{j,2}\}$ secretly to KGC.

3. The User Key Generation

   – User ID_i (pseudonym) : $x_{ID_i} \in_S \mathbb{Z}_q^*$ and computes $X_{ID_i} = x_{ID_i} \cdot P$.

4. Partial Key Extraction

   – KGC: $y_{ID_i} \in_S \mathbb{Z}_q^*$ and computes $Y_{ID_i} = y_{ID_i} \cdot P$, the partial public key.
   – KGC: Partial private key $D_{ID_i} = m_{sk} Q_i$, where $Q_i = h_0(ID_i)$ and returns (Y_{ID_i}, D_{ID_i}) to the user.

5. Private Key Verification

   – User: $D_{ID_i} \cdot P \stackrel{?}{=} h_0(ID_i) \cdot P_{pub}$
   – Public key of User ID_i: (X_{ID_i}, Y_{ID_i})
   – Private key of User ID_i: (x_{ID_i}, D_{ID_i})

6. Offline

   – User ID_i: $r_i \in_R \mathbb{Z}_q^*$. $R_i = r_i \cdot P$

7. Signcrypt (sending m_i from ID_i to ID_r)

   • $\Delta \leftarrow$ Trouble information of road conditions.
   • User ID_i computes

     – $L_i = r_i \cdot Q_r \cdot P_{pub}$
     – $h_{ic} = h_1(ID_r, R_i, X_{ID_r}, Y_{ID_r}, L_i, \Delta)$
     – Part of the ciphertext $K_i = h_{ic} \oplus (m_i || ID_i)$
     – $h_{ie} = h_2(ID_i, ID_r, R_i, K_i, X_{ID_i}, Y_{ID_i}, X_{ID_r}, Y_{ID_r})$
     – $h = h_3(\Delta)$
     – $W_i = D_{ID_i} + r_i h_{ie} + X_{ID_i} h$

   • Final ciphertext: $C_i = (R_i, K_i, W_i)$

[3] $\stackrel{?}{=}$ checks the equality on both sides of the symbol.

8. Aggregate

   - RSU computes

     - $V = h_4(X_{ID_r}W_1 || X_{ID_r}W_2 || \cdots || X_{ID_r}W_n)$
     - $W = \sum_{i=1}^{n} W_i$

   - Aggregate ciphertext $C = (V, W, R_1, R_2, \ldots, R_n, K_1, K_2, \ldots, K_n)$

9. Verification

   - The recipient computes

     - $h_{ie} = h_2(ID_i, ID_r, R_i, K_i, X_{ID_i}, Y_{ID_i}, X_{ID_r}, Y_{ID_r})$
     - $h = h_3(\Delta)$
     - Anti-collusion factor $b_i = Q_i \cdot P_{pub} + h_{ie} \cdot R_i + h \cdot X_{ID_i}$
     - $V' = h_4(x_{ID_r}b_1 || x_{ID_r}b_2 || \cdots || x_{ID_r}b_n)$
     - $V \stackrel{?}{=} V'$
     - $W \cdot P \stackrel{?}{=} \sum_{i=1}^{n} \left(Q_i \cdot P_{pub} + h_{ie} \cdot R_i + h \cdot X_{ID_i}\right)$

10. Unsigncryption

   - The recipient computes

     - $L'_i = D_{ID_r} \cdot R_i$
     - $h'_{ic} = h_1(ID_r, R_i, X_{ID_r}, Y_{ID_r}, L'_i, \Delta)$
       $m'_i || ID_i = K_i \oplus h'_{ic}$

   - Obtains $\left\{m'_i || ID_i\right\}_{i=1}^{n}$

3 Vulnerabilities of the Scheme

We expose some of the vulnerabilities of the scheme.

3.1 Master Secret Key Leakage

Consider the step 4 in Sect. 2.1 about partial key extraction. We explore the vulnerability in this scenario.

- $D_{ID_i} = m_{sk} Q_i$ where $Q_i = h_0(ID_i)$.
- ID_i can be obtained by any of the users. h_0 is publicly known. So it is direct computation to get the value of $Q_i = h_0(ID_i)$.
- Since h_0 is defined from $\{0, 1\}^* \rightarrow []\mathbb{Z}_q^*$, $Q_i \in \mathbb{Z}_q^* \implies Q_i^{-1} \in \mathbb{Z}_q^*$.
- The user with pseudonym ID_i obtains D_{ID_i} from KGC. This particular user can compute $Q_i^{-1} \cdot D_{ID_i}$ **to obtain** $\boldsymbol{m_{sk}}$.

3.2 Confidentiality Breach

The scheme in [10] states that it provides message confidentiality. We show that it is not the case.

- From Sect. 3.1, a malicious user can obtain the master secret key m_{sk}.
- An adversary with m_{sk} value and publicly available ID_r value can directly compute D_{ID_r}.
- The final ciphertext C_i contains R_i value. So $L_i = D_{ID_r} \cdot R_i$ can be directly computed.
- Having computed L_i, and with public information ID_r, R_i, X_{ID_r}, Y_{ID_r} and Δ, one can compute h_{ic}.
- $K_i \oplus h_{ic} = (m_i || ID_i)$.

Thus the message can be obtained by a malicious user. Hence the scheme does not provide confidentiality.

3.3 Impersonating KGC

The following steps show how the KGC can be impersonated by a malicious user.

- TA and KGC communicate directly to each other only until step 2 in Sect. 2.1.
- A malicious user who obtained m_{sk} as described in Sect. 3.1 can impersonate the KGC. This is since in step 5, the verification is

$$D_{ID_i} \cdot P \stackrel{?}{=} h_0(ID_i) \cdot P_{pub}$$

  in which only D_{ID_i} is created by KGC using m_{sk} and Q_i value.
- Since ID_i of any user is available to the malicious user and the user already possesses m_{sk}, this impersonation is possible.

3.4 Thwarting Entity Authentication

A malicious user who obtained m_{sk} can assign the duplicate identity ID_r and the private key value D_{ID_r} of an already existing user V_r to another user V_i. Hence we can see that entity authentication is also at stake in the protocol.

3.5 Man in the Middle Attack

Here we discuss a scenario where a malicious user V_j impersonates the TA to the user V_i and performs a Man-in-the-middle attack.

- $V_i \rightarrow$ [] TA(V_j): $PID_{i,1}$, O_i
- $V_j \rightarrow$ [] TA: $PID_{i,1}$, O_i
- TA calculates the pseudonym on V_j (instead of V_i) as $PID_{j,2} = RID_i \oplus h_5(n \cdot PID_{i,1}, T_i)$.
- TA sends $PID_j = (PID_{j,1} = PID_{i,1}, PID_{j,2})$ to KGC.
- Now the KGC has associated the user id RID_i to V_j.

The following protocol diagram illustrates the above attack.

V_i		$Adv(V_j)$		TA		KGC
$t_i \in_R \mathbb{Z}_q^*$						
$PID_{i,1} = t_i \cdot P$						
$O_i = t_i \cdot T_{pub} \oplus RID_i$						
	$\xrightarrow{PID_{i,1}, O_i}$					
			$\xrightarrow{PID_{i,1}, O_i}$			
				$\delta = h_5(n \cdot PID_{i,1}, T_i)$		
				$PID_{j,2} = RID_i \oplus \delta$		
				$PID_j = \{PID_{i,1}, PID_{j,2}\}$		
					$\xrightarrow{PID_j, T_j}$	
						$RID_i \leftrightarrow V_j$

3.6 RID leakage

There is a communication between V_i/V_r to TA where they send $PID_{i,1}/PID_{r,1}$ and O_i/O_r and requires the TA to verify as per Eq. (1). Some of the inferences here are

- TA knows RID
- There is no prior communication between V_i and TA to share RID.
- Physical exchange of RID is impossible considering the number of units involved in the communication and the time frame of the communication.

- So it must be assumed that the RID is communicated over an insecure channel or another public key encryption mechanism is involved in the process. In the first case, the vulnerability is clear as daylight and the second annuls the requirement of the complete protocol under discussion.

3.7 *Time and Space Complexities*

The vulnerabilities in Sects. 3.1, 3.2, 3.3 and 3.4 are based on finding the inverse of Q_i (mod q). The prominent methods for the same are Extended Euclidean algorithm or Fermat's method. Both are having a time complexity $\mathcal{O}(\log n)$. Since the algorithm uses a constant amount of extra space for storing intermediate results, the complexity is $\mathcal{O}(1)$. The attacks on Sects. 3.5 and 3.6 are passive adversarial attacks.

4 Conclusion

This paper exposes several critical vulnerabilities in the protocol presented in [10]. We demonstrate that the scheme fails to achieve fundamental security requirements, including confidentiality, forward secrecy, mutual authentication etc. These vulnerabilities also compromise the scheme's ability to ensure user anonymity and undeniability, as users are able to masquerade as other users. In conclusion, the scheme was proposed as an aggregate signcryption scheme secure against collusion attacks, but fell short of achieving even basic security requirements.

References

1. Anitha R, Pandiyaraju V, Muthurajkumar S, Sai Ramesh L, Rakesh R (2015) Secure data sharing in cloud storage using KAC with certificateless encryption. Adv Nat Appl Sci 9(6):134–9
2. Aravind Vishnu S, Praveen I, Sethumadhavan M (2021) An IND-CCA2 secure certificateless hybrid signcryption. Wirel Personal Commun 119(4):3589–3608
3. Barbosa M, Farshim P (2008) Certificateless signcryption. In: Proceedings of the 2008 ACM symposium on information, computer and communications security, pp 369–372
4. Chandran N, Chase M, Vaikuntanathan V (2012) Functional re-encryption and collusion-resistant obfuscation. In: Theory of cryptography: 9th theory of cryptography conference, TCC 2012, Taormina, Sicily, Italy, 19–21 Mar 2012. Proceedings 9. Springer, pp 404–421
5. Liu Z, Hu Y, Zhang X, Ma H (2010) Certificateless signcryption scheme in the standard model. Inf Sci 180(3):452–464
6. Mathi S, Nivetha R, Priyadharshini B, Padma S (2017) A certificateless public key encryption based return routability protocol for next-generation IP mobility to enhance signalling security and reduce latency. Sādhanā 42:1987–1996

7. Namboodiri K, Praveen I (2019) An efficient batch verification scheme for secure vehicular communication using bilinear pairings. In: International conference on soft computing and signal processing. Springer, pp 711–720
8. Xiong H, Guan Z, Chen Z, Li F (2013) An efficient certificateless aggregate signature with constant pairing computations. Inf Sci 219:225–235
9. Xiong W, Wang R, Wang Y, Wei Y, Zhou F, Luo X (2022) Improved certificateless aggregate signature scheme against collusion attacks for vanets. IEEE Syst J 17(1):1098–1109
10. Zhang W, Liu S, Liu Y, Cao J, Fu B, Du Y (2023) A certificateless online/offline aggregate signcryption scheme against collusion attacks based on fog computing. Electronics 12(23). https://doi.org/10.3390/electronics12234747, https://www.mdpi.com/2079-9292/12/23/4747

Author Index

A. C. Frery et al. (eds.), *Computational Intelligence and Data Analytics*, Lecture Notes on Data Engineering and Communications Technologies 236,
https://doi.org/10.1007/978-981-96-0451-7

The manufacturer's authorised representative in the EU is Springer Nature Customer Service Centre GmbH, Europaplatz 3, 69115 Heidelberg, Germany. If you have any concerns regarding our products, please contact ProductSafety@springernature.com

Printed and bound by CPI Group (UK) Ltd, Croydon, CR0 4YY
01/10/2026
02225901-0007